D1154570

The Electron

Fundamental Theories of Physics

An International Book Series on The Fundamental Theories of Physics:
Their Clarification, Development and Application

Editor: ALWYN VAN DER MERWE
University of Denver, U.S.A.

Volume 45

The Electron

New Theory and Experiment

edited by

David Hestenes

Physics Department,
Arizona State University,
Tempe, Arizona, U.S.A.

and

Antonio Weingartshofer

Laser-Electron Interactions Laboratory,
Department of Physics,
St. Francis Xavier University,
Antigonish, Nova Scotia, Canada

KLUWER ACADEMIC PUBLISHERS
DORDRECHT / BOSTON / LONDON

Library of Congress Cataloging-in-Publication Data

The Electron : new theory and experiment / edited by David Hestenes
  and Antonio Weingartshofer.
        p.   cm. -- (Fundamental theories of physics ; v. 45)
   Proceedings of the 1990 Electron Workshop held at St. Francis
Xavier University, Antigonish, Nova Scotia, Aug. 3-8.
    Includes index.
    ISBN 0-7923-1356-9 (HB : alk. paper)
    1. Electrons--Congresses. 2. Quantum electrodynamics--Congresses-
3. Dirac equation--Congresses.   I. Hestenes, David, 1933-   .
II. Weingartshofer, Antonio, 1925-   .   III. Electron Workshop (1990
: St. Francis Xavier University) IV. Series.
QC793.5.E62E42   1991
539.7'2112--dc20   91-22205
                                                                  CIP

ISBN 0-7923-1356-9

Published by Kluwer Academic Publishers,
P.O. Box 17, 3300 AA Dordrecht, The Netherlands.

Kluwer Academic Publishers incorporates
the publishing programmes of
D. Reidel, Martinus Nijhoff, Dr W. Junk and MTP Press.

Sold and distributed in the U.S.A. and Canada
by Kluwer Academic Publishers,
101 Philip Drive, Norwell, MA 02061, U.S.A.

In all other countries, sold and distributed
by Kluwer Academic Publishers Group,
P.O. Box 322, 3300 AH Dordrecht, The Netherlands.

Printed on acid-free paper

Printed in the Netherlands

This book is dedicated to the memory of
Reverend Dr. E. M. Clarke
In him there was a comfortable and congenial convergence
between Dr. Clarke the physicist and Father Clarke the priest

CONTENTS

PART II: Experiment

P R E F A C E

The importance of conducting experiments on this particle resides in our intrinsic interest in understanding the relatively simple system we call "the electron".

Robert S. Van Dyck, Jr.
Paul B. Schwinberg
Hans G. Dehmelt

This opening sentence to a comprehensive review article by scientists from the University of Washington is also a fitting opening to these Proceedings of the 1990 ELECTRON WORKSHOP for three reasons: Their history-making experiments occupy a central place in this book. (2) Their words express the spirit that brought together scientists from two continents to discuss and debate current conceptions of the electron. (3) Hopefully, this will remind the theorists that experiment has the last word.

The first word has been allotted to the eminent theorist Edwin T. Jaynes, who observes about the Electron Workshop: "It seems strange that this is the first such meeting, since for a century electrons have been the most discussed things in physics". This decade marks the centennial for both the discovery of the electron by J. J. Thomson and the invention of classical electron theory by H. A. Lorentz, but there are better reasons for the Workshop than a centennial celebration. First, new theoretical insights into the structure and self-interaction of the electron hold promise for resolving the nagging difficulties of quantum electrodynamics and achieving a deeper theory of electrons. Second, new experimental techniques provide powerful new probes for electron properties. The Electron Workshop was organized to achieve a confluence of these theoretical and experimental streams, bringing together, for mutual intellectual enrichment, physicists animated by a common will to know the electron. The success of this endeavor can be seen in the contributions to these Proceedings, which have been organized into theoretical and experimental parts.

The contributions by the theorists differ widely in approach but can be grouped into five categories. (1) Jaynes discusses the possibility of experimental tests for an old idea of Schrödinger's that the wave function really describes an extension of the electron in space. (2) A cluster of papers by Hestenes, Gull, Krüger and Boudet deals with a radical new formulation of the Dirac theory which reveals hidden geometric structure in the theory, provides powerful new computational techniques, and raises new issues of physical interpretation as well as possibilities for deepening the theory. (3) Barut contributes a comprehensive review of his own ambitious program in electron theory and quantum electrodynamics. Barut's work is rich with ingenious ideas, and the interest it provokes among other theorists can be seen in the critique by Grandy. Cooperstock takes a much different approach to nonlinear field-electron coupling which leads him to conclusions about the size of the electron. (4) Capri and Bandrauk work within the standard framework of quantum electrodynamics. Bandrauk presents a valuable review of his theoretical approach to the striking new photoelectric phenomena in high intensity laser experiments. (5) Jung proposes a theory to merge the ideas of free-free transitions and of scattering chaos, which is becoming increasingly important in the theoretical analysis of nonlinear optical phenomena.

For the last half century the properties of electrons have been probed primarily by scattering experiments at ever higher energies. Recently, however, two powerful new experimental techniques have emerged capable of giving alternative experimental views of the electron. We refer to (1) the confinement of single electrons for long term study, and (2) the interaction of electrons with high intensity laser fields. Articles by outstanding practitioners of both techniques are included in Part II of these Proceedings.

The precision experiments on trapped electrons by the Washington group quoted above have already led to a Nobel prize for the most accurate measurements of the electron magnetic moment. Sadly, there is a dearth of good theoretical ideas to be tested by this sensitive technique. Theorists, attention!

The interaction of intense laser fields with electrons and atoms is currently the subject of vast research activity. Experiment has far outstripped theory in this domain, so much of the data can be given only a qualitative explanation. This circumstance has unfortunate reper-cussions. Without sharply drawn theoretical issues, much experimental effort is wasted. Thus, there is a plethora of experiments on multiphoton ionization while more subtle, possibly more informative, phenomena are overlooked. In particular, the possibility that these phenomena might tell us something new about the electron is seldom considered. That is a point we especially wish to emphasize in these Proceedings.

In Part II, Weingartshofer provides an overview of experimental problems with intense field-electron interactions. Gallager discusses experimental results of ionization with circularly polarized microwave fields and compares them with observations in the optical regime. The outstanding research group at Saclay, represented at the Workshop by Anne L'Huillier, reviews the very informative quantitative results of their experiments on harmonics of the driving laser frequency in the radiation that accompanies ionization. The complications of multiphoton ionization are totally avoided in the next two papers discussing electron scattering techniques in the presence of intense laser fields. Wallbank describes simultaneous electron-multiphoton excitation of atoms and Morgner makes some predictions about the behavior of Penning electrons in intense laser fields. In the final paper Moorman deals with microwave ionization of hydrogen atoms as the prototype system for studying quantum chaos.

The 1990 ELECTRON WORKSHOP was held at St. Francis Xavier University in Antigonish, Nova Scotia (August 3 to 8), supported by funds from the Natural Sciences and Engineering Research Council of Canada. The Workshop organizer, Antonio Weingartshofer, wishes to thank Heinz Krüger for initiating contact with other theoreticians friendly to the electron and his colleague Barry Wallbank for invaluable help and encouragement. The enthusiastic support of University officials greatly enhanced the atmosphere of the Workshop. The Academic Vice-President, Dr. John J. MacDonald, reached deep into the empty coffers of the University to help out with expenses, and the President, Dr. David Lawless, opened the scientific sessions with a warm welcome to the University.

One of Canada's most distinguished scientists and public figures, Dr. Larkin Kerwin, returned to his Alma Mater to present the Keynote Address for this international meeting. Some thirty-six years before, his inspiring research at Laval University attracted the Reverend Dr. E. M. Clarke, and together they developed the prototype of the first practical 127° electrostatic electron selector, an instrument that soon found its way into many laboratories throughout the world as a vital component of the modern electron spectrometer. The history of the Laser-Electron Interaction Laboratory at St. Francis Xavier University began with that event while Dr. Clarke was Chairman of the Physics Department. In his address entitled "AND SO AD INFINITUM, the Continuing Evolution of the Electron", Dr. Kerwin spoke about cultural and philosophical implications as well as the scientific implications of our evolving concepts of the ultimate constituents of matter. We hope as well that this book will serve as a road sign to the physics of the 21st century.

David Hestenes
Antonio Weingartshofer

SCATTERING OF LIGHT BY FREE ELECTRONS AS A TEST OF QUANTUM THEORY

E. T. JAYNES
Arthur Holly Compton Laboratory of Physics
Washington University
St. Louis MO 63130, U.S.A.

ABSTRACT: Schrödinger and Heisenberg gave two very different views about the physical meaning of an electron wave function. We argue that Schrödinger's view may have been dismissed prematurely through failure to appreciate the stabilizing effects of forces due to Zitterbewegung, and suggest experiments now feasible which might decide the issue.

1. Introduction.

We are gathered here to discuss the present fundamental knowledge about electrons and how we might improve it. On the one hand it seems strange that this is the first such meeting, since for a Century electrons have been the most discussed things in physics. And for all this time a growing mass of technology has been based on them, which today dominates every home and office. But on the other hand, this very fact makes it seem strange that a meeting like this could be needed. How could all this marvelously successful technology exist unless we already knew all about electrons?

The answer is that technology runs far ahead of real understanding. For Centuries practical men grew better varieties of grapes and bred faster horses without any conception of chromosomes and DNA. The most easily perceived facts give sufficient knowledge to start a technology, and trial–and–error experimentation takes over from there. Because of this, the practical men who give us our technology sometimes see no need for fundamental knowledge, and even deprecate it.

This happens even within a supposedly scientific field. The mathematics of epicycles was a successful 'technology' found by trial–and–error for describing and predicting the motion of planets, and because of this success the idea that all astronomical phenomena *must* be described in terms of epicycles captured men's minds for over 1000 years. The efforts of Copernicus, Kepler, and Galileo to find the 'chromosomes and DNA' underlying epicycles were not only deprecated, but violently opposed by the practical men who, being concerned only with phenomenology, found in epicycles all they needed.

We know today that the mathematical scheme of epicycles was flexible enough (a potentially unlimited number of epicycles available, whose size and period could be chosen at will) so that however the planets moved, it could always have been 'accounted for' by invoking enough epicycles. But a mathematical system that is flexible enough to represent any phenomenology, is empty of physical content. Indeed, the real content of any physical theory lies precisely in the *constraints* that it imposes on phenomena; the stronger the constraints, the more cogent and useful the theory.

1

D. Hestenes and A. Weingartshofer (eds.), The Electron, 1–20.
© 1991 *Kluwer Academic Publishers. Printed in the Netherlands.*

In the next two Sections, we summarize the historical background of the puzzled thinking that motivates our present efforts. The reader who wants to get on with the job currently before us may turn at once to Section 4 below.

2. Is Quantum Theory a System of Epicycles?

Today, Quantum Mechanics (QM) and Quantum Electrodynamics (QED) have great pragmatic success – small wonder, since they were created, like epicycles, by empirical trial–and–error guided by just that requirement. For example, when we advanced from the hydrogen atom to the helium atom, no theoretical principle told us whether we should represent the two electrons by two wave functions in ordinary 3-d space, or one wave function in a 6-d configuration space; only trial–and–error showed which choice leads to the right answers.

Then to account for the effects now called 'electron spin', no theoretical principle told Goudsmit and Uhlenbeck how this should be incorporated into the mathematics. The expedient that finally gave the right answers depended on Pauli's knowing about the two-valued representations of the rotation group, discovered by Cartan in 1913.

In advancing to QED, no theoretical principle told Dirac that electromagnetic field modes should be quantized like material harmonic oscillators; and for reasons to be explained here by Asim Barut, we think it still an open question whether the right choice was made. It leads to many right answers but also to some horrendously wrong ones that theorists simply ignore; but it is now known that virtually all the right answers could have been found without, while some of the wrong ones were *caused by,* field quantization.

Because of their empirical origins, QM and QED are not physical theories at all. In contrast, Newtonian celestial mechanics, Relativity, and Mendelian genetics are physical theories, because their mathematics was developed by reasoning out the consequences of clearly stated physical principles which constrained the possibilities. To this day we have no constraining principle from which one can deduce the mathematics of QM and QED; in every new situation we must appeal once again to empirical evidence to tell us how we must choose our mathematics in order to get the right answers.

In other words, the mathematical system of present quantum theory is, like that of epicycles, unconstrained by any physical principles. Those who have not perceived this have pointed to its empirical success to justify a claim that all phenomena must be described in terms of Hilbert spaces, energy levels, etc. This claim (and the gratuitous addition that it must be interpreted physically in a particular manner) have captured the minds of physicists for over sixty years. And for those same sixty years, all efforts to get at the nonlinear 'chromosomes and DNA' underlying that linear mathematics have been deprecated and opposed by those practical men who, being concerned only with phenomenology, find in the present formalism all they need.

But is not this system of mathematics also flexible enough to accommodate any phenomenology, whatever it might be? Others have raised this question seriously in connection with the BCS theory of superconductivity. We have all been taught that it is a marvelous success of quantum theory, accounting for persistent currents, Meissner effect, isotope effect, Josephson effect, etc. Yet on examination one realizes that the model Hamiltonian is phenomenological, chosen not from first principles but by trial–and–error so as to agree with just those experiments.

Then in what sense can one claim that the BCS theory gives a *physical explanation* of superconductivity? Surely, if the Meissner effect did not exist, a different phenomenological model would have been invented, that does not predict it; one could have claimed just as great a success for quantum theory whatever the phenomenology to be explained.

This situation is not limited to superconductivity; in magnetic resonance, whatever the observed spectrum, one has always been able to invent a phenomenological spin–Hamiltonian that "accounts" for it. In high–energy physics one observes a few facts and considers it a big advance – and great new triumph for quantum theory – when it is always found possible to invent a model conforming to QM, that "accounts" for them. The 'technology' of QM, like that of epicycles, has run far ahead of real understanding.

This is the grounds for our suggestion (Jaynes, 1989) that present QM is only an empty mathematical shell in which a future physical theory may, perhaps, be built. But however that may be, the point we want to stress is that the success – however great – of an empirically developed set of rules gives us no reason to believe in any particular physical interpretation of them. No physical principles went into them.

Contrast this with the logical status of a real physical theory; the success of Newtonian celestial mechanics does give us a valid reason for believing in the restricting inverse–square law, from which it was deduced; the success of relativity theory gives us an excellent reason for believing in the principle of relativity, from which it was deduced. Theories need not refer specifically to physics: the success of economic predictions made from the restricting law of supply and demand gives us a valid reason for believing in that law.

3. But What is Wrong With It?

Of course, finding a successful empirical equation can be an important beginning of real understanding; perhaps even the necessary first step. In this sense, the mathematics of QM does contain some very important and fundamental truth; but the process by which it was found reveals nothing about its meaning, and it remains not only logically undefined, but pragmatically incomplete. It can, for example, predict the relative time of decay of two Co^{60} nuclei only with a probable error of about five years; but the experimentalist can measure this interval to a fraction of a microsecond.

Contemplating this, we understand why Bohr once remarked that the 'deep truths' are ones for which the opposite is also true; repeatedly, the attempt to present a unified front on questions of interpretation forces QM into schizoid positions. In spite of the fact that the experimenter can measure details about individual decays that the theory cannot predict, those who speculate about the deeper immediate cause of each individual decay are considered incompetent, and the currently taught physical interpretation claims that QM is already a complete description of reality.

Then it contradicts itself by its inability to describe reality at all. For example, we are not allowed to ask: "What is really happening when an atom emits light?" We may ask only: "What is the probability that, if we make a measurement, we shall find that a photon has been emitted?" As Bohr emphasized repeatedly, the Copenhagen interpretation of quantum theory cannot, as a matter of principle, answer any question of the form: "What is really happening when - - - ?" Yet we submit that such questions are exactly the ones that a physicist ought to be trying to answer; for the purpose of science is to understand the real world. If there were no such thing as a reality that exists independently of human knowledge, then there could be no point to physics or any other science.

It is not only in radioactivity that QM is pragmatically incomplete; for example, the data record from a Stern-Gerlach experiment can tell not only the number of particles in each beam, but the time order in which 'spin up' and 'spin down' occurred. Indeed, as we noted long ago (Jaynes, 1957), in every real experiment the experimenter can observe things that the theory cannot predict. Always, official quantum theory takes a schizoid position, admitting that the theory is observationally incomplete; yet persisting in the claim that it is logically complete.

Even the EPR paradox failed to force retraction of this claim, and so currently taught quantum theory still contains the schizoid elements of local acausality on the one hand – and instantaneous action at a distance on the other! We find it astonishing that anyone could seriously advocate such a theory.

In short, the currently taught physical interpretation has elements of nonsense and mysticism which have troubled thoughtful physicists, starting with Einstein and Schrödinger, for over sixty years. The more deeply one thinks about these things, the more troubled he becomes and the more convinced that the present interpretive scheme of quantum theory is long overdue for a drastic modification. We want to do everything we can to help find it.

Not surprisingly, there has been no really significant advance in basic understanding since the 1927 Solvay Congress, in which this schizoid mentality became solidified into physics. Theoretical physics can hardly hope to make any further progress in such understanding until we learn how to separate the permanent mathematical truths from the physical nonsense that now obscures them.

To be fair, we should add that some of these contradictions disappear when we note that "currently taught" quantum theory is quite different from the "Copenhagen theory", defined as the teachings of Niels Bohr. The latter is much more defensible than the former if we recognize that Bohr's intention was never to describe reality at all; only our information about reality. This is a legitimate goal in its own right, and it has a useful – indeed, necessary – role to play in physics as discussed further in Jaynes (1986, 1990).

The trouble is that this is far from the *only* legitimate goal of physics; yet for 60 years Bohr's teachings have been perverted into attempts to deprecate and discourage any further thinking aimed at finding the causes underlying microphenomena. Such thinking is termed 'obsolete mechanistic materialism', but those who hurl such epithets then reveal their schizoid mentality when they ascribe unquestioning ontological reality – independent of human information – to things such as 'quantum jumps' and 'vacuum fluctuations'. This does violence to Bohr's teachings; yet those who commit it claim to be disciples of Bohr.

For a time we were optimistic because it appeared that the new thinking of John Bell (1987) might show us the way. It was refreshing to see from his words that he was not brainwashed by the conventional muddled thinking and teaching, but was able to discern the real difficulty. But his recent work (Bell, 1990) shows him apparently at the end of his rope, reduced to destructive criticism of the ideas of everybody else but offering nothing to replace them. Therefore it is up to us to find the new constructive ideas that theoretical physics needs.

4. Our Job for Today

Theoretical work of the kind presented at this meeting is sometimes held to be "out of the mainstream" of current thinking; but that is quite mistaken. There is no mainstream today; it has long since dried up and our vessel is grounded. We are trying rather to start a new stream able to carry science a little further. Indeed, our efforts are much closer to the traditional mainstream of science than much of what is done in theoretical physics today. Talk of tachyons, superstrings, worm holes, the wave function of the universe, the first 10^{-40} second after the big bang, etc., is speculation vastly more wild and far–fetched than anything we are doing.

In the present discussion we want to look at the problems of QM from a very elementary, lowbrow physical viewpoint in the hope of seeing things that the highbrow mathematical viewpoint does not see. I want to suggest, in agreement with David Hestenes, that Zitterbewegung (ZBW) is a real phenomenon with real physical consequences that underlie all of quantum theory; indeed, such important consequences that without ZBW the world would be very different and we would not be here. But my ZBW differs from his in some basic qualitative respects, and so our first order of business is to describe this difference and see whether it could be tested experimentally. Then we can proceed to some speculations about the role of ZBW in the world.

5. The Puzzle of Space-Time Algebra

It is now about 25 years since I started trying to read David Hestenes' work on space-time algebra. All this time, I have been convinced that there is something true, fundamental, and extremely important for physics in it. But I am still bewildered as to what it is, because he writes in a language that I find indecipherable; his message just does not come through to me. Let me explain my difficulty, not just to display my own ignorance, but to warn those who work on space–time algebra: nearly all physicists have the same hangup, and you are never going to get an appreciative hearing from physicists until you learn how to explain what you are doing, in plain language that makes physical sense to us.

Physicists go into a state of mental shock when we see a single equation which purports to represent the sum of a scalar and a vector. All of our training, from childhood on, has ground into us that one must never even dream of doing such an absurd thing; the sin is even worse than committing a dimensional inhomogeneity. How can David get away with this when the rest of us cannot?

If u and v are vectors, then in Hestenes' equation $(uv = u \cdot v + u \wedge v)$ the symbol '+' must have a different meaning than it does in conventional mathematics. But then the symbol '=' must also have some different meaning, and he does not choose to enlighten us, so the above equation remains incomprehensible to me. The closest I can come to making sense out of it is to note that we cannot speak of a sum of apples and oranges; yet we may place an apple and an orange side by side and contemplate them together. Perhaps this is something like the intended meaning.

There is another possibility. Perhaps '+' and '=' have their conventional meanings after all, but u and v do not. In my view it simply does not make sense to speak of the sum of a scalar and a vector, any more than of the proverbial square circle; but it makes perfectly good sense to speak, as Cartan does, of the sum of two *matrices*, interpreted as abstract mathematical *representations* of a scalar and a vector. If this is what David really means, he could have prevented decades of confusion by a slight change in verbiage. Physicists are very touchy about the distinction between an physical or geometrical object and a mathematical representation of that object, because the mathematical representation usually holds only in some restricted domain that does not apply to the object itself.

But I suspect that what he "really means" is something more abstract than either of these two suggestions, and neither his writings nor his talks provide enough clues for me to decide what it is. I have never been able to get past that equation, with any comprehension of what is being said.

Passing on to the next part of Hestenes' work, we encounter a physical difficulty. In discussing the Dirac equation we read that some symbol "stands for a rotation". Now it is evident that from a mathematical standpoint there is an abstract correspondence with rotations (composition law of the rotation group). But from a physical standpoint, rotation of *what*? When I look at the Dirac equation, I see just what Schrödinger did: a four-component wave function $\{\psi_\sigma(x,t)\}$ representing a continuous distribution of something (perhaps probability or charge); but that something itself has no further internal directional structure that could 'rotate'.

But just at this meeting, I finally picked up the first clue as to what David is talking about here. It seems that when he looks at the Dirac equation, he sees not a continuous distribution of anything, but *a tangle of all the different possible trajectories of a point particle*! Presumably, these are the things that are rotating. This must be the most egregious example of a hidden–variable theory ever dreamt of, and nothing of his that I have ever read prepared me for this revelation.

I would never, in 1000 years, have thought of looking at the Dirac equation in that way. For, in any theory where the underlying reality is conceived to be a single particle hiding somewhere in the wave packet, the behavior of the packet is determined, not by what that particle actually does, but by the range of *possible* things that it might have done, but did not. Thus the wave packet cannot itself describe any reality; it represents only a state of knowledge about reality. Indeed, this is the view that Heisenberg (1958) stated very explicitly.

You can, of course, account for many facts by such a picture – namely, those so unsharp in time and space that what the experimentalist observes can be regarded as some kind of average over an ensemble of many different trajectories. But when the experimental result ought to depend on *one* such motion, I think that the point particle trajectory picture will surely fail, because there is no such thing in Nature as a point particle.

Put more constructively, a point particle theory can be confirmed only by an experiment which actually sees that particle, removed from its ensemble or wave packet, doing something *as a particle*, all by itself. In an exactly similar way, Louis Pasteur's microbe theory of diseases could be confirmed only by developing the instruments by which a single microbe could be seen and its behavior observed.

Of course, we must be as demanding as Pasteur about the resolving power of our instruments. A continuous structure that is small compared to our resolution distance will appear to us as a point particle. That a microbe is not a point particle, but a continuous structure with definite shape and internal moving parts, can be learned only with a microscope that has a resolution distance small compared to the dimensions of the microbe.

We must accomplish something like this, in order to check the reality of those tangled trajectories which were supposed to be the 'chromosomes and DNA' hiding in the Dirac wave function. But is there any such experiment where our observation is so sharp in both time and space that it depends on a single trajectory? It seems to me that there are simple (in concept) experiments now on the borderline of feasibility, which are capable of testing this rather fundamental issue. In fact, they do not differ in principle – or even in the relevant dimensions and resolving power – from Pasteur's microscopes.

6. What is a Free Electron?

We have long been intrigued with the fact that in applications of quantum theory, in our equations we write only wave functions ψ, either explicitly or implicitly (as matrix elements between such wave functions). But in the interpretive words between those equations we use only the language of point particles. Even the Feynman diagrams are a part of that inter–equation language, depicting particles rather than waves.

Thus the wave–particle duality is partly an artifact of our own making, signifying only our own inability to decide what we are talking about. But the predictions of observable facts come entirely from wave functions $\psi(x,t) = r(x,t)\exp[i\phi(x,t)]$; and not merely the magnitudes $|\psi|^2 = r^2$, but even more from the phases $\phi(x,t)$. Then if anything in the mathematics of QM could be held to represent some kind of reality, it is surely the complex wave function itself, not that point particle imagined to be hiding somewhere in it, but which plays no part in our calculations. This is just Schrödinger's original viewpoint.

The idea that a free electron is something more like an amoeba than a point particle was suggested by David Bohm (1951); but there he was only trying to help us visualize the mathematics of wave packets. Here we want to endow that suggestion with physical reality and suppose, with Schrödinger, that a wave packet $\psi(x,t)$ is not merely a representation of a state of information about an electron; but a physically real thing in its own right, with a shape and internal moving parts that are capable of being changed by external interactions and observed by us. The arguments that were raised against this picture long ago (spreading of the wave packet, etc.) are easily answered today, as we shall see. The spreading wave packet solution does not, in our theory, describe the physical free electron; among all solutions of the Dirac equation it represents a set of measure zero.

This thinking – long on the back burner, so to speak – was moved to the front burner by an incident that occurred at the 1977 meeting on free electron lasers at Telluride, Colorado. I gave a talk (Jaynes, 1978) on the general principles for generating light from electrons, which contained some speculation about possible explanations of the then much discussed 'blue electron' effect reported by Schwartz & Hora (1969). Here 50 kev free electrons are irradiated by blue light from an Argon laser, then drift 20 cm and were reported to emit the same color blue light on striking an alumina screen which is normally not luminescent.

It requires very little thought to see that such an effect cannot be accounted for by mutual coherence properties of different electrons, which would require impossible collimation of the electron beam; somehow, each individual electron must be made to carry the information about the light wavelength, for a million light cycles after irradiation.

My calculation showed that interaction with laser light can perturb the wave packet of a free electron, $\psi_0(x,t) \rightarrow \psi_0 + \psi_1$, so that it is partially separated into a linear array of smaller lumps, making the transverse density profile

$$\rho(x) = |\psi_0 + \psi_1|^2 \propto |\psi_0|^2 \left[1 + A\cos(2\pi x/\lambda)\right]$$

look somewhat like a comb, with teeth [those internal moving parts] separated by the light wavelength λ; here A is proportional to the light amplitude and the cosine of a polarization angle. Then when all these lumps, moving in the z-direction, strike a screen simultaneously, there is in effect a pulse of simultaneous current elements with that separation, which can act like an end–fire array and radiate light in the x-direction with the original wavelength λ (but, of course, with a much broader spectrum, whose width indicates the lateral size of the electron wave packet $|\psi_0|^2$ in the x-direction).

To my astonishment, Willis Lamb objected strongly to this, saying "You don't under-stand quantum theory. The electron is not broken up into many little electrons; the lumps are only lumps of probability for one electron. There is no interference effect in the radiation emitted when the electron is suddenly decelerated, because the electron is in reality in only one of those lumps."

This is a beautiful example of that schizoid attitude toward reality that believers in QM are obliged to develop; for he believed at the same time that if an electron wave packet is broken up into separated lumps by passing through the standard two slit apparatus, there is interference between those lumps, making the standard electron diffraction pattern (and showing, at least to me, that the electron *was* in both lumps simultaneously).

I was so taken aback by Willis's objection that I then did a conventional QED calculation, which showed to my satisfaction that standard QED *does* predict the interference effect that I had obtained so much more easily from a semiclassical picture. In fact, my calculation predicted just the dependence on polarization and drift distance that Schwartz and Hora reported seeing.

Other experimentalists have insisted that the effect does not exist (or at least could not have been observed because of impossible requirements for collimation of the electron beam) and it is not considered respectable to mention it today. But according to our wave packet theory, the collimation should not matter; perhaps their failure to confirm the effect was due just to their concentration on the irrelevancy of collimation, while the essential thing was that Schwartz's electron gun, with its small holes, produced electrons in wave packets of the right size.

If the effect does not exist, this would seem to be a major embarrassment for quantum theory; surely, predicting so easily an effect that does not exist is just as bad as failing to predict an effect that does exist [as someone put it at the time, if the effect is confirmed it will become known as *der Schwartz–Hora Effekt*; if not, it will be *die schwarze Aura*]. In any event, if the effect was not reproducible at the time – for whatever reason – the technology available today might overcome the old difficulties.

What has remained from this incident is the picture of the wave packet of a free electron as something with a physical meaning, its size and shape in principle measurable. For example, an electron in a wave packet ten microns long is physically different from one in a wave packet two microns long; a spherical wave packet electron is physically different from one a cigar–shaped one, and those differences should be observable in experiments now becoming feasible.

This issue now suggests our proposed experiment. Eventually we shall return to Zitterbewegung, but for the time being the only issue before us is: *Does interference exist between light waves scattered from different parts of a free electron wave packet?* If it does not, then something like the Lamb–Hestenes picture must be correct; if it does, then a whole new world of observable physical phenomena is opened up. As the old 1977 calculation showed in one case, quantum theory predicts a mass of very detailed, observable, new phenomena. They would make a free electron, when prepared in various sizes and shapes by previous irradiation, even more versatile in behavior than an amoeba.

In fact, that technology noted in the Introduction has already indicated something of the possibilities here. The electron diffraction microscope is able to reveal to us an amazing variety of fine detail. All this information is contained somehow in the electron wave functions; but it is surely not in coherence properties of different electrons, which could never be collimated well enough for that. Each individual electron must be carrying, in its phases (*i.e.* the distortion of its wave fronts) a vast amount of information about what it has passed through. It seems to us that if electrons lacked this plastic, amoeba–like character, electron microscopes would not work.

Consider, then, the exact analog of Pasteur's observation of a microbe: scattering of light of wavelength λ by a free electron which we represent by a wave packet of dimensions perhaps $2\lambda - 10\lambda$. If we are to see any interference between different parts of the wave packet, then that hypothetical point particle must be in two different places at the same time (or at least, at two places with spacelike separation). But it is easy for a continuous wave structure to give such interferences.

7. Relativistic Basis of the Nonrelativistic (NR) Schrödinger Equation.

To define the proposed experiment it will be sufficient to use the non–relativistic spinless Schrödinger equation, if we note first how and under what circumstances it can arise from the relativistic Klein–Gordon equation satisfied by each Dirac wave component:

$$\nabla^2 \phi - \frac{1}{c^2} \frac{\partial^2 \phi}{\partial \phi^2} = \left(\frac{mc}{\hbar}\right)^2 \phi \tag{1}$$

This has plane wave solutions of the form

$$\phi(x, t) = \exp(ik \cdot x - \omega t) \tag{2}$$

with

$$\omega \equiv c\sqrt{k^2 + \left(\frac{mc}{\hbar}\right)^2} \tag{3}$$

and thus $|\omega| \geq mc^2/\hbar$. The possible frequencies of propagating waves lie into two ranges separated by $2mc^2/\hbar$, which is just the ZBW frequency. Frequencies $|\omega| < mc^2/\hbar$ correspond to waves evanescent in space. Now given any solution of (1) which has only frequencies in one of those propagating ranges, say $\exp(-i\omega t)$ with $mc^2/\hbar \leq \omega$, we can view the solution as a rapid oscillation at frequency $mc^2/\hbar$, modulated by an envelope function which may be slowly varying. To separate them, make the substitution

$$\phi(x, t) = \psi(x, t) \exp\left(-\frac{imc^2}{\hbar} t\right) \tag{4}$$

whereupon $\psi(x, t)$ is found to satisfy the equation

$$i\hbar\dot{\psi} = -\frac{\hbar^2}{2m} \nabla^2 \psi + \frac{\hbar^2}{2mc^2} \ddot{\psi} \tag{5}$$

which is exact. Now if $\psi(x, t)$ contains frequencies up to order ν, the two time derivative terms are in approximately the ratio

$$\frac{\hbar^2 \ddot{\psi}/2mc^2}{i\hbar\dot{\psi}} \approx \frac{\hbar\nu}{2mc^2} \tag{6}$$

and if this is small compared to one, we have the NR Schrödinger equation

$$i\hbar\dot{\psi} = -\frac{\hbar^2}{2m} \nabla^2 \psi \tag{7}$$

but we now understand when it applies and what it means. It describes the slow 'secular' variations in the envelope when the relativistic wave function has only components $\exp(-i\omega t)$ with no admixture of terms $\exp(+i\omega t)$.

The point we stress is that solving the NR Schrödinger equation does not give us an approximation to an arbitrary relativistic solution; but only to those particular solutions *which contain no ZBW effects*. But the solutions we have discarded in making this approximation comprise in a sense the 'vast majority' of all possible relativistic solutions!

With this perhaps 'new' understanding, we may reexamine the conventional solutions of (7) for our problem. It has the general initial-value solution

$$\psi_0(x, t) = \int G(x - x') \, \psi_0(x', 0) \, d^3 x' \tag{8}$$

with the Green's function

$$G(x - x'; t) = \left(\frac{m}{2\pi i \hbar t}\right)^{3/2} \exp\left\{\frac{imr^2}{2\hbar t}\right\} \tag{9}$$

corresponding to diffusion with an imaginary diffusion coefficient. In particular, with an initial packet of RMS radius a,

$$\psi_0(x, 0) = \left(\frac{1}{2\pi a^2}\right)^{3/4} \exp\left(-\frac{r^2}{4a^2}\right) \tag{10}$$

we find the canonical spreading wave packet solution of our textbooks:

$$\psi_0(x, t) = \left(\frac{a}{\sqrt{2\pi}\left(a^2 + \frac{i\hbar t}{2m}\right)}\right)^{3/2} \exp\left\{-\frac{r^2}{4\left(a^2 + \frac{i\hbar t}{2m}\right)}\right\} \tag{11}$$

for which the probability density or normalized charge density is given by

$$|\psi_0(x, t)|^2 = \left(\frac{1}{2\pi\sigma^2}\right)^{3/2} \exp\left\{-\frac{r^2}{2\sigma^2}\right\} \tag{12}$$

with

$$\sigma^2(t) \equiv a^2 + \frac{\hbar^2 t^2}{4m^2 a^2} \tag{13}$$

so the packet grows with an ultimate spreading velocity which 'remembers' its initial size compared to the Compton wavelength:

$$\left(\frac{v}{c}\right)_{final} = \frac{(\hbar/mc)}{2a} \tag{14}$$

Doubtless all of us were, as students, assigned the homework problem of calculating from (13) how long it would require for some object like a marble to double its size (which now seems to me a totally wrong conception of what quantum theory is and says).

8. Non–Aristotelian Scattering of Light by Free Electrons

We want to look at the wave packets of free electrons in the same way that Pasteur looked at microbes. But conventional quantum mechanical scattering theory does not contemplate this, being based on an older idea, namely Aristotle's theory of Dramatic Unity. This prescribes that "The action must be complete, having

(1) a beginning which implies no necessary antecedent, but is itself a natural antecedent of something to come,
(2) a middle, which requires other matters to precede and follow, and
(3) an end, which naturally follows upon something else, but implies nothing following it."
                                                                     (Taylor, 1913)

Conventional scattering theory follows this plan perfectly, presupposing an initial state in which the particles involved are propagating as free particles but we ask not whence they come, an intermediate state in which the action takes place, and a final state in which they again propagate as free particles and we ask not where they go.

Note that the conventional textbook arguments (the Heisenberg γ–ray microscope, etc.) warning us that there is an uncertainty principle making it impossible to do so many things, always presuppose Aristotelian scattering and draw those conclusions from overall momentum conservation, coupled with a naïve 'buckshot' picture of a photon and ascribing a separate ontological reality to the scattering of each individual photon. But the actual mathematical formalism of QED, as developed afterward, does not have anything in it corresponding to the concept of 'a given photon'.

It seems to us that these things should be pointed out in elementary QM courses; the Heisenberg conclusion, far from being a firmly established foundation principle of physics, is not an experimental fact at all, only an unverified conjecture which presupposes just the things that we want to test here. It is high time that we found out whether it is true, and our technology is just now coming to the point where the experiments are possible.

But the scattering theory that we and Pasteur need is non–Aristotelian, in that the action does not have a beginning or an end at any times that are relevant to what we observe; rather we have something akin to a slowly changing nearly steady state, the incident light constantly bathing the electron or microbe and the scattered radiation constantly proceeding from it.

This makes an important technical difference, in that principles like overall momentum conservation, although we do not deny them, do not have the same application that they have in conventional scattering theory. This is not a handicap; on the contrary, it means that we, like electron microscopists, shall be able to see details that Aristotelian scattering theory does not describe. Requiring that the action be complete before we observe anything greatly restricts what one could see.

In the above we have used the notation ψ_0 to denote solutions of the free-particle equation (7). In a transverse EM field, the minimal coupling *ansatz* makes the replacement $p \rightarrow p - (e/c)A$, giving in first order

$$i\hbar\dot{\psi} = \frac{p^2}{2m} - \frac{e}{mc}\left(A \cdot p\right)\psi \qquad (15)$$

To see what this interaction term means physically, consider a classical point electron in the same EM field:

$$m\dot{v} = eE = -\frac{e}{c}\dot{A} \qquad (16)$$

or, in a gauge where $A = 0$ when $v = 0$,

$$v(t) = -\frac{e}{mc} A(t) \tag{17}$$

as in the London theory of superconductivity (where this equation accounts for a great deal of that phenomenology noted above). Therefore, we have

$$\frac{e}{mc} A \cdot p\psi = -v \cdot p\psi = i\hbar\,(v \cdot \nabla)\,\psi \tag{18}$$

and two terms of (15) combine thus

$$i\hbar\,\dot\psi + \frac{e}{mc} A \cdot p\psi = i\hbar \left(\frac{\partial}{\partial t} + v \cdot \nabla\right)\psi \tag{19}$$

which we recognize as the 'convective derivative' of hydrodynamics, so we have simply

$$i\hbar\,\frac{D\psi}{Dt} = -\frac{\hbar^2}{2m} \nabla^2\,\psi\,, \tag{20}$$

which is exactly equivalent to (15). The change in ψ as seen by a local observer moving at the classical electron velocity is just the free particle spreading! The external perturbation, in first order, merely translates the wave function locally by the classical motion. Recognizing this, we can write the solution and the current response immediately without any need to work out perturbation solutions of (15). A fairly good solution of the initial value problem for (15) is simply

$$\psi(x,t) = \psi_0(x - \int vdt,\ t) \tag{21}$$

which we may visualize as in (Fig. 1).

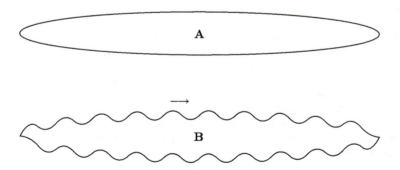

Fig. 1. (A) A cigar-shaped wave packet 5 microns long, unperturbed. (B) The same packet with light of wavelength $\lambda = 0.5$ micron incident from the left, polarized vertically. The undulations are moving to the right at the velocity of light.

To fix orders of magnitude, think of a wave packet a few microns in size, optical wave-lengths of perhaps a half micron. Whenever the optical wavelength is small compared to the size of the wave packet, then the perturbation converts an initially smooth wave packet into something more like a caterpillar than an amoeba, with undulations moving forward at the speed of light (of course, the size of the undulations is greatly exaggerated in Fig. 1; they are the same size as the displacement ($x_0 = eA/mc\omega$) of a classical point electron, and amount only to perhaps 10^{-9} cm in rather intense laser light at optical frequencies).

Then the incident wave $A_{inc}(x,t)$ induces a local current response

$$J(x,t) = \frac{e}{c}|\psi(x,t)|^2\, v(t) = -\frac{e^2}{mc^2}|\psi|^2\, A_{inc}(x,t) \qquad emu \qquad (22)$$

and we note that the original Klein-Nishina derivation of the Compton cross–section, and the Dyson (1951) calculation of vacuum polarization started from this same local current response. To calculate the field scattered by it they used standard classical EM theory (thus providing two early examples of the fact that QED does not actually use field quantization); and we shall do the same.

Resolving the current and field into time fourier components:

$$J(x,t) = \int \frac{d\omega}{2\pi}\, J(x,\omega)\, \exp(-i\omega t) \qquad (23)$$

etc., the scattered field at position x' is

$$A_{sc}(x',\omega) = \int d^3x\, J(x,\omega)\, \left(\frac{e^{i\omega r/c}}{r}\right) \qquad (24)$$

where $r \equiv |x - x'|$. Using (23) and making the far-field approximation this becomes

$$A_{sc}(x',\omega) = -\frac{e^2}{mc^2}\, \frac{e^{ikR}}{R} \int d^3x\, |\psi(x)|^2\, e^{-ik\cdot x}\, A_{inc}(x,\omega) \qquad (25)$$

where R is the distance from the center of gravity x_0 of the wave packet to the field point x', and k is a vector of magnitude ω/c pointing from x_0 to x'. Finally, noting that the coefficient is the classical electron radius $r_0 = e^2/mc^2$ and taking the incident field as a plane wave with propagation vector k_0:

$$A_{inc}(x,\omega) = A_0 e^{ik_0\cdot x} \qquad (26)$$

we have the scattered wave in the direction of k:

$$A_{sc}(x',\omega) = -r_0 A_0\, \frac{e^{ikR}}{R}\, \rho(k_0 - k) \qquad (27)$$

where

$$\rho(k_0 - k) \equiv \int d^3x\, |\psi(x)|^2\, e^{i(k_0-k)\cdot x} \qquad (28)$$

is the space fourier transform of the wave packet density. Note that $\rho(0) = 1$ is the statement that the wave function is normalized. The differential scattering cross–section into the element of solid angle $d\Omega$ is then

$$\frac{d\sigma}{d\Omega} = r_0^2 |\rho(k_0 - k)|^2 \sin^2 \theta \qquad (29)$$

where $\theta = (A_0, k)$ is a colatitude polarization angle. As a check, if the wave packet size $a << \lambda$, then $\rho(k_0 - k) \rightarrow 1$, and we have for the total scattering cross–section

$$\sigma = \oint d\Omega\, r_0^2 \sin^2 \theta = \frac{8\pi}{3} r_0^2 , \qquad (30)$$

the usual classical Thompson cross-section. Put most succinctly, the scattering experiment measures

$$\left[\frac{Cross\ section\ of\ wave\ packet}{Cross\ section\ of\ classical\ electron} \right] = |\rho(k_o - k)|^2 \qquad (31)$$

so if the experiment is feasible and the wave function is something physically real, one should get information about the size and shape of the wave packet from the directional properties of the scattered light. If the true object is only a point particle and the wave function represents only a state of knowledge about its possible positions, then at optical frequencies one should see only the classical cross–section at all scattering angles, whatever the size of the wave packet.

Scattering of laser light by free electrons has been observed experimentally – as long ago as 1963 experimenters were reporting it – but to the best of our knowledge they have not looked for this effect. The first order of business is simply to verify whether these interference effects are or are not observable; but we have not yet specified what size wave packets are to be expected. One experimental clue is provided by electron interference downstream from a fine Wollaston wire; how thick can the wire be while we still see interference? The answer appears to be about 4 microns; thus we would conclude that the wave packet must be about this size.

The result is interpreted differently by those experimentalists who talk in terms of collimation of electron beams instead of wave packets; they represent an electron by a plane wave and view the 4 microns as the coherence range of plane waves with slightly different directions. Of course, a wave packet $\psi(x)$ can be fourier analyzed and the components $\Psi(k)$ would have that meaning; but the spread of k–values in a wave packet has nothing to do with the spread of velocities of the different electrons in the beam; it appears to us that they fail to see this distinction. In our view, the 4 microns must be seen as a property of a single electron, not an indication of lack of parallelism of trajectories of different electrons.

9. Back to Zitterbewegung

We propose that, while ZBW is such a high-frequency effect that it does not play any great role in the scattering of optical frequency light, nevertheless ZBW is the origin of forces that modify the electron wave packet, so that the 'spreading wave packet' solution (11) does not describe the real free electron. Indeed, we have seen from the above derivation that the conventional spreading wave packet solution describes only the NR approximation to a very special relativistic motion in which ZBW effects are absent. But the description of a real free electron must use the relativistic Dirac equation, and include the effects of the interaction of the electron with its own electromagnetic field.

Mathematically, in early QED this interaction diverged, and in no theory can it be considered a negligibly small perturbation. Conceptually, as Einstein warned us long ago (Jaynes, 1989), neither the electron nor the field can exist without the other and their interaction is never turned off, so it is not possible to describe either correctly if we ignore the other. But whenever ZBW oscillations are present this represents a current oscillating at frequency $\omega \sim 2mc^2/\hbar$, whose electromagnetic field reacts back on the electron and modifies its behavior.

A current $J(x',t')$ generates a field $A(x,t) = \int D(x-x',t-t')J(x',t')d^3x'dt'$. This in turn exerts a local force density on the current $J(x,t)$ given by $F = J(x,t) \times [\nabla \times A(x,t)]$. If the currents at both positions have the same ZBW frequency, then there is a time–average secular force that depends on their relative phases and can be either attractive or repulsive.

We emphasize again that these speculations are quite modest and respectable compared to those utterly wild ones noted above, which dominate present theoretical physics. The effect we are proposing is not strange or new; it is predicted by standard relativistic Quantum Theory (which does not forbid the use of wave packets in our calculations); only it was not heretofore overtly mentioned. We are only trying to anticipate, by physical reasoning, what observable effects this secular force might have.

One reason why the effect was not seen clearly before is that the current–current interaction $\int \int J(x)D(x-y)J(y)$ was perceived as only an *energy* term. That it also represents a *force* could, of course, have been found by carrying out a variation δa of the size of the wave packet, but here the physical outlook of the calculators prevented them from doing this. One did not think of a free electron wave packet as a physically real thing that might be distorted by local forces; scattering theories calculated only matrix elements between those Aristotelian plane–wave states. Thus the actual size and shape of a wave packet never got into the calculations.

Why do we include only transverse fields here? We think the answer is that, because of fine features of the Dirac equation not presently in view, these are the only fields actually generated by the Dirac current. In any event, to include a longitudinal interaction in the present calculations would introduce a Coulomb repulsion between different parts of the wave packet, which would probably be much stronger than the transverse forces studied here. That would cause the wave packet to explode in a time far shorter than the Gaussian spreading time of (13).

But we knew from the start that we must never include a coulomb interaction of an electron with itself. For example, if in the hydrogen atom we interpret $\rho(x,t) = e|\psi(x,t)|^2$, we must still use the Hamiltonian $H = p^2/2m + V(x)$ including only the coulomb field $V_{prot} = -e^2/r$ of the proton. We must not include a term $V_{int} = e \int (\rho(x')/r)d^3x'$, or our hydrogen atom would be completely disrupted into something qualitatively different from the atom that we know experimentally. In quantum theory, longitudinal and transverse fields have quite different properties (and, perhaps, different physical origins). This issue requires more study, both in our theory and in conventional QED (where we have also managed only to get around it in a pragmatic sense, not actually resolve it theoretically).

10. Forces due to Zitterbewegung

Our calculation is not different in principle from those that Asim Barut and Tom Grandy do, trying to find solutions of the coupled Maxwell and Dirac equations. However, we are looking at a different phenomenon for which specific solutions may be much harder to find, so in the present work we concentrate on getting a clear picture of the physical mechanisms at work; only after this is well understood would we be ready to tackle the explicit solutions that we shall demand eventually.

Our current comes from the Dirac equation in the standard way: $j^\mu = e\,\bar\psi\gamma^\mu\psi$. But if the wave function is an admixture of what are usually called positive and negative energy solutions [which we think should be called only positive and negative frequency solutions], this has high frequency oscillations at the ZBW frequency $\omega = 2mc^2/\hbar$. For a monochromatic component of the current, the time component of j^μ is determined by charge conservation from the ordinary vector components $J(x,t) = \{j^1, j^2, j^3\}$, which therefore determine the entire radiated field. In the present case, we consider the oscillating current solenoidal: $\nabla \cdot J = 0$, so only transverse fields are generated.

Consider now two small regions of space d^3x_1, d^3x_2 both inside the wave packet but separated by a distance $r \equiv |x_2 - x_1|$ large compared to the ZBW wavelength $\hbar/2mc \simeq 10^{-10}$ cm. Denote the current elements in these by

$$J(x_1, t)\, d^3x_1 = I_1 \cos(\omega t + \phi_1)$$
$$J(x_2, t)\, d^3x_2 = I_2 \cos(\omega t + \phi_2) \tag{32}$$

The current element I_1 generates at the position x_2 an EM field

$$A_1(x_2, t) = f(r, t, \phi_1)\, I_1 \tag{33}$$

where the 'propagator' is

$$f(r, t, \phi_1) \equiv \frac{\cos(\omega t - kr + \phi_1)}{r}. \tag{34}$$

and, as usual, $k \equiv \omega/c$. This exerts on d^3x_2 an element of force

$$dF_2(t) = J(x_2, t) \times [\nabla \times A_1(x_2, t)]\, d^3x_2 = \cos(\omega t + \phi_2)\, I_2 \times [\nabla f \times I_1] \tag{35}$$

But

$$\nabla f = \frac{\partial f}{\partial r}\nabla r = k\,\frac{\sin(\omega t - kr + \phi_1)}{r}\, n \tag{36}$$

where $n \equiv \nabla r$ is a unit vector pointing from $x_1 \to x_2$, and we used the aforementioned long distance condition $kr >> 1$ to discard a near field term that is appreciable only at points within a ZBW wavelength of the current.

Now the force density at x_2 takes the form

$$F(x_2, t) = \frac{k\,\sin(\omega t - kr + \phi_1)\,\cos(\omega t + \phi_2)}{r}\, I_2 \times (n \times I_1) \tag{37}$$

This has terms oscillating at frequency 2ω and constant ones: to get the time average over a ZBW cycle, note that

$$\overline{\sin(\omega t - kr + \phi_1)\,\cos(\omega t + \phi_2)} = (1/2)\,\sin(\phi_1 - \phi_2 - kr) \tag{38}$$

so the time average force density seen at x_2 due to the current element at x_1 is

$$\overline{F} = \frac{k \, \sin(\phi_1 - \phi_2 - kr)}{2r} \, [n(I_1 \cdot I_2) - I_1(n \cdot I_2)] \tag{39}$$

Its component along n depends only on the product of transverse components of the currents:

$$n \cdot \overline{F} = \frac{k \, \sin(\phi_1 - \phi_2 - kr)}{2r} \, [(I_1 \cdot I_2) - (n \cdot I_1)(n \cdot I_2)] \tag{40}$$

which can be either positive (repulsive) or negative (attractive) depending on the relative phases. Thus ZBW currents generate secular force terms which must modify the wave packet of a free electron, and give it a tendency to expand or contract, in addition to the conventional spreading in (11).

There is also a secular term from the ZBW electric field; but this is purely transverse to n and does not contribute to attraction or repulsion. Indeed, since it is unnatural to suppose that two isolated current elements can produce a net torque on themselves, the time average of the electric field forces must just cancel the transverse component of (39): $n \times \overline{F}_{total} = 0$, and (40) gives the entire force. Note that the phases ϕ can vary with position along r in such a way that there would be virtually no external radiation in the n direction, while the total phase term $(\phi_1 - \phi_2 - kr)$ remains mostly at values which give attractive internal forces; this hints at a more general stability property.

But this is of no interest unless the ZBW forces are large enough to compete with the spreading tendency exhibited in (11). Now it is evident already that these forces are orders of magnitude stronger than the corresponding ones due to the same currents at optical frequencies, because from (36) the magnetic field in the radiation zone generated by a given current is proportional to the frequency. Let us estimate their general magnitude, very crudely.

Supposing a wave packet with dimensions a, a normalized wave function will have a magnitude indicated by $|\psi|^2 a^3 \simeq 1$. Therefore the current densities $J = e\bar{\psi}\gamma^\mu\psi$ might conceivably be as large as about e/a^3; let us suppose they are a tenth of that, about $e/10a^3$. Then consider the wave packet broken up into two volume elements $d^3x \simeq a^3/2$, separated by an average distance of about $a/2$. The current coefficients I_1, I_2 above are of the order of $(e/10a^3)(a^3/2) = e/20$. If phases are optimal, the attractive force (40) between them could be as large as $F \simeq kI^2/a = ke^2/400a$. The binding energy of these parts to each other would then be of the order of $Fa/2 = (1/800) \, mce^2/\hbar = mc^2/(800 \cdot 137)$, or about 4.5 ev. The parts would also have some internal binding energy of their own.

We see that the ZBW forces are easily strong enough to do the job we require of them; one can depart considerably from the 'optimum' conditions and still have 0.5 ev of binding energy, enough to stabilize the packet. We might put it thus: in a kind of bootstrap operation, the wave packet digs its own potential well and is confined in it (unable to spread), while remaining free to move about and carry the well along with it, like a turtle trapped in its own shell.

There is a close analogy – perhaps more than an analogy – to some well known things in solid–state theory. As in Fig 2, the propagating frequency ranges are the 'conduction bands' of space, while the 'forbidden band' in which solutions oscillatory in time are spatially evanescent, lies between them. But in the solid–state case, any additional potential which perturbs the periodic lattice potential, can make additional solutions possible in the forbidden band; the localized bound states lying just outside the conduction band, caused by donor or acceptor impurity atoms. This is depicted io Fig 2; and having seen it, we find it hard to avoid supposing that free positrons must be represented by short lines just above the lower conduction band. Some physical arguments in support of this view of things are given below.

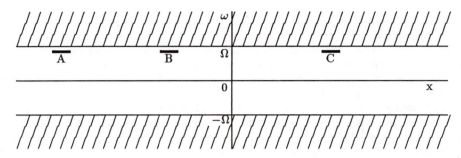

Figure 2. The solid-state analogy. The 'conduction' or propagating regions have frequencies $\omega > \Omega$, $\omega < -\Omega$, while the 'forbidden' region lies between them. Then the physical free electron is thought of as corresponding to localized but mobile states (A, B, C), with wave function oscillating at just below the propagating frequency $\Omega \equiv mc^2/\hbar$.

Given this picture, let us close by indulging in some perhaps wild and free speculation. Evidently, there is a bewildering variety of possible solutions here; presumably, many different initial conditions can be realized, for which the subsequent motions do almost every conceivable thing. But attractive forces always win over repulsive ones because they lower the energy, and then there is no way to escape from them; the energy needed to expand the packet against attractive forces has been radiated away. So after some time the solutions must, inexorably, settle down to some final stable nonradiating limit cycle.

This, we suggest, means that the original arguments against Schrödinger's interpretation of the wave function no longer hold; the real free electron wave packet does not spread indefinitely, but settles down into some steady state of definite size (perhaps about 4 microns) in which the attractive ZBW forces just balance the spreading tendency in (11). This would be the experimental 'aged' free electron, loosely analogous to the 'dressed' particle of current theory, but quite different in essential properties.

The final stable solution will be easier to find than any particular transient one because we have some guiding principles for this. Firstly, the balancing of attractive and spreading tendencies calls for an analysis rather like Einstein's original argument relating mobility and diffusion constant. Secondly, mechanical stability against arbitrary small deformations gives us a variational principle. Thirdly, the condition of no net external radiation imposes conditions on the possible stable current distributions. Finally, all this must satisfy the Dirac equation, in such a way as to just maintain this steady current oscillation.

In fact, a very general electro–mechanical theorem tells us that the second and third principles are closely related, and gives us the additional information that the actual oscillation frequencies $\pm\Omega$ in the final steady packet will be slightly less than $mc^2/\hbar$, the difference indicating the 'binding energy' of the wave packet relative to a plane wave as depicted in Fig. 2.

The condition for mechanical stability is the same as the condition for no net external radiation. To state this, let a current distribution $j(x, t)$ be confined to a finite spatial volume, and take the fourier transform

$$J(k, \omega) \equiv \int d^3x dt \, j(x, t) \exp[(i(\omega t - k \cdot x)]$$

Then from standard EM theory, the necessary and sufficient condition for no external radiation is $k \times J(k, \omega) = 0$ whenever $c|k| = \omega$. When this condition is not met, the radiation can exert relatively strong forces on external objects, inducing changes in their states; thus stability is a joint property of a system and its surroundings.

Why do we not see fragments of electrons (leptoquarks?) flying about, produced by collisions, slits, etc? Perhaps we do, but are not mentally prepared to recognize them. But in those experiments there are always very many electrons involved, and in those fragments the nonradiation condition is far from satisfied; they interact resonantly with neighboring fragments to bring them back together into the stable 'aged' electrons before they arrive at our detectors (and indeed, very few electron detectors actually measure the charge of the things they are detecting). Again, even though the initial phases are random, attractive forces always win out because their effects reinforce, drawing the pieces together and thus increasing the attractive forces. Note also that the ZBW forces are long range compared to Coulomb forces, falling off only as $1/r$; thus fairly distant fragments can interact strongly.

On any slow perturbation of an oscillating system, there is a principle of stationary action, which makes changes of energy proportional to changes of frequency. This was known to Lord Rayleigh in acoustical systems, and was used by Wien in the theory of black-body radiation, and by Einstein and Born as the 'adiabatic principle' in the early days of quantum theory.

In the writer's Neoclassical theory of electrodynamics (Jaynes, 1973) we showed a classical Hamiltonian for which this result is not merely an approximation for slow perturbations, but is an exact conservation law; and our present equations of motion can be derived from a Hamiltonian of the same form. In effect, it is a 'new' constant of the motion never contemplated in classical statistical mechanics because it is not a conservation of energy or momentum. It is a law of conservation of *action* E/ω and much of the phenomenology of quantum theory (which Bohr saw as revealing 'the inadequacy of classical concepts') is explained by it, very easily.

11. Conclusions

The full story that we have started here is much too long to tell in a single article, but let us take a glimpse at what lies ahead, if these speculations prove to have some truth in them. Of course, it is too much to expect that every detail of our present thinking will be correct; but that is unnecessary. Indeed, even if all our speculations prove to be wrong, these ideas may stimulate new constructive thought in the right direction, which would not have happened without them.

We have developed a physical picture of ZBW as performing two essential functions in the world. Its transient solutions provide the "dither" that initiates changes of state;

while its limit cycles provide the stabilizing forces that hold particles together. As early as 1904 Poincaré perceived the need for these forces, but to the best of our knowledge every previous attempt to find them has sought them in static rather than oscillatory models. The great 'advantage' of high frequency oscillations is that a given current generates much stronger EM fields, so that quite large energies and forces result.

Once one has seen this much, a great variety of new effects can be seen, which begin to suggest simple causal explanations for many of those mysterious "quantum effects" previously thought to defy explanation in classical terms. But it is evident that a very large amount of hard work remains to be done before much of this picture can be realized.

12. References

Bell, J. (1987), *Speakable and Unspeakable in Quantum Mechanics*, Cambridge University Press, U.K.

Bell, J. (1990), "Against 'Measurement'", Physics World, Aug. 1990, pp. 33–40.

Bjorken, J. D. & Drell, S. D. (1964), *Relativistic Quantum Mechanics*, Vol. 1, McGraw–Hill Book Co., N. Y.

Bohm, D. (1951), *Quantum Theory*, Prentice–Hall, Inc., Englewood Cliffs, N. J.

Cartan, E. (1966), *Theory of Spinors*, Hermann, Paris.

Dyson, F. J. (1951), Notes on QED, Cornell University

Heisenberg, W. (1958), Daedalus, **87**, 100.

Hestenes, D. (1966), *Space–Time Algebra*, Gordon & Breach, N. Y.

Jaynes, E. T. (1957), "Information Theory and Statistical Mechanics II", Phys. Rev. **118**, pp. 171–190.

Jaynes, E. T. (1973) "Survey of the Present Status of Neoclassical Radiation Theory", in *Coherence and Quantum Optics*, L. Mandel and E. Wolf, Editors, Plenum Publishing Company, N. Y., pp. 35–81.

Jaynes, E. T. (1978) "Ancient History of Free–Electron Devices", in *Novel Sources of Coherent Radiation*, S. F. Jacobs, Murray Sargent II, & M. O. Scully, Editors, Addison–Wesley Publishing Company, Reading MA; pp. 1–39.

Jaynes, E. T. (1986) "Predictive Statistical Mechanics", in *Frontiers of Nonequilibrium Statistical Physics*, G. Moore & M. Scully, Editors, Plenum Press, N. Y., pp. 33–55.

Jaynes, E. T. (1989) "Clearing up Mysteries: the Original Goal", in *Maximum Entropy and Bayesian Methods*, J. Skilling, Editor, Kluwer Academic Publishers, Dordrecht–Holland, pp. 1–27.

Jaynes, E. T. (1990) "Probability in Quantum Theory", in *Complexity, Entropy, and the Physics of Information*, W. H. Zurek, Ed., Addison–Wesley, Redwood City CA, pp. 38–403.

Schwartz, H. & Hora, H. (1969); App. Phys. Lett. **15**, 349. For a summary and extensive bibliography of later developments, see H. Hora, Il Nuovo Cimento **26**, 295 (1975).

Taylor, H. O. (1913), *Ancient Ideals*, MacMillan, London; p. 288.

ZITTERBEWEGUNG IN RADIATIVE PROCESSES

David Hestenes
Physics Department
Arizona State University
Tempe, Arizona 85287
USA

ABSTRACT. The zitterbewegung is a local circulatory motion of the electron presumed to be the basis of the electron spin and magnetic moment. A reformulation of the Dirac theory shows that this interpretation can be sustained rigorously, with the complex phase factor in the wave function describing the local frequency and phase of the circulatory motion directly. This reveals the zitterbewegung as a mechanism for storing energy in a single electron, with many implications for radiative processes.

1. INTRODUCTION

Schrödinger was never satisfied with quantum mechanics. He was especially disturbed by the absence of a clear physical mechanism for radiative processes. This issue is of renewed interest today, for powerful new experimental techniques make it possible to investigate radiative processes with unprecedented resolution and precision. Indeed, the discovery of multiphoton ionization and related phenomena has already upset the conventional wisdom about the photoelectric effect [1]. It seems though, that the initial confusion has been cleared up to the satisfaction of most theorists in the field, and extensive theoretical work has produced explanations for most of the new phenomena brought to light in high intensity laser-atom interactions. All this has been accomplished without any fundamental changes in theory, so some regard it as another triumph of quantum electrodynamics. However there are good reasons to be doubtful.

Most explanations in quantum optics are *phenomenological* in the sense that each is based on some *ad hoc* hamiltonian tailored to the problem at hand. A truly *fundamental* explanation, of course, must be derived from the Dirac equation. To be sure, that is not always possible or practical. But it is essential if anything truly new about radiating electrons is to be learned. Phenomenological models for laser-electron interactions are incapable of distinguishing collective activity from the radiative behavior of single electrons. Consequently, I believe, opportunities for discovering fundamentally new knowledge about radiative processes have been missed.

In this note I will argue that a generally overlooked feature of the Dirac theory, the *zitterbewegung* (ZBW), is the key to understanding radiative processes, and genuinely new physics is to be expected from studying its implications theoretically and experimentally. The argument has three main steps. The first step is a purely mathematical reformulation of

21

D. Hestenes and A. Weingartshofer (eds.), The Electron, 21–36.

the Dirac theory, so it should be uncontroversial. Nevertheless, the results are so surprising and unfamiliar that most physicists are taken aback. Briefly, the reformulation eliminates superfluous degrees of freedom and reveals a hidden geometric structure in the Dirac theory. The imaginary factor $i\hbar$ in Dirac's equation automatically becomes identified with the electron spin, and the electron wave function has a geometrical interpretation wherein the spin is the angular momentum of a local circulatory motion, that is, the ZBW. This raises serious questions about the interpretation of quantum mechanics which are discussed in the second step of the argument. The final step is concerned with a qualitative discussion of the ZBW in radiative processes and the prospects for new physics.

The key to recognizing the geometric structure of the Dirac theory is reformulating it in terms of *spacetime algebra,* a Clifford algebra providing the optimal encoding of spacetime geometry in algebraic form. Only a "bare bones" account of the spacetime algebra and the ZBW structure of the Dirac theory can be given here. Much more is provided in [2] and the many references therein. Hopefully, the present account can serve as an intelligible introduction to the other articles in these proceedings which employ the spacetime algebra. However, there is no getting around the fact that genuine insight into any mathematical system requires a good deal of time and effort.

2. SPACETIME ALGEBRA.

The spacetime algebra (STA) is generated from spacetime vectors by introducing a suitable rule for multiplying vectors. We begin with the usual Minkowski model of spacetime as a 4-dimensional vector space $\mathcal{M}^4$. In mathematical parlance, STA is the real Clifford algebra of the Minkowski metric on $\mathcal{M}^4$. More specifically, STA is a real associative algebra generated from $\mathcal{M}^4$ by defining an associative product on $\mathcal{M}^4$ with the special property that the square of every vector is scalar-valued. I call this product the *geometric product* to emphasize the fact that it has a definite geometric interpretation which fully characterizes the geometrical properties of spacetime. The geometric product uv of vectors u and v can be interpreted by decomposing it into symmetric and skewsymmetric parts; thus,

$$uv = u \cdot v + u \wedge v, \tag{1}$$

where two new products have been introduced and defined by

$$u \cdot v = \tfrac{1}{2}(uv + vu) = v \cdot u, \tag{2}$$

$$u \wedge v = \tfrac{1}{2}(uv - vu) = -v \wedge u. \tag{3}$$

It follows from the definition of the geometric product that $u \cdot v$ is scalar-valued; indeed, it is the usual inner product defined on Minkowski space. The quantity $u \wedge v$ is called a *bivector,* and it represents a directed plane segment in the same way that a vector represents a directed line segment.

In these proceedings, Ed Jaynes [3] confesses to a long-standing "hang-up" over Eq. (1) which has prevented him from getting into STA. As I have the greatest respect for Ed's intellect and other physicists may suffer the same hang-up, I shall attempt a cure forthwith. But first, some further discussion will be helpful in preparation.

Note that for orthogonal vectors (as defined by $u \cdot v = 0$), Eq. (1) gives $uv = u \wedge v = -vu$. Thus, the geometric relation of orthogonality is expressed algebraically by an anticommutative geometric product. Similarly, collinearity is expressed by a commutative geometric product. For in that case $u \wedge v = 0$, Eqs. (1) and (2) gives $uv = u \cdot v = vu$. In general, (1) shows that the geometric product represents the relative direction of any two vectors by a combination of commutative and anticommutative parts.

To facilitate comparison with the Dirac matrix algebra, it is convenient to characterize the structure of STA in terms of a basis. Let $\{\gamma_\mu; \mu = 0, 1, 2, 3\}$ be a righthanded orthonormal basis for $\mathcal{M}^4$ with timelike vector γ_0 in the forward (future) lightcone. In terms of this basis the spacetime metric is expressed by the equations

$$\gamma_0{}^2 = 1 = -\gamma_k{}^2 \quad \text{for } k = 1, 2, 3, \tag{4}$$

and

$$\gamma_\mu \cdot \gamma_v = 0 \quad \text{for } \mu \neq v. \tag{5}$$

Other basis elements of STA, each with a definite geometric interpretation, can be generated from the γ_μ by multiplication. For example, $\gamma_2 \gamma_1 = \gamma_2 \wedge \gamma_1$ is a bivector of unit magnitude, as expressed by

$$(\gamma_2 \gamma_1)^2 = -1. \tag{6}$$

Returning now to Ed's hang-up, he believes that the validity of Eq. (1) requires some new concept of addition. On the contrary, the concept of addition in Eq. (1) is identical to the one physicists are familiar with in complex numbers. Indeed, Eq. (1) can be read as a separation of a complex number $z = uv$ into real and imaginary parts. To make that obvious, suppose that u and v are spacelike unit vectors subtending an angle θ, so that $u \cdot v = -\cos\theta$, where the minus sign is due to the negative signature. The area of the parallelogram determined by u and v is given by $\sin\theta$, so we can write $u \wedge v = i \sin\theta$, where i is the unit bivector for the spacelike plane containing u and v. If γ_1 and γ_2 compose an orthonormal basis for that plane, then $i = \gamma_2 \gamma_1$ and $i^2 = -1$. Thus, Eq. (1) assumes the familiar form

$$z = uv = -\cos\theta + i\sin\theta = -e^{-i\theta}. \tag{7}$$

Of course, this gives a much richer concept of complex numbers than the ordinary one. The i has a twofold geometrical meaning: It is the generator of rotations in the plane, as can be seen by solving (7) for v; thus,

$$v = -uz = ue^{-i\theta}. \tag{8}$$

It is also the unit directed area element for the plane. Of great importance to us later on will be the fact that all Lorentz rotations are generated by the bivectors of STA. Although STA enriches the concept of complex number, it employs the same old concept of addition.

Formally, addition is defined by the associative and commutative rules. When adding complex numbers, these rules enable us to collect and concatenate real and imaginary parts separately. The same is true when adding combinations of scalars and vectors in STA. I suspect that the underlying cause of Ed's hang-up is the worry that scalars and vectors will get inextricably mixed-up under addition. But addition doesn't mix-up real and imaginary parts of complex numbers. Why? Because they are *linearly independent*! That, I believe, is the concept that Ed overlooked in this context. Scalars and vectors don't get mixed-up under addition because they are linearly independent. Indeed, the addition of scalars to vectors is equivalent to augmenting the vectors with an additional component. But if scalars and vectors cannot be concatenated, why add them at all? The answer is the same as for complex numbers: Because multiplication intermixes them, creating valuable new entities such as the spin representations of the rotation group.

There is a certain historical irony that a discussion like this should be necessary in this day and age. The matter was already cleaned up more than a century ago. More than 150 years ago, William Rowan Hamilton worried that a complex number written as the sum of a real and imaginary parts can have no meaning, because unlike things cannot be added. The concept of linear independence had not been invented yet, but it was implicit in his resolution of the problem: He showed that complex algebra is equivalent to a system of operations relating pairs of real numbers. That insight helped gain general acceptance for complex numbers. A decade later, when Hamilton invented the quaternions, the adding of unlike things didn't bother him any more. Our terms *scalar* and *vector* were coined by Hamilton to denote the two unlike parts of a quaternion (though Hamilton's vectors actually correspond to bivectors in STA). Thus, as originally conceived, scalars and vectors were added. Most physicists became familiar with quaternions from Maxwell's great *Treatise on Electricity and Magnetism* (1873). This included J. Williard Gibbs, who developed the standard vector calculus of today primarily by dismantling quaternions into separate scalar and vector parts. A few generations later the physics community had forgotten about quaternions, and young physicists like Ed were inculcated with the dogmatic proscription against adding scalars and vectors. Trained incapacity! Fortunately, the true relation of vector algebra to quaternions (which Gibbs and everyone else at the time had failed to see) is perfectly clear within the broader perspective of STA. Indeed, as demonstrated in detail elsewhere, both these algebraic systems are fully encompassed and integrated by STA.

Now let us return to discussing the structure of STA. The unit *pseudoscalar* for spacetime is so important that the special symbol i will be reserved to represent it. Its generation by the vector basis is expressed by

$$i = \gamma_0 \gamma_1 \gamma_2 \gamma_3. \tag{9}$$

Geometrically, it represents the unit oriented 4-volume element for spacetime. Its algebraic properties

$$i^2 = -1, \tag{10}$$

$$\gamma_\mu i = -\gamma_\mu i \tag{11}$$

make it easy to manipulate. Multiplication of (9) by γ_0 yields the pseudovector

$$\gamma_1\gamma_2\gamma_3 = \gamma_0 i. \tag{12}$$

Geometrically, this is the (directed) unit 3-volume element for a hyperplane with normal γ_0.

By forming all distinct products of the γ_μ we obtain a complete basis for STA consisting of the $2^4 = 16$ linearly independent elements

$$1, \quad \gamma_\mu, \quad \gamma_\mu \wedge \gamma_\nu, \quad \gamma_\mu i, \quad i. \tag{13}$$

It follows that a generic element M in STA, called a *multivector,* can be written in the *expanded form*

$$M = \alpha + a + B + bi + \beta i, \tag{14}$$

where α and β are scalars, a and b are vectors and (with summation over repeated indices)

$$B = \tfrac{1}{2} B^{\mu\nu} \gamma_\mu \wedge \gamma_\nu \tag{15}$$

is a bivector with scalar components $B^{\mu\nu}$. Ed should note that (13) implies that the STA is a 16-dimensional linear space, so (14) is equivalent, with respect to addition, to a vector with 16 components. But multiplication is a different story.

The multivector M in (14) can be decomposed into an even part M_+ and an odd part M_-, as expressed by

$$M = M_+ + M_-, \tag{16a}$$

$$M_+ = \alpha + B + \beta i, \tag{16b}$$

$$M_- = a + bi. \tag{16c}$$

A multivector is said to be even (odd) if its odd (even) part vanishes.

For M in the expanded form (14), the operation of reversion in STA is defined by

$$\tilde{M} = \alpha + a - B - bi + \beta i. \tag{17}$$

It follows that for any multivectors M and N,

$$(MN)^{\tilde{}} = \tilde{N}\tilde{M}. \tag{18}$$

Essentially, reversion amounts to reversing the order of geometric products.

The relation of STA to the Dirac algebra is now easy to state. The Dirac matrices, commonly denoted by the symbols γ_μ, can be put into one-to-one correspondence with the basis vectors denoted by the same symbols above. Then the algebra generated by the Dirac matrices *over the reals* is isomorphic to STA. It follows that the geometric meaning attributed to the vectors γ_μ and their products above is inherent in the Dirac algebra, though it is scarcely recognized in the literature. This isomorphism completely defines the geometric content of the Dirac algebra with respect to spacetime. It suggests also that the representation of the γ_μ by matrices is irrelevant to their function in physical theory. This suggestion is confirmed in the next Section by casting the Dirac theory in terms of STA with no reference at all to matrices.

The full Dirac algebra is generated by the γ_μ over a complex instead of a real number field. The fact that the real field suffices to express the full geometric content of the algebra suggests that the 16 additional degrees of freedom introduced by employing a complex field instead are physically irrelevant. This suggestion is also confirmed in the next Section by formulating the Dirac theory without them. Elimination of the irrelevant $\sqrt{-1}$ in the complex number field opens up the possibility of discovering a geometric meaning for the $\sqrt{-1}$ which occurs so prominently in the equations of quantum mechanics. Indeed, equations (4), (6) and (8) show that STA contains *many different roots of minus one*, including three geometrically different types. Each type plays a different role in the Dirac theory.

3. GEOMETRY OF THE DIRAC THEORY.

In the language of STA, the *Dirac equation* can be written in the form

$$\Box \, \psi i\hbar - \frac{e}{c} A\psi = mc\psi\gamma_0, \tag{19}$$

where

$$\Box = \gamma^\mu \partial_\mu, \tag{20}$$

$A = A_\mu \gamma^\mu$ is the usual electromagnetic vector potential, and $\mathbf{i}$ is the unit bivector

$$\mathbf{i} = \gamma_2 \gamma_1 = i\gamma_3 \gamma_0. \tag{21}$$

The Dirac wave function $\psi = \psi(x)$ at each spacetime point $x = x^\mu \gamma_\mu$ is an even multivector with the invariant canonical form

$$\psi = (\rho e^{i\beta})^{\frac{1}{2}} R, \tag{22}$$

where i is the unit pseudoscalar, ρ and β are scalars and R satisfies

$$R\tilde{R} = \tilde{R}R = 1. \tag{23}$$

A brief proof that the above STA representation of the Dirac equation and wave function is mathematically equivalent to the conventional matrix representation is given in the appendix to Gull's article [4].

Equation (19) is Lorentz invariant, despite the explicit appearance of the constants γ_0 and $\mathbf{i} = \gamma_2\gamma_1$ in it. These constants are arbitrarily specified by writing (19). They need not be identified with the vectors of a particular coordinate system, though it is often convenient to do so. The only requirement is that γ_0 be a fixed timelike unit vector, while $\mathbf{i}$ is a spacelike unit bivector which commutes with γ_0. Of course, the γ_0 and $\mathbf{i} = \gamma_2\gamma_1$ in

(19) are the same constants that appear in the expressions (25) and (27) below for the Dirac current and the spin.

The most striking thing about (19) is that the role of the unit imaginary in the matrix version of the Dirac equation has been taken over by the unit bivector **i**, and this reveals that it has a geometric meaning. Indeed, equations (27) and (28) below show that $i\hbar$ is to be identified with the spin.

Equation (19) may look more complicated than the conventional matrix form of the Dirac equation, but it actually simplifies and enriches the analysis of solutions by making their geometric structure manifest, as is shown in the detailed calculations of Krüger [5]. The key result of the STA formulation is the invariant decomposition (22) of the Dirac wave function. Its geometrical and physical significance is determined by its relation to observables of the Dirac theory, which we specify next.

At each point x, the function $R = R(x)$ in (22) determines a Lorentz rotation (i.e. a proper, orthochronous Lorentz transformation) of a given fixed frame of vectors $\{\gamma_\mu\}$ into a frame $\{e_\mu = e_\mu(x)\}$ given by

$$e_\mu = R\gamma_\mu \tilde{R}. \tag{24}$$

In other words, R determines a unique frame field on spacetime. This frame field has a physical interpretation.

First, the vector field

$$\psi\gamma_0\tilde{\psi} = \rho e_0 = \rho v \tag{25}$$

is the *Dirac current,* which according to the Born interpretation, is to be interpreted as a probability current. Accordingly, at each point x, the timelike vector $v = v(x) = e_0(x)$ is interpreted as the probable (proper) velocity of the electron, and $\rho = \rho(x)$ is the relative probability (i.e. proper probability density) that the electron actually is at x.

Second, the vector field

$$\frac{\hbar}{2}\psi\gamma_3\tilde{\psi} = \rho\frac{\hbar}{2}e_3 = \rho s \tag{26}$$

is the *spin (or polarization) vector* density. The *spin angular momentum* $S = S(x)$ is actually a bivector quantity, related to the spin vector s by

$$S = isv = \frac{\hbar}{2}ie_3e_0 = \frac{\hbar}{2}e_2e_1 = \frac{\hbar}{2}R\gamma_2\gamma_1\tilde{R} \tag{27}$$

Multiplying this on the right by (22) and using (23), one easily obtains

$$S\psi = \tfrac{1}{2}\psi\gamma_2\gamma_1\hbar, \tag{28}$$

which relates the spin S to the bivector $\gamma_2\gamma_1\hbar$.

In general, six parameters are needed to specify an arbitrary Lorentz rotation. Five of the parameters in the Lorentz rotation (24) are needed to specify the direction of the electron velocity v and spin s. This also determines the plane containing e_1 and e_2, as shown in (27). The remaining parameter ϕ determines the orientation of e_1 and e_2 in the e_2e_1, plane. This can be expressed by factoring R into the form

$$R = R_0 e^{i\phi}, \tag{29}$$

where R_0 is determined by the first 5 parameters just mentioned. The parameter ϕ is the *phase* of the wave function, and here we have a geometrical interpretation of the phase. The vectors e_1, and e_2 are not given a physical interpretation in the conventional formulation of the Dirac theory, because the matrix formalism suppresses them completely. But they will be given a kinematical interpretation when the ZBW interpretation is introduced below.

The factorization (22) of the wave function ψ can now be seen as a decomposition into a 6-parameter *kinematical factor* R and a 2-parameter *statistical factor* $(\rho e^{i\beta})^{\frac{1}{2}}$. The parameter ρ is clearly a probability density. The physical interpretation of β raises problems which are yet to be fully resolved. Important insights into this issue are supplied by other articles in these Proceedings. Boudet [6] describes formal properties of β in the geometry of the Dirac theory. Krüger [5] finds new solutions for the hydrogen atom with $\beta = 0$, in sharp contrast to the strange properties of β in the Darwin solution. Gull [4] discusses the *essential* role of β and its relation to spin in matching boundary conditions at a potential step, including the Klein paradox. My own expectation is that a full understanding of β will come only from elaborating the statistical interpretation of the the Dirac theory discussed below. That is why I have relegated β to the statistical factor in the wave function.

The physics (in contrast to the statistics) in the wave function appears to be in the kinematical factor R. Support for this assertion comes from examining the "free particle" solutions of the Dirac equation. There are two distinct types of plane wave solutions with momentum $p = mcv$, an *electron* solution and a *positron* solution. The electron solution has the form

$$\psi = \rho^{\frac{1}{2}} R_0 e^{i\phi}, \tag{30}$$

where ρ and R_0 are constant, but the phase ϕ has the spacetime dependence

$$\hbar\phi = p \cdot x = mcv \cdot x = mc^2 \tau. \tag{31}$$

Here τ is the *proper time* along "*streamlines*" of the Dirac current, which are straight lines with tangent v orthogonal to the 1-parameter family of hyperplanes with constant phase constituting a moving plane wave. According to (25) and (26) the electron velocity v and spin s are constant everywhere. But along a streamline ϕ increases uniformly, so the phase factor in (30) rotates e_1 and e_2 in the plane of the spin S with the circular *zitterbewegung* *frequency*

$$\omega_0 = 2mc^2/\hbar = 1.6 \times 10^{21} \mathrm{s}^{-1}. \tag{32}$$

A similar rotation takes place along the streamlines of every solution of the Dirac equation, though, in general, with a variable frequency. Indeed, the decomposition (29) tells us that generally the phase $\phi = \phi(x)$ at each spacetime point x determines a well-defined rotation, not just in some abstract complex plane, but in a definite physical plane, the plane of the spin S at x.

Jaynes [3] tells us that when he looks at the standard Dirac wave function he doesn't see anything that could rotate. This is a striking illustration of how crucially the interpretation of a theory depends on the form of its mathematical representation. The STA formulation makes the rotation inherent in the wave function absolutely explicit. But, as physicists, we are not satisfied with a "mere" mathematical rotation. With Jaynes, we demand to know "What physically is rotating?" Here again the plane wave solution helps us gain a vital insight.

The kinematical factor in (29) can be written in the form

$$R = e^{\frac{1}{2}\Omega\tau}R_0,$$
(33)

where Ω is the constant bivector

$$\Omega = mc^2 S^{-1} = \frac{2mc^2}{\hbar}e_1 e_2,$$
(34)

with $e_1 e_2 = R\gamma_1\gamma_2\tilde{R} = -R_0 i \tilde{R}_0$. Accordingly, Ω is the angular velocity of the frame $\{e_\mu = e_\mu(x(\tau))\}$ as it moves along a streamline. Both $e_0 = v$ and $e_3 = \hat{s}$ are constants of the motion, but

$$e_1(\tau) = e^{\Omega\tau}e_1(0) = e_1(0)\cos\omega_0\tau + e_2(0)\sin\omega_0\tau,$$

(35)

$$e_2(\tau) = e^{\Omega\tau}e_2(0) = e_2(0)\cos\omega_0\tau - e_1(0)\sin\omega_0\tau,$$

where $\omega_0 = |\Omega|$ is the ZBW frequency. These equations describe the rotation of the frame explicitly.

Now, it has often been suggested on heuristic grounds the electron spin and magnetic moment may be generated by some kind of local circular motion of the electron. This idea cannot be maintained if the electron velocity is identified with the streamline velocity v of the Dirac current, because v is orthogonal to the spin. If the idea is physically correct, the true electron velocity must have a component in the spin plane. Our geometrical representation of the electron plane wave presents us with an obvious choice. We suppose that the velocity of the electron can be identified with the null vector

$$u = e_0 - e_2.$$
(36)

Of course, this means that the electron moves with the speed of light, as in Schrödinger's original ZBW model. This hypothesis defines what I call the *zitterbewegung interpretation of the Dirac theory* [2]. It is more general than Schrödinger's idea of the ZBW, for (36) is obviously applicable to any solution of the Dirac equation. In the plane wave case, however it is easy to integrate.

With the time dependence of e_2 given by (35) and $u = c^{-1}\dot{z}$, Eq. (36) is easily integrated to get the history $z = z(\tau)$ of the electron; thus,

$$z(\tau) = vc\tau + \left(e^{\Omega\tau} - 1\right)r_0 + z_0. \tag{37}$$

This is a parametric equation for a lightlike helix $z(\tau) = x(\tau) + r(\tau)$ centered on the streamline $x(\tau) = vc\tau + z_0 - r_0$ with radius vector

$$r(\tau) = e^{\Omega\tau}r_0 = -\frac{c}{\omega_0}e_1 = -\frac{c}{\omega_0^2}\dot{u}. \tag{38}$$

The radius of the helix is half the electron Compton wavelength

$$\lambda_0 = c/\omega_0 = \hbar/2mc = 1.9 \times 10^{-13}\,\text{m.}. \tag{39}$$

The Dirac current describes the mean velocity over a zbw period:

$$\bar{u} = e_0 = v, \tag{40}$$

so the Compton wavelength is the diameter of ZBW fluctuations about this mean.

From (34) and (38) , which imply $\dot{r} = \Omega r$, we find

$$S = mc^2\Omega^{-1} = mr^2\Omega = m\dot{r}r. \tag{41}$$

Thus, the spin angular momentum can be regarded as the angular momentum of ZBW fluctuations.

With τ expressed as a function of spacetime position by (31), Eq. (37) describes a spacetime-filling congruence of lightlike helixes centered on Dirac streamlines, with exactly one helix through each spacetime point. In accord with the statistical interpretation of the Dirac wave function, each helix is a possible worldline for the electron, and the modulus of the wave function determines the probability that the electron traverses any particular helix. All these conclusions about the geometry of the plane wave solutions apply generally to every solution of the Dirac equation, though, of course, in the presence of external fields the helixes are bent and distorted. Jaynes [3] has described this view of Dirac solutions aptly as a "*tangle of all the different possible trajectories of a point particle,*" and he exclaimed "I would never in 1,000 years have thought of looking at the Dirac equation in that way!" Never without the STA formulation! Mark again how crucial the mathematical representation is to physical interpretation! The tangled geometry of helixes has been inherent in the Dirac theory all the time; only a suitable representation and definition was necessary to reveal it. Mark that the *ZBW interpretation attributes a purely kinematical meaning to the phase factor,* so the entire factor R in (29) and (22) has a purely kinematical interpretation. This gives the interpretation of the Dirac wave function a maximum degree of coherence.

Berry [7] has given the quantum phase factor a general geometrical interpretation. According to the ZBW interpretation, the phase factor is more literally geometrical than anyone had imagined.

IV. WHAT IS AN ELECTRON, REALLY?

Is the electron a particle always, sometimes, or never? Theorists have come down on every side of this question. A definitive answer is essential to any sort of objectivity attributed to quantum mechanics. I am pleased that Ed Jaynes [3] has come down on the side opposite mine, for the comparison of contrasting interpretations helps highlight the critical issues. I am equally pleased that he has placed Willis Lamb on my side.

The contrasting interpretations that Ed and I defend should not be regarded as dogmatic stances, nor should it overshadow the great extent to which we agree. The participants at this conference know that Ed is well established as one of the world's leading practitioners of quantum mechanics, especially in the domain of quantum optics. All the while, though, he has been one of the most astute critics of quantum mechanics. His criticism has always been based, not on sterile philosophical speculation or mathematical formalism, but on cogent physical reasoning born of his intimate knowledge of both classical and quantum electrodynamics and how they relate to real experimental data. The criticism he presents in these proceedings is only part of the extensive critical evaluation he has presented on other occasions. I find myself in whole-hearted agreement with the entire body of his criticism, and I commend it to any serious student of the foundations of quantum electrodynamics. Ed and I agree that there is great truth in standard quantum mechanics, but the problem is to separate the truth from the fiction. We also agree that the interpretation of the electron wave function is a critical issue. We part company on what to do about it. Though I am sure that Ed agrees that the Dirac theory is somehow more fundamental, like most other theorists in quantum optics, he is content to base his analysis on the Klein-Gordon and Schrödinger approximations to it. I regard that as a grave mistake, for the ZBW structure of the Dirac theory could never be discovered in these approximate theories, even though it is inherent in the phase factor of the wave function, and they thereby inherit a ZBW interpretation from the Dirac theory. With this understood, let us return to the particle issue.

Ed Jaynes, like Asim Barut [8], wants to interpret the electron wave function as describing a real physical entity, rather than just a state of knowledge about the electron as I wist. While I believe that that viewpoint faces insuperable difficulties, I applaud Ed's objective to bring the matter to decisive experimental test, and I agree that this is feasible. I note that the main reason for Ed's stance is that he believes, along with most other physicists, that electron diffraction can *only* be explained as due to " interference of the electron with itself," so with admirable consistency, Ed maintains that the electron must be extended in space like the wave function. This issue of how to explain diffraction is one of the great bugaboos of quantum mechanics, so I will address it from the particle perspective below.

Ed seems to have a hang-up about point particles as well as STA. Let me attempt some therapy. The question "Is the electron a particle?" can and must be addressed at different levels, where different physical issues are at stake. Let me call the first level the *interpretation level*. Here the question is "Does the Dirac theory admit to a coherent particle interpretation which is superior to alternative interpretations?" My answer to this question is, of course, yes! Indeed, I maintain that the ZBW interpretation is the only one which comes close to giving a coherent account of all details of the Dirac theory. It is not maintained at this level that the electron *really* is a point particle, but only that the Dirac theory says it is, in the sense that it ascribes to the electron no internal structure and no finite dimensions. The electron spin and magnetic moment are features of electron kinematics rather than internal structure.

In the spirit of Jaynes, it might be suggested that the electron is an extended body and the helixes are world lines of its component parts. This suggestion faces difficulties which seem to rule it out. First, there is an absence of evidence for any interaction among the parts which would be needed to make the body cohere. Second, the dimensions of the body would have to be on the order of a Compton wavelength ($\sim 10^{-13}$m). But this is much too big! Scattering experiments limit the size of the electron (i.e. the size of the domain in which momentum transfer takes place) to less than 10^{-18}m [9]. Only the particle interpretation appears to be consistent with this experimental evidence. Additional evidence for the particle interpretation ([2], [10]) is less direct. For example, the explanation for Van der Waals forces requires that atoms are fluctuating dipoles, which they certainly are if electrons are particles orbiting the nucleus rather than Jaynesian amoebas enveloping the nucleus in static charge clouds. Moreover, the time dilatation in the decay of μ^- particles captured in atomic s-states indicates that they *really are moving with the Bohr velocity* in those states [11]. So must electrons move also.

It seems to me that the Born statistical interpretation is essential for understanding scattering data, and this demands the particle interpretation. How then do we explain the structure in a diffraction pattern? "Interference" is the standard answer! But there is a lesser known alternative which has been propounded vigorously by David Bohm [12] and others for years. This puts Bohm firmly on my side, though Jaynes cites him as a precursor to his amoebic viewpoint. Bohm maintains that the electron is a particle with a definite trajectory and that the wave function determines a family of possible trajectories, just as I do in my ZBW interpretation. On this point, we differ only in details of how the trajectories are determined by the wave function. Bohm uses Schrödinger theory rather than Dirac theory. The trajectories have actually been calculated from the Schrödinger equation for the double slit experiment [13], and the Dirac equation would surely yield essentially the same result. The trajectories flow uniformly through both slits, but thereafter they spread out, bunching up at diffraction maxima and thinning out at minima. When a single electron has been detected on the "diffraction screen," one can (in principle to any desired precision) determine which of the trajectories it *actually* followed and trace the trajectory backwards to determine where the electron passed through one of the slits. In this sense, quantum mechanics allows us to measure definite electron trajectories.

This *description* of electron diffraction is a self-consistent interpretation of the equations of quantum mechanics. It has the great advantage of preserving a consistent particle interpretation, allowing us to maintain that every electron has a continuous (albeit indirectly observable) trajectory. But physicists want more. They want an *explanation* of diffraction, not just a description. They want to identify a causal mechanism underlying diffraction. I don't believe that standard quantum mechanics has achieved that, but I suggest below where the missing mechanism might be found. On the contrary, standard quantum mechanics purports to explain diffraction as a consequence of interference. The possibility of such an explanation is a *mathematical consequence* of the fact that the QM wave equation is linear, so it can be argued in the double slit experiment that the diffraction pattern is caused by interference in the superposition of particular solutions with each slit as source. Accepting this mathematical possibility as physical reality has the strange consequence that the electron must somehow pass through both slits in order to interfere with itself. I maintain that this interpretation buys nothing but trouble, since it is obviously inconsistent with the factually grounded particle interpretation, but it has no greater predictive power. It is as awkward as it is unnecessary. There is actually only one valid solution of the wave equation which matches the boundary conditions in a diffraction experiment, and only that solution is used in the above particle interpretation of the experiment. The subdivision of that solution into interfering particular solutions which separately do not satisfy the boundary conditions can therefore be safely dismissed as a mere mathematical artifice. Accordingly, *the interference explanation of particle diffraction can be dismissed as an artifice introduced in an attempt to manufacture an explanation out of a description.*

Now let us address the particle question at a second, more *fundamental level*. At this level, I agree with Einstein, Rosen and Cooperstock [14] that the electron, as a particle, must not be treated independently of the electromagnetic field but as part of it. The electron in the Dirac theory is an emasculated charged particle, stripped of its own electromagnetic field, like a classical test charge. The central problem of quantum electrodynamics, as recognized by Barut [8] and many others, is to restore the electron's field and deduce the consequences. This is *the self-interaction problem.* Whether, in the ultimate solution to this problem the electron will emerge as a true singularity in the field or some kind of soliton [14] is anybody's guess. One thing is certain, though, the problem is nonlinear. And if quantization is a consequence of this nonlinearity, as I have suggested elsewhere [10], then the self-interaction problem can never be solved with standard quantum mechanics; a more fundamental starting point must be found.

Though the Dirac theory omits the electron's field, it appears to contain vestiges of self-interaction which are valuable clues to a deeper theory. It is widely believed that the electron mass and spin are consequences of self-interaction. But these are properties of the ZBW, so the ZBW itself must derive from self-interaction. Already this suggests [2] that the electron self-field is of magnetic type to produce the spin, and the electron mass comes from a kind of self-inductance of the circular motion.

Interpreted literally, the ZBW motion should be reflected in the electron's electromagnetic field. Specifically, the electron should be the seat of a nonradiating field that oscillates with the ZBW frequency. Call it the *ZBW field.* The usual Coulomb and magnetic dipole fields of the electron are then averages of the ZBW field over a ZBW period. The ZBW frequency is much too high to detect experimentally. However, it has been suggested [10] that many familiar quantum phenomena might be explained as consequences of ZBW resonances. Here are three examples:

(1) *Electron Diffraction.* The ZBW field broadcasts the electron's deBroglie frequency and wavelength to the environment. I submit that in diffraction it is the ZBW field, rather than the electron itself, that feels out the topology of the target and by feedback produces a shift in the phase of the ZBW motion which alters the electron's trajectory. In crystal diffraction, the Bragg angles must then be conditions for resonance between the *broadcasted ZBW wave* and the *feedback wave* scattered off the crystal. They are thus conditions for resonant momentum transfer between the electron and the crystal. An attractive feature of this explanation is that it includes a mechanism for momentum transfer which is missing from conventional explanations of diffraction.

(2) *Atomic States.* An electron bound in an atom is in a ZBW resonant state, wherein the frequency of the orbital motion is a harmonic of the ZBW frequency. The principal quantum number indexes the harmonics. Now, if the above explanation of diffraction is correct, the electron must be broadcasting a ZBW wave which is scattered resonantly off the nucleus and back to the electron. An atomic state is thus a state of

resonant momentum exchange between the electron and itself. This is to say that an electron accelerated by the field of the nucleus is always *radiating continuously, but it is also continuously absorbing its own radiation.* In the ground state, all the radiated energy must be absorbed, since the state is stable. However, in an excited state the *radiation rate* must exceed the *absorption rate* so the state decays. It should therefore be possible to *calculate the lifetime* of the state from the mismatch between these two rates, thus to explain spontaneous emission. All this is just another example of diffraction, with atomic states corresponding to diffraction peaks and "quantum conditions" corresponding to the Bragg law. The main difference between *transient diffraction* by a crystal and *continuous diffraction* within an atom is that the momentum transfer is between two different objects in the first case but between an object and itself in the second.

(3) *Pauli Principle.* Two electrons in the same atomic state will certainly have resonant ZBW frequencies, so momentum exchange via their ZBW fields is to be expected. Thus we have here a natural mechanism for explaining the Pauli principle, Evidently, then, if the electrons have antiparallel spins the ZBW interaction produces a stable two electron state, while if the spins are parallel the state is unstable and so never seen.

Though suggested by the Dirac theory, all this goes well beyond it. It is conceivable that, besides spontaneous emission, the ZBW is responsible for other phenomena, such as the Lamb shift and the anomalous magnetic moment, which are attributed to quantization of the electromagnetic field. Clearly the ZBW idea is pregnant with possibilities for new physics.

V. RADIATIVE PROCESSES.

Ed Jaynes is quite right to assert that if the electron really is a particle but quantum mechanics describes only the behavior of an ensemble, then it must be possible to extract the particle from the ensemble and study it all by itself. The problem with such an extraction, of course, is to ascertain suitable equations of motion for the particle alone, because they might differ significantly from equations for the particle behavior within the ensemble. Nevertheless, as a first approach to the problem, I propose to extract a single ZBW worldline from the Dirac theory and interpret it literally as the worldline of an individual electron. Even without equations of motion, we can reason qualitatively about the electron's behavior from what we know about solutions of the Dirac equation. Such reasoning can be quite provocative. Even if it cannot be refined by calculations from exact equations of motion, it may prove useful in guiding the solving and interpreting of the Dirac equation.

When an electron is placed in an external field, energy can be absorbed by the ZBW field, producing an increase in ZBW frequency and hence a decrease in the ZBW radius. We know that from solutions of the Dirac equation where binding energies appear in the complex phase factor. Indeed, the *Minimal Coupling Ansatz* can be interpreted as specifying that external fields produce shifts in the ZBW frequency. As Steve Gull puts it, *the electron is a parametric oscillator* with frequency modulated by external fields. A shift in the ZBW frequency ω, is also a shift in electron mass m, because $\hbar\omega = mc^2$ holds generally. The so-called electron *rest mass* is therefore only a lower bound to the electron mass. The electron mass is actually variable and changing all the time in interactions. However, if no external field is present to induce radiation, it may be that the electron can retain a mass greater than its empirical rest mass. In other words, it may be that *energy can be stored in the ZBW of a single free electron.* This possibility can surely be put to experimental test. Indeed, the basic mechanism may have been probed already by recent experiments in quantum optics.

For example, the ZBW mechanism can be deployed to explain *multiphoton ionization* [1]. When a bound atomic electron is irradiated by an intense laser field, the ZBW may absorb a harmonic of the laser frequency, with an attendant increase of electron mass and shrinking of its atomic orbit. Evidently this excited ZBW state is metastable and may persist for some time after the laser field is off. Then the stored energy is liberated either by reradiation or ionization. The phenomenon of *above threshold ionization* [1] shows that the electron (ZBW) may absorb much more than the minimum necessary for ionization. If, indeed, the ZBW is the mechanism for multiphoton and above-threshold ionization, then it must be possible to demonstrate these phenomena in experiments with single atoms. According to the standard explanations, such experiments should not work. It may be added that final state interactions in ionization should be significantly affected by ZBW mass shifts.

This ZBW explanation for the new photoelectric phenomena may appear to be incompatible with conventional explanations. An excellent and accessible explanation grounded in standard quantum electrodynamics is given by André Bandrauk [15]. The idea is that embedding a molecule in a laser alters the effective electronic potential to create a new set of bound states which can be observed with electron probes. This is not necessarily inconsistent with the ZBW explanation, but the putative physical mechanism is quite different. Other experiments will probably be necessary to distinguish between the two possibilities.

Evidence that irradiated single free electrons can absorb harmonics of the laser frequency exists already in the pioneering "stimulated bremsstrahlung" experiments of Tony Weingartshofer [16]. These experiments have been regarded as anomalous in the high intensity laser field, because they cannot be explained by standard arguments. However, I submit that they are just further examples of the ZBW mechanism at work.

To establish unequivocally that energy can be stored in the ZBW of a single free electron, we need cleaner experiments on single electrons. The prediction is that an electron can absorb an n-th order harmonic to put it in a metastable state with mass m given by $mc^2 = m_0 c^2 + n\hbar\omega_\ell$, where m_0 is the rest mass and ω_ℓ is the laser frequency. Then, under suitable conditions, the electron can be released in this excited state to transport the additional energy until the electron is induced to release it by a collision or some other means. This phenomenon may actually have been observed already in the infamous *Schwartz-Hora effect* described briefly by Jaynes [3]. I hold with Jaynes that this effect is probably real and the possibility deserves to be investigated thoroughly. Our explanations for the effect may appear to be quite different, but remember, I attribute the standard QM phase factor to the ZBW, and the phase factor plays the key role in Jaynes' argument. The main difference is that Jaynes sees the effect as due to coherent action of parts of the electron spread out over a wave packet. The issues are clear. The truth will be found out.

Acknowledgement. The idea that energy can be stored in the ZBW of a single free electron was developed jointly with Heinz Krüger in several conversations.

REFERENCES

[1] P. Agostini and G. Petite (1988), Photoelectric effect under strong irradiation, CONTEMP. PHYS. **1**, 57-77.

[2] D. Hestenes (1990), The Zitterbewegung Interpretation of Quantum Mechanics, *Found. Phys.* **20**, 1213-1232.

[3] E. T. Jaynes (1991), Scattering of Light by Free Electrons as a Test of Quantum Theory, (these Proceedings).

[4] S. F. Gull (1991), Charged Particles at Potential Steps, (these Proceedings).

[5] H. Krüger, New Solutions of the Dirac Equation for Central Fields, (these Proceedings).

[6] R. Boudet (1991), The Role of Duality Rotation in the Dirac Theory, (these Proceedings).

[7] A. Shapere & F. Wilczek (1989), **GEOMETRIC PHASES IN PHYSICS**, World Scientific.

[8] A. O. Barut (1991), Brief History and Recent Developments in Electron Theory and Quantum Electrodynamics, (these Proceedings).

[9] D. Bender *et. al.*(1984), Tests of QED at 29 GeV center-of-mass energy, *Phys. Rev.* **D30**, 515 .

[10] D. Hestenes (1985), Quantum Mechanics from Self-Interaction, *Found. Phys.* **15**, 63-87 .

[11] M. Silverman (1982), Relativistic time dilatation of bound muons and Lorentz invariance of charge, *Am. J. Phys.* **50**, 251-254.

[12] D. Bohm & B. Hiley (1985), Unbroken Quantum Realism, from Microscopic to Macroscopic Levels, *Phys. Rev. Letters* **55**, 2511.

[13] J.-P. Vigier, C. Dewdney, P.R. Holland & A. Kypriandis (1987), Causal particle trajectories and the interpretation of quantum mechanics. In **Quantum Implications**, B.J.Hiley & F.D. Peat (eds.), Routledge and Kegan Paul, London.

[14] F. I. Cooperstock (1991), Non-linear Gauge Invariant Field Theories of the Electron and other Elementary Particles, (these Proceedings).

[15] A. D. Bandrauk (1991), The Electron and the Dressed Molecule, (these Proceedings).

[16] A. Weingartshofer, J. K. Holmes, G. Caudle, E. M. Clarke & H. Krüger (1977), Direct Observation of Multiphoton Processes in Laser-Induced Free-Free Transition, Phys. Rev. Let., **39**, 269-270. A. Weingartshofer, J. K. Holmes, J. Sabbagh & S. L. Chin (1983), Electron scattering in intense laser fields, J. Phys B **16**, 1805-1817.

CHARGED PARTICLES AT POTENTIAL STEPS

S. F. Gull
Mullard Radio Astronomy Observatory
Cavendish Laboratory, Madingley Road
Cambridge CB3 0HE, U.K.

Abstract. The behaviour of charged particles at electromagnetic steps is analysed using the powerful mathematical tools provided by the SpaceTime Algebra. Currents predicted in the evanescent region of a Dirac wavefunction strongly suggest that the electron "zitterbewegung" (ZBW) represents a real circulation. At higher potentials, the Klein paradox reveals a crucial difficulty of interpretation of "positronic" wavefunctions that must be overcome before Hestenes' ZBW model can be taken seriously. The problem of radiation reaction is still not solved. Solutions of the Lorentz-Dirac equation for a potential step show crazy teleological features: certain input velocities have no possible future output states. The prospects for realistic electron models are briefly discussed.

1. Introduction

I can safely say to this audience that, as a radio astronomer, I have observed more electrons than anyone else present. A medium-sized radio galaxy displays radio synchrotron emission from about 10^{63} electrons as jets of relativistic electrons and positrons shoot out into space from a black hole deep inside a galactic nucleus. Our interest in these, the most violent objects in the Universe, requires a corresponding understanding of the humble lepton and its radiation mechanisms, particularly at the highest energies.

Like some of the other participants here, I have for a long time been deeply unhappy about the accepted theories of this little particle and I have to admit that I have been unable to heed Feynman's (1967) excellent advice not to ask oneself

> " 'But how can it be like that?' because you will get 'down the drain',
> into a blind alley from which no one has yet escaped.".

Two years ago, however, an unexpected event brightened up my view of this particular drain, when I became aware of the work by David Hestenes on Geometric Algebra. He claimed (Hestenes 1986) that our mathematical language is seriously incomplete, because physicists do not know how to multiply vectors together, and that they are thereby missing the geometrical content of the equations of physics, particularly the Dirac equation. I rapidly became convinced that Geometric Algebra, which includes the algebra of our spacetime, the SpaceTime Algebra (STA), as a special case, is an essential ingedient in correcting our misconceptions about the nature of quantum mechanics.

As a beginner in a strange new field, I have modest aims in this paper and focus attention on the behaviour of a charged particle encountering an electromagnetic potential step, all

37

D. Hestenes and A. Weingartshofer (eds.), The Electron, 37–48.
© 1991 *Kluwer Academic Publishers. Printed in the Netherlands.*

the time using the powerful mathematical tools provided by the STA. This situation forms a convenient "theoretical laboratory" for the electron, allowing us to examine critically the predictions of presently-available models. After a brief review of the STA and the Dirac equation, I consider the riddle of the charge current in the Dirac theory and the extent to which the behaviour of an electron wave at an electromagnetic potential step throws light upon the new "zitterbewegung" (ZBW) interpretation (Hestenes 1990). Klein's paradox is briefly mentioned, but *not* resolved. I believe that the difficulties of the Dirac current are clearly exposed by this paradox and that our present ways of overcoming it (essentially due to Feynman) are unsatisfactory. "The cure is worse than the disease", as Ed Jaynes is fond of saying.

In the second part of this paper we return to the potential step to examine the mystery of the radiation reaction. This is still an important problem, because it is extremely desirable to have a self-consistent equation of motion for an accelerated charge that takes into account its own radiation. The case of a finite-sized charge distribution is tractable if sufficient care is used in any approximations, but the difficulties for a point charge seem insuperable. The most famous attempt, the Lorentz-Dirac equation (Dirac 1938), fails miserably when confronted with the problem of the potential step, showing unacceptable overall scattering states that prohibit certain ranges of input condition. The STA suggests an alternative equation.

Inspired by the ZBW interpretation of quantum mechanics, we have a brief look at realistic electron models (Barut & Zhanghi 1984). Translated into the STA, these models begin to look very interesting, but certain defects become obvious.

2. The SpaceTime Algebra and the Dirac equation

This paper embraces the ideas and notations developed by Hestenes over the last 30 years (Hestenes 1966, 1986). The SpaceTime Algebra (STA) is a real, Geometric (Clifford) algebra developed on a 4-dimensional flat spacetime with a standard Minkowski metric. The basic ingredients of this algebra are an orthonormal frame of vectors $\{\gamma_0, \gamma_1, \gamma_2, \gamma_3\}$, where $\gamma_0^2 = -\gamma_k^2 = 1$. The time-like vector γ_0 defines a Lorentz frame, that we can think of as representing the *laboratory frame*. The $\{\gamma_\mu\}$ satisfy the same algebraic relations as the Dirac γ-matrices, but it must be stressed that they here represent 4 unit vectors in spacetime and *not* the 4 components of a single vector. We shall have no use for a matrix representation of the $\{\gamma_\mu\}$ here, but a translation table is given in the Appendix.

From this basic set of vectors we build up the 16 ($= 2^4$) geometric elements of the STA:

$$\begin{array}{ccccc} 1 & \{\gamma_\mu\} & \{\sigma_k, i\sigma_k\} & \{i\gamma_\mu\} & i \\ \text{1 scalar} & \text{4 vectors} & \text{6 bivectors} & \text{4 pseudovectors} & \text{1 pseudoscalar} \end{array} \cdot$$

The time-like bivectors $\sigma_k \equiv \gamma_k\gamma_0$ obey the same algebraic relations as the Pauli spin-matrices, but in the STA they represent an orthonormal frame of vectors in space *relative* to the laboratory time vector γ_0. The unit pseudoscalar of spacetime is defined as

$$i \equiv \gamma_0\gamma_1\gamma_2\gamma_3 = \sigma_1\sigma_2\sigma_3.$$

The STA is the 16-dimensional *real* linear space formed from these geometric objects. We will not need to consider the coefficients to be *complex* scalars, because the STA already contains 10 geometrically distinct square roots of -1, which is plenty enough for our purposes.

Indeed, the STA representation of the complex imaginary is different from one application to another, so that the use of complex numbers as *scalars* may be unnecessary in physics.

A particularly important subalgebra of STA is the set of *spinors*, which, for spacetime, is simply the even subalgebra, comprising the scalar, bivectors and pseudoscalar. We can write a general spinor in the form

$$\psi = a_0 + \mathbf{a} + i(b_0 + \mathbf{b}),$$

where $\mathbf{a} \equiv a_k \sigma_k$ and $\mathbf{b} \equiv b_k \sigma_k$ are *relative vectors*. A very important point of interpretation is the fact that spinors of the STA are geometrical objects in their own right, rather than (as in matrix versions) residing in a complex "spin-space". The notion of "spin-space" is not needed here.

The spinors are important because they define the transformation properties of frames in spacetime. A spinor ψ defines a Lorentz rotation through the transformation

$$\rho e_\mu = \psi \gamma_\mu \tilde{\psi},$$

where the $\{e_\mu\}$ are a new frame of orthogonal vectors and $\tilde{\psi}$ is the *reverse* of ψ, formed by reversing the order of all geometric products. To see this explictly we write a Dirac spinor ψ in canonical form:

$$
\begin{array}{cccc}
\psi & = & (\rho e^{i\beta})^{\frac{1}{2}} & L(\mathbf{u}) & R(\theta, \phi, \chi). \\
\text{spinor} & & \text{complex} & \text{Lorentz} & \text{Rotation by} \\
& & \text{amplitude} & \text{transformation} & \text{Euler angles} \\
& & & \text{to velocity } v = e^{\mathbf{u}}\gamma_0 & (\theta, \phi, \chi)
\end{array}
$$

Explicit forms for $L(\mathbf{u})$ and $R(\theta, \phi, \chi)$ are

$$L(\mathbf{u}) = e^{\mathbf{u}/2} \quad \text{and} \quad R(\theta, \phi, \chi) = e^{-i\sigma_3\phi/2}\, e^{-i\sigma_2\theta/2} e^{-i\sigma_3\chi/2}.$$

Writing the Dirac equation in the STA (see Appendix) we have

$$\nabla \psi i \sigma_3 - eA\psi = m\psi\gamma_0.$$

This admits plane-wave solutions for 4-momentum p (with $p^2 = m^2$),

$$
\begin{aligned}
\psi \quad &= \quad L(\mathbf{p})R(\theta, \phi, \chi)e^{-i\sigma_3 p \cdot x} \\
\text{or} \quad & \quad L(\mathbf{p})R(\theta, \phi, \chi)i\, e^{+i\sigma_3 p \cdot x}.
\end{aligned}
$$

The Dirac current $\psi\gamma_0\tilde{\psi} \equiv \rho v$ and the spin current $\frac{1}{2}\psi\gamma_3\tilde{\psi} \equiv \rho s$ can then be interpreted using the STA. The Dirac wavefunction represents a change of frame, providing a Lorentz boost of γ_0 to the comoving velocity v and a rotation of γ_3 to align it with the spin axis of the electron. The phase of the wavefunction gives the rotation of the $\gamma_2\gamma_1$ plane about the spin axis. Of the two other factors contained in $\psi\tilde{\psi} \equiv \rho e^{i\beta}$, ρ can be taken to represent the probability of finding the electron at position x, but the rôle of β is uncertain. We can see that it distinguishes between positive ($\beta = 0$) and negative ($\beta = \pi$) frequency (energy) states.

In the ZBW interpretation of quantum mechanics (Hestenes 1990), the Dirac current is redefined as $\rho v \equiv \psi \gamma_- \tilde{\psi}$, where $\gamma_- \equiv \gamma_0 - \gamma_2$. This would make the worldline of a particle into a light-like helix, making manifest a transverse ZBW as the source of the electron's spin and magnetic moment. The positive frequency solutions above are interpreted as electrons, and the negative frequencies as positrons, so that β measures the extent to which we have a pure particle/antiparticle state. It is safe to say, however, that no entirely satisfactory interpretation of β is yet available.

3. The Dirac electron at a potential step

An elementary application of the Dirac equation is to consider an electron wave incident normally upon a simple electromagnetic potential step of magnitude ϕ, so that $(A = 0, z < 0)$ and $(A = \phi \gamma_0, z > 0)$. In the STA we write the incident, reflected and transmitted waves as follows:

Incident	Reflected	Transmitted
$\psi_I = e^{u\sigma_3/2} \Phi e^{-i\sigma_3 p_I \cdot x}$	$\psi_r = r\, e^{-u\sigma_3/2} \Phi e^{-i\sigma_3 p_r \cdot x}$	$\psi_t = t\, e^{u'\sigma_3/2} \Phi e^{-i\sigma_3 p_t \cdot x}$
$p_I = E\gamma_0 + p\gamma_3$	$p_r = E\gamma_0 - p\gamma_3$	$p_t = E\gamma_0 + p'\gamma_3,$

where Φ is a Pauli spinor describing the spin state, $\Phi = 1$ corresponding to longitudinal spin "up" and $\Phi = -i\sigma_2$ to spin "down". Matching the spinors at $z = 0$ we find

$$\cosh(u/2)(1 + r) = \cosh(u'/2)\, t,$$

$$\sinh(u/2)(1 - r) = \sinh(u'/2)\, t,$$

$$r = \frac{\sinh(u - u')/2}{\sinh(u + u')/2}.$$

This is plausible, the reflection coefficient increasing to unity as the step height approaches the classical reflection point at $e\phi = E - m$.

For steps higher than this, the reflection coefficient is unity and the wavefunction inside the step is evanescent. We have to take some care when the solutions are written in terms of the STA, because we cannot just let p' become imaginary. The evanescent wave is

$$\psi_t = \psi_0 e^{-\kappa \cdot x} e^{-i\sigma_3 p' \cdot x},$$

where ψ_0 is a constant Dirac spinor. We can find a matching condition if $\kappa \cdot p' = 0$ and $(p')^2 - \kappa^2 = m^2$. The solution displays very interesting general features that may give us a clue about the nature of the ZBW. For the case of longitudinal spin we find

$$\psi_0 = e^{i\beta/2} e^{-\kappa z} \Phi e^{-i\sigma_3 Et},$$

where $E - e\phi = m\cos\beta$ and $\kappa = \pm m\sin\beta$ for the two polarisation states, spin "up" taking the positive sign. The evanescent wave has a non-zero β, which flips to $\pm\pi$ as $e\phi$ approaches $E + m$.

If the spin of the incident wave is tranverse (polarised in the $x - y$ plane), then there is a non-zero Dirac current in the evanescent region, the Dirac "velocity" having the value κ/m. The direction of this velocity is perpendicular both to the incident spatial momentum and to the spin. Similar effects can be seen in solutions of the Dirac or the non-relativistic Pauli equations whenever the amplitude of ψ varies with position. Other examples are the hydrogen wavefunction or a Gaussian wave-packet (Hestenes 1979). It is tempting the interpret this phenomenon in terms of the non-cancelling of "Amperian" currents due to the circulation of the ZBW. It seems that the evanescent wave is giving us a "tomographic" view of the ZBW, implying again that the spin of the electron represents a real circulation. There is a need, however, for a sensible interpretation of the longitudinal behaviour, which shows a non-zero value of β.

The Klein paradox and Feynman's resolution

When the size of the potential step exceeds $E + m$ the solution becomes propagating. Defining $E' = e\phi - m$, $(p')^2 = (E')^2 - m^2$ and $\tanh u' = p'/E'$, we find

$$r = \frac{\cosh(u - u')/2}{\cosh(u + u')/2}; \quad r < 1: \quad \beta = \pi.$$

This case needs some care in order to match on to a solution with a positive group velocity inside the step. Many standard books give a solution with a negative group velocity, a notable example being Bjorken & Drell (1964).

This result for the reflection coefficient is actually very odd indeed, and I suspect that the reason some books give the wrong solution (with $r > 1$) is simply wishful thinking. We could perhaps understand $r > 1$ as representing the creation of electron/positron pairs at the step, with the positrons continuing into the step. In fact we seem to have an entirely different situation, and the transmitted wave looks more like a "positronic hole", having negative mass and negative charge.

The central problem is, as everybody knows, that the Dirac current $e\psi\gamma_0\tilde{\psi}$ is positive definite, and thus cannot represent a positronic current without some modification. The problem does not arise for the (non-STA) Klein-Gordon equation

$$\left(D^2 + m^2\right)\psi = 0, \quad \text{where } D \equiv \nabla + ieA$$

(the i is now just the uninterpreted $\sqrt{-1}$ of common usage). The Klein-Gordon reflection coefficients are

$$r = \frac{p - p'}{p + p'} < 1 \quad (e\phi < E - m),$$

$$r = \frac{p + p'}{p - p'} > 1 \quad (e\phi > E + m).$$

The Klein-Gordon charge current behaves sensibly, showing a strong resonance at $p = p'$, which one could interpret as stimulated emission of pairs.

The reason for the striking difference between these solutions is easy to find, because the Dirac equation does not lead to the minimally-coupled Klein-Gordon equation:

$$\nabla\psi i\sigma_3 - eA\psi = m\psi\gamma_0 \Longrightarrow$$

$$\nabla^2\psi + m^2\psi - e^2A^2\psi + 2emA\nabla\psi i\sigma_3 + eF\psi i\sigma_3 = 0,$$

where $F = \nabla A = \mathbf{E} + i\mathbf{B}$ ($\nabla \cdot A = 0$). There is an extra spin term $eF\psi i\sigma_3$, which couples into the very strong $\mathbf{E}$ field in the step, effectively changing the matching conditions. We see that the *spin* is crucial to this longitudinal ZBW phenomenon, which again cries out for a better physical interpretation.

Why should we bother with the Klein paradox of the Dirac equation when we are going to use field theory? The very simple case of an electromagnetic step has revealed a stark contradiction about the nature of the Dirac current. In almost any other branch of science we would conclude at this point that we have *already* done something wrong, and that we should go back and correct it. It seems, however, that in quantum theory we conclude that we have to do something else as well. This is more than a little odd and, in any case, it is unreasonable to suppose that such a contradiction will go away just by making the theory more complicated.

I am not able to understand or willing to repeat the arguments that lead to the standard resolution of the Klein paradox; a representative reference is Nikishov (1970). It appears that somewhere in the quantum fog the ideas of "exclusion principle" and "no stimulated emission of fermions" are invoked. The conclusion, on the other hand, is crystal-clear: $r = 1$ and there is no longer any paradox.

> I know an old lady who swallowed a fly,
>
> I don't know why she swallowed a fly, perhaps she'll die
>
> (Traditional nursery rhyme).

There seems to be a fundamental truth in the Dirac equation; we can interpret it geometrically. I believe that the same is probably true of the Weinberg/Salam electroweak model (Abers & Lee 1973, Hestenes 1982). But the Dirac theory *doesn't quite work* even when, with apologies to David Hestenes, it is translated into the STA. How should we patch it up? QED? Families of leptons? Quarks? Higgs particles? I suspect that we are in the same unfortunate position as the the old lady in the nursery rhyme who swallowed a fly and, finding it not to her liking, followed it with a series of increasingly indigestible remedies. My diagnosis is that we physicists have now reached the stage where we are attempting to swallow a "goat"and should beware the "horse" that must be waiting for us soon. I honestly believe that most "unifications" (except Weinberg/Salam) are headed in the wrong direction and that we (or at least some of us) should try to turn the clock back and start again using the proper mathematical tools. The SpaceTime Algebra is an essential ingredient in this programme.

Pushing the analogy a little further, the place to start is, I suppose, at the point we swallowed the fly. My guess is that we went wrong when the first infinity appeared in physical theory: the self-energy of a charge. To this end I am sympathetic to the view expressed by Jaynes (1990) about the reality or otherwise of the electromagnetic field, following the lead given by Wheeler & Feynman (1949). In this view the electromagnetic field does not actually exist in spacetime, but represents instead an *information storage device*. The electromagnetic field at any point might be a *summary* of what we need to know about distant charges in order to predict the behaviour of a charge *if there should be one* at that point. But I have already been sufficiently radical in this paper and I shall defer any further heresy. There is, however, a closely related problem concerning *radiation reaction.* Is it possible to have a self-consistent equation of motion for a particle that takes account of its electromagnetic radiation?

4. Radiation reaction at a step

The radiation reaction on an accelerated charge is a century-old problem that is still not resolved, despite the valiant efforts of generations of theoretical physicists. The Lorentz-Dirac equation (Dirac 1938) is "self-consistent" in the sense that it correctly accounts for the energy budget of a radiating particle, but it suffers from strange pre-acceleration effects and admits runaway solutions. Dirac derived the equation by expanding the field near a point charge as a power-series in retarded time and collecting the parts that did not diverge at the origin. The standard form of this equation (in SI units) is

$$m\dot{v} - \frac{e^2}{6\pi\epsilon_0 c^3}\left(\ddot{v} + \dot{v}^2 v\right) = eF \cdot v.$$

The derivation of this equation is badly flawed: the power-series in retarded time has only a finite radius of convergence. Burke (1970) shows that, if the same mathematical techniques are applied to the case of spherical oscillations of a rubber ball in air, pre-acceleration and runaway solutions again appear. Careful analysis of *finite-sized* charged distributions do not show such peculiar effects, and *do not* yield the Lorentz-Dirac equation (Jaynes 1980 (unpublished notes), Grandy & Aghazadeh 1983). Although these analyses are fine, I do not believe that the electron has a finite size (though see Jaynes' paper in this volume for a different point of view), and it is still very attractive to have an equation of motion that self-consistently accounts for the radiation of an accelerated point charge. Consequently, the Lorentz-Dirac equation is still used (see, for example, Barut 1988, Barut 1990), usually with the added boundary condition (due to Dirac) that the velocity remains finite as $t \to \infty$. Unfortunately, even with Dirac's condition, the solutions of this equation are particularly strange and physically unacceptable for the apparently innocuous electromagnetic potential step, which we now study.

The usual form of the Lorentz-Dirac equation does not fully reveal its geometrical meaning. Expressing the equation in the STA, we note that $v^2 = 1$, $\dot{v} \cdot v = 0$ implies

$$\ddot{v} + \dot{v}^2 v = \ddot{v} - (v \cdot \ddot{v})v = \frac{1}{2}\left(\ddot{v} - v\ddot{v}v\right) \equiv \ddot{v}_\perp,$$

which is the component of $\ddot{v}$ projected perpendicular to v. Because the acceleration $\dot{v}$ and the Lorentz force $eF \cdot v$ are both perpendicular to v, the reactive term must also be perpendicular. Multiplying by v we find

$$\frac{1}{2}\left(\ddot{v}v - v\ddot{v}\right) = \frac{1}{2}\frac{d}{d\tau}\left(\dot{v}v - v\dot{v}\right) = \frac{d}{d\tau}\left(\dot{v}v\right),$$

where we have again used $v \cdot \dot{v} = 0$. Defining $e^2/(6\pi\epsilon_0 c^3) \equiv \tau_0$, we can then rewrite the equation in terms of the rest-frame acceleration bivector $\Omega_v \equiv \dot{v}v = \dot{v} \wedge v$:

$$\frac{d\Omega_v}{d\tau} - \frac{\Omega_v}{\tau_0} = -\frac{e}{2m\tau_0}\left(F - vFv\right) = -\frac{eE_v}{m\tau_0},$$

where $E_v \equiv \frac{1}{2}(F - vFv)$ is the electric field in the rest frame.

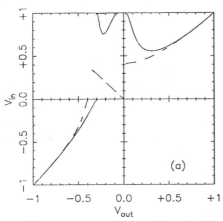 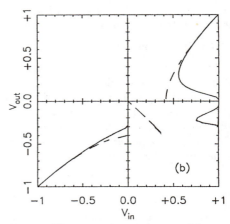

Figure 1. Input and output velocities for the Lorentz-Dirac equation at a potential step of $e\phi = 0.1m$. (a) $v_{\rm in}$ is a single-valued function of $v_{\rm out}$. (b) $v_{\rm out}$ is a multi-valued function of $v_{\rm in}$. The dashed curves show the behaviour of a non-radiating particle.

The field-free solutions can be expressed using the STA as $\Omega_v = B\exp(\tau/\tau_0)$ for any simple time-like bivector B, so that $v(\tau) = \exp[B\exp(\tau/\tau_0)]v(0)$. The solutions clearly have very undesirable properties as $\tau \to \infty$ unless we employ Dirac's boundary condition. To apply the Lorentz-Dirac equation to the potential step problem we search all possible finite output velocities $v_{\rm out}$ and integrate the equation *backwards in time* to see what the appropriate input velocities $v_{\rm in}$ must have been.

For a particle incident on a step having higher potential for $x > 0$ there are three distinct cases to consider.

1. Transmission left to right ($v_{\rm in}$ and $v_{\rm out}$ both positive). Note that, if $v_{\rm out}$ is large, then the particle, which has pre-accelerated up to $v_{\rm out}$ before it reached the step, goes through the step quickly, thereby receiving a small influence from it, so that $v_{\rm in}$ must have been large. However, if $v_{\rm out}$ is small the particle spends a longer time in the step and the change in velocity due to the step (not *at* it, though) is also large, so that $v_{\rm in}$ is again large.

2. Transmission right to left ($v_{\rm in}$ and $v_{\rm out}$ both negative). The particle, which has $v_{\rm out} < v_{\rm in}$, has pre-accelerated by an amount that increases as $v_{\rm out}$ is reduced. When $v_{\rm out}$ is sufficiently small, the long-term value of $v_{\rm in}$ is positive, so that the step is re-encountered.

3. Anomalous reflection ($v_{\rm in} > 0$; $v_{\rm out} < 0$). The particle approaches the step from the left, pre-decelerates towards the step and crosses it. On the other side the pre-deceleration of its next step-crossing changes the sign of its velocity and drags it back into the step, from which it emerges, having been reflected.

In Figure 1 the values of $v_{\rm in}$ and $v_{\rm out}$ are plotted for the particular case $e\phi = 0.1\,m$. The dashed curve shows, for comparison, the behaviour of a non-radiating particle. Even if the pre-acceleration properties of the equation are forgiven, the solutions are totally crazy. Figure 1(a) plots v_{in} as a singled-valued function of v_{out}, but a time-dweller's view of Figure 1(b) shows that for $v_{in} > 0$ there are ranges of v_{in} with either 2 or 4 values of v_{out}. In case that is not bad enough, for low values of v_{in} (corresponding in the non-radiating case to simple reflection) there are *NO* allowed values of v_{out}. We have, apparently, lost control of the input velocity! That is the sort of thing which is almost bound to happen when complicated equations are integrated backwards in time.

The firm conclusion of this little investigation is that Dirac's suggested boundary condition is not correct; merely making v finite at $\tau \rightarrow \infty$ is not sufficient to rescue the Lorentz-Dirac equation. It should also be noted that there are no mathematical difficulties associated with the idealised case of a sharp step, which can be approached as a well-behaved limit of a finite-width step of any shape. An extra condition on the Lorentz-Dirac equation (Sawada, Kawabata & Uchiyama 1983), which apparently prohibits such steps, cannot remove the difficulty, therefore.

A recent suggestion by Barut (1990) of "renormalising" the Lorentz-Dirac equation, taking solutions without pre-acceleration and removing the runaway part by decree is, I believe, even worse, because it does not conserve energy. Applying it to the finite-width step problem, and letting the step width go to zero whilst keeping the height constant, we find $v_{out} \rightarrow v_{in}$. That is extremely unfortunate if the step is a decelerating one as we have supposed, because the particle is now in a region of higher potential, which we could use to accelerate the particle in another, wider, downward step. This leads to an interesting new design of linear accelerator!

Modified Lorentz-Dirac equations

We can throw some light on the problems of the Lorentz-Dirac equation by writing the equation of motion as $m\,\dot{v} = e(F + F_s) \cdot v$, where F_s is a "self-field" representing the inner workings of the particle. It would be natural to suppose that these internal mechanisms would introduce a time lag, so that the effective force might be due to some retarded average of the applied field, for example $F_{ret} \equiv \frac{1}{\tau_0} \int_{-\infty}^{\tau} e^{(\tau'-\tau)/\tau_0} F(\tau')\mathrm{d}\tau' \approx F(\tau - \tau_0)$. A simple example shows why this suggestion is unsatisfactory.

Imagine a classical electron orbiting a nucleus. The instantaneous force is always towards the nucleus, so that the particle is kept in a circular orbit. The retarded force is directed *ahead* of the nucleus so that it would tend to increase the radius of the orbit, and hence the energy. Dirac solved this problem by using the *advanced* force $F_{adv} \equiv \frac{1}{\tau_0} \int_{\tau}^{\infty} e^{(\tau-\tau')/\tau_0} F(\tau')\mathrm{d}\tau'$, so that the response to any impulse precedes its cause. This can be easily seen by rewriting the Lorentz-Dirac equation in terms of F_s:

$$\dot{F}_s - \frac{1}{\tau_0} F_s = -\frac{\mathrm{d}E_v}{\mathrm{d}\tau}.$$

In this form the equation looks extremely dangerous, as we have already shown it to be in practice. An obvious causal modification is

$$\dot{F}_s + \frac{1}{\tau_0} F_s = \frac{\mathrm{d}E_v}{\mathrm{d}\tau},$$

which uses $(2F - F_{ret})$ instead of F_{adv}. It responds to an impulse like a mass of $m/2$ and then remembers the rest of its mass over a time $\approx \tau_0$. Applied to the step problem, the modified equation has properties very similar to the sensible branches of the Lorentz-Dirac solutions in Figure 1(b). The disallowed values of v_{in} now have $v_{\text{out}} = 0$, and we take the upper branch for large v_{in}. However, although the modification produces results indistinguishable to the accuracy of plotting from those of Figure 1(b), they are *not* identical, and the suggestion is *ad hoc*. To make further progress we must have a more detailed model of electron behaviour, but I conclude from this study that we probably have to revise our notions about the rate of radiation from accelerated charges.

5. Realistic classical models

Making classical models of particles with spin is an interesting game, all the more so when the STA is available. An example of such a model has been given by Barut & Zhanghi (1984); in addition to its position x, the particle has an internal mechanism or "clock" determined by a Dirac spinor ψ. Using the STA, and returning to natural units ($\hbar = c = 1$), we translate their Lagrangian as

$$\mathcal{L} = \frac{1}{2} \langle \, \dot{\psi} i \sigma_3 \tilde{\psi} \, \rangle + \langle \, p(\dot{x} - \psi \gamma_0 \tilde{\psi}) \, \rangle + e \langle \, A(x) \psi \gamma_0 \tilde{\psi} \, \rangle \,,$$

where $\langle \, \rangle$ means "scalar part of ". The equations of motion are

$$\dot{\psi} = \pi \psi \gamma_0,$$

$$v \equiv \dot{x} = \psi \gamma_0 \tilde{\psi},$$

$$\dot{\pi} = eF \cdot v,$$

where $\pi \equiv p - eA$. The Hamiltonian is $\pi \cdot v = m$ and the angular momentum bivector is $\frac{1}{2} \psi i \sigma_3 \tilde{\psi} + x \wedge p$.

The solution of these equations for zero field can be written in terms of

$$\psi_- \equiv e^{\mathbf{u}/2} \Phi e^{-i\sigma_3 m \tau} \quad ; \quad \psi_+ \equiv e^{\mathbf{u}/2} \Phi i e^{i\sigma_3 m \tau}.$$

The model can display longitudinal ZBW through interference of positive and negative frequencies but, unlike solutions of the Dirac equation, the negative frequency solution has the same charge/mass ratio as the positive energy one. To reverse the charge/mass ratio, it is necessary for the momentum π to point into the backward light-cone, thereby making the Hamiltonian negative. When the particle encounters a potential step there is a sudden change in its momentum π, but the spinor ψ is continuous. Energy is conserved; the time component of π changing by $e\phi$. The subsequent motion inside the step shows longitudinal ZBW and analysis of the mean velocity shows that the apparent rest mass has decreased.

The behaviour of the model in a magnetic field $F = Bi\sigma_3$ shows that it has no magnetic moment ($g = 0$), because a stationary solution ($v = \gamma_0$, $\pi = m\gamma_0$, $\psi = e^{-i\sigma_3 m \tau}$) can always be found. Following the suggestion by Hestenes (1990) for the Dirac current, we can make the ZBW of this model manifest by redefining the velocity $v = \psi \gamma_- \tilde{\psi}$, where $\gamma_- \equiv \gamma_0 - \gamma_2$ again. Examination of the stationary solutions shows that the negative frequencies have been eliminated and the positive frequencies moved to $2m$. There is now a magnetic moment; computer studies show that $g = 1$ for this modification, with the ZBW orbit plane precessing at one half the gyro frequency eB/m.

I conclude that, whilst Barut's little model is not yet satisfactory to describe the electron, there is still great promise for realistic models, particularly when the STA is employed.

6. Conclusions

We now have a natural language for spacetime physics that simplifies manipulations and gives equations which are fully Lorentz invariant and coordinate-free. The SpaceTime Algebra is able to provide important insights into the geometrical content of Dirac theory. Hestenes' ZBW interpretation of quantum mechanics seems promising, and there are important clues suggesting that the electron has a helical motion associated with its spin, but the longitudinal ZBW remains a mystery.

The problem of radiation reaction will not go away, but we can at least hope that the Lorentz-Dirac equation will finally be laid to rest. The crazy behaviour of this equation is well illustrated by its disastrous failure for the potential step problem.

Acknowledgements

I have greatly benefited from discussions with Anthony Lasenby and David Hestenes.

References

Abers, E. & Lee, B. (1973). Gauge Theories. *Phys. Reports*, **9**, 1–141.

Barut, A. O. (1988). Lorentz-Dirac Equation and Energy Conservation for Radiating Electrons. *Phys. Lett.*, **131**, 11–12.

Barut, A. O. (1990). Renormalisation and Elimination of Preacceleration and Runaway Solutions of the Lorentz-Dirac Equation. *Phys. Lett.*, **A131**, 11–12.

Barut, A. O. & Zhanghi, N. (1984). Classical Model of the Electron. *Phys. Rev. Lett.*, **52**, 2009–2012.

Bjorken, J. D. & Drell, S. D. (1964). *Relativistic Quantum Mechanics*, Vol. 1. McGraw-Hill, New York.

Burke, W. L. (1970). Runaway Solutions: Remarks on the Asymptotic Theory of Radiation Damping. *Phys. Rev.*, **A2**, 1501–1505.

Dirac, P. A. M. (1938). Classical Theory of Radiating Electrons. *Proc. Roy. Soc. Lond.*, **A47**, 148–169.

Feynman, R. P. (1967). *The Character of Physical Law*. MIT Press, Cambridge MA.

Grandy, W. T., Jr. & Aghazadeh, A. (1983). Radiative Corrections for Extended Charged Particles in Classical Electrodynamics. *Ann. Phys.*, **142**, 284–298.

Hestenes, D. (1966). *Space-Time Algebra*. Gordon & Breach, New York.

Hestenes, D. (1979). Spin and Uncertainty in the Interpretation of Quantum Mechanics. *Am. J. Phys.*, **47**, 399–415.

Hestenes, D. (1982). Space-Time Structure of Weak and Electromagnetic Interactions. *Found. Phys.*, **12**, 153–168.

Hestenes, D. (1986). A Unified Language for Mathematics and Physics. In *Clifford Algebras and Their Application in Physics*, (ed. J. S. R. Chisholm & A. K. Common), pp. 1–23. D. Reidel, Dordrecht.

Hestenes, D. (1990). The Zitterbewegung Interpretation of Quantum Mechanics. *Found. Phys.*, **20**, 1213–1232.

Jaynes, E. T. (1990). Probability in Quantum Theory. In *Complexity, Entropy, and the Physics of Information*, (ed. W. H. Zurek), pp. 33–55. Addison-Wesley, Redwood City CA.

Nikishov, A. I. (1970). Barrier Scattering in Field Theory and Removal of the Klein Paradox. *Nuclear Phys.*, **B21**, 346–358.

Sawada, T., Kawabata, T. & Uchiyama, F. (1983). Derivation of an extra condition on the Lorentz-Dirac equation from a conservation law. *Phys. Rev.*, **D27**, 454–455.

Wheeler, J. A. & Feynman, R. P. (1949). Classical Electrodynamics is Terms of Direct Interparticle Action. *Rev. Mod. Phys.*, **21**, 425–433.

Appendix. A translation table for Dirac spinors

In this Appendix the conventional and STA versions of the Dirac spinor are written down explicitly, so that one can translate freely between them. For the standard column spinor we employ the form of the Dirac matrices given by Bjorken & Drell (1964), the components of which are themselves 2-component Pauli spin matrices:

$$\hat{\gamma}_0 = \begin{pmatrix} 1 & 0 \\ 0 & -1 \end{pmatrix}; \quad \hat{\gamma}_k = \begin{pmatrix} 0 & \sigma_k \\ -\sigma_k & 0 \end{pmatrix}.$$

We identify a 4-component complex column vector Ψ in the standard treatment with an even multivector ψ of the STA. To make it quite clear that Ψ is a column vector in "spin-space", we will use Dirac notation. Thus

$$\Psi \equiv |\psi\rangle \quad \leftrightarrow \quad \psi.$$
$$\text{conventional} \qquad \text{STA}$$

The STA form of the Dirac spinor can be written in terms of its components:

$$\psi = a_0 + a_k\sigma_k + b_0 i + b_k i\sigma_k.$$

This translates into a matrix as above, and can be converted into the conventional spinor by taking its first column:

$$|\psi\rangle = \begin{bmatrix} a_0 + ib_3 \\ -b_2 + ib_1 \\ a_3 + ib_0 \\ a_1 + ia_2 \end{bmatrix}.$$

We translate the operators of matrix theory as follows:

$$\hat{\gamma}_\mu|\psi\rangle \leftrightarrow \gamma_\mu\psi\gamma_0,$$

$$i|\psi\rangle \leftrightarrow \psi i\sigma_3.$$

We are now in a position to translate the Dirac equation from the conventional form

$$i\hat{\gamma}^\mu\partial_\mu|\psi\rangle - e\hat{\gamma}^\mu A_\mu|\psi\rangle = m|\psi\rangle \leftrightarrow \gamma^\mu\partial_\mu\psi\gamma_0 i\sigma_3 - e\gamma^\mu A_\mu\psi\gamma_0 = m\psi.$$

Multiplying by γ_0 and re-assembling the vectors $A = \gamma^\mu A_\mu$ and $\nabla = \gamma^\mu\partial_\mu$ we obtain the STA version of the Dirac equation

$$\nabla\psi i\sigma_3 - eA\psi = m\psi\gamma_0.$$

NEW SOLUTIONS OF THE DIRAC EQUATION
FOR CENTRAL FIELDS

HEINZ KRÜGER

Fachbereich Physik der Universität
Postfach 3049
6750 Kaiserslautern

West Germany

ABSTRACT. A modified form of Hestenes's space-time version of Dirac's equation is separated in spherical polar coordinates. The angular part of the solution spinor is derived in its most general form in terms of Gegenbauer polynomials and exponentials. One- and two-valued (multi-valued) spinors are obtained as a generalization of Weyl's spherical harmonics with spin. The radial spinor equation defines a reduced real space-time algebra with one time and two space dimensions and displays symmetries which are hidden in the conventional matrix form. The superiority of this real valued Clifford algebraic formulation of the radial problem is demonstrated for the bound states of hydrogen-like atoms. The single-valued solutions turn out to be the ones of Darwin in a different representation.

One- and two-valued solutions together imply the unexpected discovery that the hydrogen atom in its states of lowest energy may realize the following values of its magnetic moment: $\mu = \mu_D$ and $\mu = \mu_D/\pi$, where $\mu_D = \mu_{D\,arwin} = \frac{1}{3}(1+2\gamma)\mu_B$, $\gamma = \sqrt{1-\alpha^2}$, $\alpha = \frac{q^2}{hc}$, $\mu_B = \mu_{B\,ohr} = \frac{|q|h}{2mc}$. Presumably it is an effect of the selfinteraction which seems to prefer the single-valued Darwin solution of the Dirac equation in the normal state of the hydrogen atom.

1. Introduction

When I presented my new $\beta = 0$ solutions for the first time at the Antigonish workshop 1990, I was asked the following question:

Can these $\beta = 0$ solutions be generalized to also include magnetic quantum numbers different from zero?

This article gives an answer in terms of a new and complete separation of the real Dirac equation with respect to both angles of spherical polar coordinates.

49

D. Hestenes and A. Weingartshofer (eds.), The Electron, 49–81.

The separation is carried through in section 2. In section 3 the general solution of the angle equations is derived, leading to multi-valued, square-integrable spherical harmonics with spin. Section 4 is reserved for the study of algebraic properties of the radial spinor equation and their symmetries. The special $\beta = 0$ solutions mentioned above are identified within the general context in section 5. The investigation of the transformation properties of the general central field spinor with respect to the space inversion - or parity operation is the content of section 6. It is shown that the repeated application of the parity operation P provides a simple coordinate-free criterion for the multi-valuedness of a spinor on the 3-dimensional euclidean subspace $I\!E^3$ in the Minkowski space $I\!M(1,3)$: A spinor is single valued iff (if and only if) the twofold application of P is the identity, i.e. $P^2 = 1$. So, a spinor is two-valued on $I\!E^3$, iff $P^2 = -1$, four-valued, iff $P^4 = -1$, etc..

Hydrogen-like bound states and their normalization are derived in section 7 for nuclear charge numbers $Z \leq 137$. The intent of this section only is to demonstrate in detail the practical feasibility of this representation-free Clifford algebraic spinor calculus. It is evident, that for instance the case $Z > 137$ (Klein paradox) and Coulomb scattering can be treated with the same efficiency. These investigations as well as their applications to radiative transitions between bound- and scattering states (bremsstrahlung and pair creation) are subject to forthcoming publications.

In section 8, a brief formulation of the anomalous Zeeman effect by means of space-time algebra is presented for the first time. The states of lowest energy are discussed in detail and the magnetic moments announced in the abstract are calculated.

2. Separation of Variables

The real Clifford algebraic formulation of Dirac's equation developed by David Hestenes [1]

$$\hbar \partial \psi_H\, i_3 - \frac{q}{c} A \psi_H = mc\psi_H\, \gamma_0 \quad , \quad i\psi_H = \psi_H\, i \quad , \quad i = \gamma_0 \gamma_1 \gamma_2 \gamma_3 \qquad (2.1)$$

explicitely depends on the basis vector γ_0, $\gamma_0^2 = 1$ defining the time axis and on the spacelike bivector $i_3 = \gamma_2 \gamma_1 = i\vec{\sigma}_3$, which at the first view may seem to prefer the 3-axis in $I\!E^3$. That this preference in fact is as spurious as the dependence of a matrix formulation of Dirac's theory on the choice of a special representation has been discussed in [1]. We are free to rotate the axis $\vec{\sigma}_3$ with the help of a unitary spinor U to an arbitrary axis $\vec{n}$, $\vec{n}^{\,2} = 1$ by

$$\psi_H = \psi U \quad , \quad \psi = \psi_H U^\dagger \quad ,$$
$$U^\dagger = \tilde{U}^* = \gamma_0 \tilde{U} \gamma_0 = U^{-1} \quad , \quad U\vec{\sigma}_3 = \vec{n} U \quad . \tag{2.2}$$

The sign $\tilde{M}$ means the reverse of a multivector M and is obtained by reversing the order of all vector factors in all the products of which M is composed. From (2.2) and (2.1) we find the Dirac equation

$$\hbar \partial \psi i \vec{n} - \frac{q}{c} A\psi = mc\psi\gamma_0 \quad , \quad i\psi = \psi i \quad . \tag{2.3}$$

Just as the 3-axis in Hestenes's equation (2.1), the direction $\vec{n}$ may be interpreted as the unit normal on the spin plane in which the "Zitterbewegung" of the lepton takes place. For the standard chart of the spherical polar coordinates discussed below there is a special choice of $\vec{n}$, which greatly simplifies the task of separating the variables, namely

$$\vec{n} = \vec{\sigma}_2 \quad , \quad U = e^{i_1 \pi/4} = \frac{1+i_1}{\sqrt{2}} \quad , \quad i_1 = \gamma_3\gamma_2 = i\vec{\sigma}_1, \tag{2.4}$$

$$i_k = i\vec{\sigma}_k \quad , \quad \vec{\sigma}_k = \gamma_k\gamma_0 \quad , \quad k = 1,2,3 \quad , \quad i = \gamma_0\gamma_1\gamma_2\gamma_3. \tag{2.5}$$

The standard chart of spherical polar coordiates

$$\vec{r} = x\vec{\sigma}_1 + y\vec{\sigma}_2 + z\vec{\sigma}_3 = r\left[\sin\vartheta\,(\vec{\sigma}_1\cos\varphi + \vec{\sigma}_2\sin\varphi) + \vec{\sigma}_3\cos\vartheta\right] \quad ,$$
$$0 < r < \infty \quad , \quad 0 < \vartheta < \pi \quad , \quad 0 < \varphi < 2\pi \quad , \tag{2.6}$$

can be factorized in spinor form according to

$$\vec{r} = r\left(\sin\vartheta\,\vec{\sigma}_1 e^{i_3\varphi} + \vec{\sigma}_3\cos\vartheta\right) = re^{-i_3\varphi/2}\left[\vec{\sigma}_1\sin\vartheta + \vec{\sigma}_3\cos\vartheta\right]e^{i_3\varphi/2}$$
$$= re^{-i_3\varphi/2}\,e^{-i_2\vartheta/2}\,\vec{\sigma}_3\,e^{i_2\vartheta/2}\,e^{i_3\varphi/2} \tag{2.7}$$
$$= rS\vec{\sigma}_3\tilde{S} = \vec{r} \quad ,$$

$$S = e^{-i_3\varphi/2}e^{-i_2\vartheta/2} \quad , \quad S^{-1} = \tilde{S} = S^\dagger \quad , \quad S^* = S. \tag{2.8}$$

This factorization of $\vec{r}$ induces a spinor factorization of the invariant derivative (gradient operator) both on $I\!\!E^3$

$$\vec{\partial} = \vec{\sigma}_1 \frac{\partial}{\partial x} + \vec{\sigma}_2 \frac{\partial}{\partial y} + \vec{\sigma}_3 \frac{\partial}{\partial z} = \Omega \vec{\partial}\,' \Omega^{-1} \quad , \quad \Omega r \sqrt{\sin \vartheta} = S \quad , \qquad (2.9)$$

$$\vec{\partial}\,' = \vec{\sigma}_3 \partial_r + \frac{\vec{\sigma}_1}{r} \partial_\vartheta + \frac{\vec{\sigma}_2 \partial_\varphi}{r \sin \vartheta} \quad , \quad \partial_r \equiv \frac{\partial}{\partial r} \quad , \quad \partial_\vartheta \equiv \frac{\partial}{\partial \vartheta} \quad , \quad \partial_\varphi \equiv \frac{\partial}{\partial \varphi} \quad , \quad (2.10)$$

and on the Minkowski space $I\!\!M(1,3)$

$$\partial = \Omega \partial' \Omega^{-1} \quad , \quad \partial' = \gamma_0 (\partial_0 + \vec{\partial}\,') \quad , \quad \partial_0 = \frac{\partial}{\partial r_0} \quad , \quad r_0 = ct \quad . \qquad (2.11)$$

For a central field

$$qA = \gamma_0 V(r) \quad , \quad r = |\vec{r}| \quad , \qquad (2.12)$$

equation (2.3) with the choice (2.4) and (2.11) becomes

$$\hbar \partial' \Omega^{-1} \psi i_2 - \frac{1}{c} \gamma_0 V \Omega^{-1} \psi = mc \Omega^{-1} \psi \gamma_0 \quad , \qquad (2.13)$$

which permits a separation of the r_0- and φ-dependence with the two separation constants E and λ

$$\psi = \psi(r_0, r, \vartheta, \varphi) = \Omega \psi_1(r, \vartheta) \exp \left\{ i_2 \left(\lambda \varphi - \frac{E r_0}{\hbar c} \right) \right\} \quad , \quad \lambda, E \in I\!\!R \quad , \quad (2.14)$$

$$-\hbar \left(\gamma_3 \partial_r + \frac{\gamma_1}{r} \partial_\vartheta \right) \psi_1 i_2 + \frac{\hbar \lambda \gamma_2}{r \sin \vartheta} \psi_1 + \frac{1}{c} \gamma_0 (E - V) \psi_1 = mc \psi_1 \gamma_0 \quad . \qquad (2.15)$$

Separation of r and ϑ however is less trivial because of the noncommutativity of the Clifford product and the post multiplication factors i_2 and γ_0 in (2.15). For solving this problem I put

$$\psi_1(r, \vartheta) = \eta(r) g(\vartheta) + i_3 \eta(r) f(\vartheta) \quad , \quad i\eta = \eta i \quad , \qquad (2.16)$$

where f and g have to fulfill the conditions

$$f = f^* = \gamma_0 f \gamma_0 \quad , \quad g = g^* \quad , \quad i_2 f = f i_2 \quad , \quad i_2 g = g i_2 \quad . \tag{2.17}$$

Equation (2.15) then leads to

$$
0 = - \hbar \gamma_3 \eta' i_2 g - \frac{\hbar \gamma_1}{r} \eta \left(g' i_2 + \frac{\lambda f}{\sin \vartheta} \right) + \gamma_0 \frac{1}{c}(E - V)\eta g - m c \eta g \gamma_0
$$

$$
- \hbar \gamma_3 i_3 \eta' i_2 f + \frac{\hbar i_3 \gamma_1}{r} \eta \left(f' i_2 - \frac{\lambda g}{\sin \vartheta} \right) + \gamma_0 \frac{1}{c}(E - V)i_3 \eta f - m c i_3 \eta f \gamma_0 = 0 \quad ,
$$

where the primes stand for the derivatives with respect to the corresponding variables r and ϑ. Introducing the separation constant $\kappa \in I\!R$ according to

$$
f' i_2 - \frac{\lambda g}{\sin \vartheta} = \kappa f \quad , \quad g' i_2 + \frac{\lambda f}{\sin \vartheta} = -\kappa g \quad ,
$$
$$
f = f(\vartheta) \; , \; g = g(\vartheta) \; , \; \kappa \in I\!R \; , \tag{2.18}
$$

I achieve a complete separation of the angle spinors $f(\vartheta)$ and $g(\vartheta)$ from the radial spinor $\eta(r)$, which has to solve the radial equation

$$
-\hbar \gamma_3 \eta' i_2 + \left(\gamma_0 \frac{E - V}{c} + \gamma_1 \frac{\hbar \kappa}{r} \right) \eta = m c \eta \gamma_0 \quad , \quad \eta = \eta(r) \; , \; i \eta = \eta i \; . \tag{2.19}
$$

3. Spherical Harmonics with Spin

Before proceeding to derive the general solution of the angle equations (2.18), it is worth to study the special case $\lambda = 0$ first. For $\lambda = 0$ the spinors f and g are decoupled and one finds two types of linear independent solutions

$$
f = 0 \quad , \quad g \pi \sqrt{2} = e^{i_2 \kappa \vartheta} \tag{3.1}
$$

$$
g = 0 \quad , \quad f \pi \sqrt{2} = e^{-i_2 \kappa \vartheta} . \tag{3.2}
$$

The general solution then is a linear combination of (3.1) and (3.2). In perhaps more familiar physical terms, (3.1) and (3.2) comprise a complete set of two normal-vibration solutions into which any $\lambda = 0$ solution of (2.18) may be decomposed. In section 5 it is shown that each of (3.1) and (3.2) separately, when inserted into (2.16), lead to a solution of the Dirac equation with a vanishing Yvon-Takabayasi parameter $\beta = 0$. Just these solutions I have presented at the Antigonish workshop 1990.

A first step towards the general solution of (2.18) for $\lambda \neq 0$ is to decouple f and g. For this sake let me introduce

$$f_1 = i_2 f \tag{3.3}$$

instead of f, which leads to

$$f_1' + i_2 \kappa f_1 = \frac{\lambda g}{\sin \vartheta} = \frac{\lambda (\tilde{g})^\sim}{\sin \vartheta} \quad , \tag{3.4}$$

and after a reversion of the equation (2.18) for g, one notes the suggestive form

$$\tilde{g}\,' + i_2 \kappa \tilde{g} = \frac{\lambda \tilde{f}_1}{\sin \vartheta}. \tag{3.5}$$

Thus $\tilde{g}$ is up to a constant spinor factor equal to f_1, viz.,

$$\tilde{g} = f_1 \varepsilon e^{-2 i_2 \delta} \quad , \quad \varepsilon > 0 \quad , \quad \delta \in I\!\!R \ . \tag{3.6}$$

Inserting (3.6) into (3.5) and (3.4), one notes that $\varepsilon = 1$ and that

$$\omega = f_1 e^{-i_2 \delta} \tag{3.7}$$

has to fulfill

$$\omega' + i_2 \kappa \omega = \frac{\lambda \tilde{\omega}}{\sin \vartheta} \ . \tag{3.8}$$

Choosing the phase $\delta = 0$, I obtain

$$f = -i_2 \omega \quad , \quad g = \tilde{\omega} \quad , \tag{3.9}$$

and from (2.16)

$$\psi_1(r, \vartheta) = \eta(r)\tilde{\omega}(\vartheta) - i_3 \eta(r) i_2 \omega(\vartheta) \ . \tag{3.10}$$

There is a close connection between the spinor equation (3.8) and the spherical harmonics with spin $P(\vartheta)$ and $Q(\vartheta)$, P, $Q \in \mathbb{R}$, defined by Hermann Weyl eventually before 1929 [2]. In fact,

$$\omega = \sin^\lambda \vartheta \left[Q \cos \frac{\vartheta}{2} + i_2 P \sin \frac{\vartheta}{2} \right] , \tag{3.11}$$

and the replacement of κ by k and λ by $-m - \frac{1}{2}$ lead to equations (7.13) in [2] for the variable $z = \cos \vartheta$.

Since I am interested in the general solution of (3.8), *I make no use of Weyl's single-valued special solutions.* As the discussion of the case $\kappa = 0$ already shows, there are square-integrable but multi-valued solutions of (3.8) for $|\lambda| < \frac{1}{2}$

$$\sin \vartheta \, \partial_\vartheta \omega = \lambda \tilde{\omega} = (z^2 - 1)\partial_z \omega \quad , \quad z = \cos \vartheta \quad , \tag{3.12}$$

given by

$$\omega = c_1 \left(\frac{1-z}{1+z} \right)^{\lambda/2} + i_2 c_2 \left(\frac{1+z}{1-z} \right)^{\lambda/2} \quad , \quad |\lambda| < \frac{1}{2} \quad . \tag{3.13}$$

Inspection of (2.14) however shows that the factor $e^{i_2 \lambda \varphi}$ for $|\lambda| < \frac{1}{2}$ gives rise to a solution (2.14) whose valuedness is greater than 2! Therefore, *the finally adopted restriction [3] to one- and two-valued spinors (2.14) leads to the conclusion, that $\kappa = 0$ is no admissible eigenvalue of (3.8).*

In order to derive the general solution of (3.8) in the case $\lambda \neq 0$, I put

$$\omega = \sin^{|\lambda|} \vartheta \, e^{i_2 \vartheta/2} (a + b\partial_\vartheta) C(\vartheta) \quad ,$$
$$C \in \mathbb{R} \quad , \quad a = a_0 + i_2 a_2 \quad , \quad b = b_0 + i_2 b_2 \quad , \tag{3.14}$$

where the constant spinors a and b may be adjusted such that

$$\lambda \tilde{b} = -|\lambda| b \quad , \quad a = -i_2 b \left(\kappa + \frac{1}{2} - |\lambda| \right) \quad , \tag{3.15}$$

whence the real function $C(\vartheta)$ has to solve the differential equation of Gegenbauer functions

$$\left[\partial_\vartheta^2 + 2|\lambda| \cot \vartheta \, \partial_\vartheta + \left| \kappa + \frac{1}{2} \right|^2 - |\lambda|^2 \right] C(\vartheta) = 0. \tag{3.16}$$

With the scalar integration constants c_1, c_2 the general solution of (3.16) is [4]

$$C(\vartheta) = c_1 C^{|\lambda|}_{|\kappa + \frac{1}{2}| - |\lambda|}(\cos\vartheta)$$

$$+ c_2 \,_2F_1\left(\frac{|\kappa + \frac{1}{2}| + |\lambda|}{2}, \frac{|\kappa + \frac{1}{2}| + |\lambda| + 1}{2}; \left|\kappa + \frac{1}{2}\right| + 1; \cos^2\vartheta\right), \qquad (3.17)$$

$$\lambda \neq 0 \quad , \quad \lambda \in I\!R \quad ,$$

from which a square-integrable spinor ω results if $c_2 = 0$ and only if

$$\left|\kappa + \frac{1}{2}\right| - |\lambda| = p = 0, 1, 2, \ldots \in I\!N_0. \qquad (3.18)$$

Under these conditions (3.17) degenerates to a Gegenbauer polynomial $C^{|\lambda|}_p(\cos\vartheta)$ of degree p in $z = \cos\vartheta$, and from (3.15), (3.14) one infers for $\lambda > 0$

$$\omega = \omega_+ = |b| \sin^{|\lambda|}\vartheta \; e^{i_2\vartheta/2}\left(i_2\partial_\vartheta + \kappa + \frac{1}{2} - |\lambda|\right) C^{|\lambda|}_p(\cos\vartheta) \quad , \qquad (3.19)$$

while for $\lambda < 0$

$$\omega = -i_2\omega_+. \qquad (3.20)$$

Before proceeding to calculate the normalization constant $|b|$ in (3.19), it is necessary to study the case $p = 0$ in (3.18) first. For $\kappa + \frac{1}{2} = |\lambda|$ one notes the result $\omega = \omega_+ = 0$ and for $\kappa + \frac{1}{2} = -|\lambda|$, $\omega_+ = -2|\lambda||b| \sin^{|\lambda|}\vartheta \; e^{i_2\vartheta/2}$. Imposing the normalization condition

$$4\pi \int_0^\infty d\vartheta \; \omega\tilde{\omega} = 1 \; , \qquad (3.21)$$

one finds

$$(4\pi|b|)^2 \; \Gamma(2|\lambda| + 1) = 2^{2|\lambda|} \; \Gamma^2(|\lambda|) \; . \qquad (3.22)$$

For $p = 1, 2, \ldots \in I\!N$, the relation

$$\partial_\vartheta C_p^{|\lambda|}(\cos \vartheta) = -2|\lambda| \sin \vartheta \, C_{p-1}^{|\lambda|+1}(\cos \vartheta) \quad , \tag{3.23}$$

allows to write (3.19) in the form

$$\omega_+ = |b| \sin^{|\lambda|} \vartheta \; e^{i_2 \vartheta/2} \left[\left(\kappa + \frac{1}{2} - |\lambda| \right) C_p^{|\lambda|}(z) - 2i_2 |\lambda| \sqrt{1 - z^2} \; C_{p-1}^{|\lambda|+1}(z) \right] \quad ,$$

$$z = \cos \vartheta \quad ,$$

$$\tag{3.24}$$

such that the necessary integrals in (3.21) are reduced to normalization integrals of Gegenbauer polynomials. By use of $\Gamma(x + 1) = x\Gamma(x)$ and comparison with (3.22), one obtains the normalization conditions

$$|b| = b_< \quad , \quad b_<^2 \, (4\pi)^2 \, \Gamma(p + 2|\lambda| + 1) = p! \, 2^{2|\lambda|} \, \Gamma^2(|\lambda|) \quad ,$$

$$\kappa + \frac{1}{2} = -p - |\lambda| \quad , \quad p \geq 0 \tag{3.25}$$

$$|b| = b_> \quad , \quad b_>^2 \, (4\pi)^2 \, \Gamma(p + 2|\lambda|) = (p - 1)! \, 2^{2|\lambda|} \, \Gamma^2(|\lambda|) \quad ,$$

$$\kappa + \frac{1}{2} = p + |\lambda| \quad , \quad p \geq 1 . \tag{3.26}$$

Equation (3.8) shows that $\tilde{\omega}$ also is a solution if κ is replaced by $-\kappa$, and (3.25) shows a shift $p \to p + 1$ with respect to (3.26). So, $\omega_>$ obtained from (3.26), (3.24) and (3.20) after $p \to p+1$ has to be proportional to $\tilde{\omega}_<$, as derived from (3.26), (3.24) and (3.20), and, the factor of proportionality must be scalar. Thus for $p \to p+1$ in $\omega_>$, the relation must hold

$$\tilde{\omega}_> = c\omega_< \quad , \quad c = \tilde{c} . \tag{3.27}$$

In fact, making use of

$$2|\lambda|(1 - z^2)C_p^{|\lambda|+1}(z) + (p + 1)zC_{p+1}^{|\lambda|}(z) = (p + |\lambda|)C_p^{|\lambda|}(z)$$

and

$$2|\lambda|zC_p^{|\lambda|+1}(z) - (p + 1)C_{p+1}^{|\lambda|}(z) = 2|\lambda|C_{p-1}^{|\lambda|+1}(z) \quad ,$$

one notes the result $c = -1$, provided the Gegenbauer polynomials $C_p^\alpha(z)$ for $\alpha \neq 0$ are defined according to

$$C_{-1}^\alpha(z) = 0 \quad , \quad C_0^\alpha(z) = 1 \quad ,$$
$$(p+1)C_{p+1}^\alpha(z) = 2(p+\alpha)zC_p^\alpha(z) - (p+2\alpha-1)C_{p-1}^\alpha(z) \quad , \tag{3.28}$$
$$p = 0, 1, \ldots \in I\!N_0 \ .$$

Without loss of generality, the factor $c = -1$ can be absorbed into a convenient choice of phase, such that, for $\lambda \neq 0$ the normalized solutions ω of (3.8) can be written in the following final form:

$$\omega = \omega_< = \omega_{p\lambda}(\vartheta)\, e^{\frac{i_2\pi(|\lambda|-\lambda)}{4|\lambda|}} \quad ,$$
$$\kappa = -k \quad , \quad k = p + |\lambda| + \frac{1}{2} \quad , \quad p = 0, 1, 2, \ldots \in I\!N_0 \quad , \tag{3.29}$$

$$\omega = \omega_> = \tilde{\omega}_< \quad , \quad +\kappa = k = p + |\lambda| + \frac{1}{2} \quad , \quad p \in I\!N_0 \quad , \tag{3.30}$$

$$\omega_{p\lambda}(\vartheta) = \frac{2^{|\lambda|}\,\Gamma(|\lambda|)}{4\pi} \left(\frac{p!}{\Gamma(p+2|\lambda|+1)} \right)^{1/2} \sin^{|\lambda|}\vartheta\ e^{i_2\theta/2}$$
$$\left[(p+2|\lambda|)C_p^{|\lambda|}(\cos\vartheta) + 2i_2|\lambda|\sin\vartheta\ C_{p-1}^{|\lambda|+1}(\cos\vartheta) \right] \quad ,$$
$$C_{-1}^\alpha(\cos\vartheta) = 0 \quad , \quad C_0^\alpha(\cos\vartheta) = 1 \quad , \quad \lambda \neq 0 \quad , \tag{3.31}$$

$$\omega_{p\lambda}(\pi - \vartheta) = \tilde{\omega}_{p\lambda}(\vartheta)\, e^{i_2\pi(p+\frac{1}{2})} \quad , \tag{3.32}$$

$$4\pi \int_0^\pi d\vartheta\, [\omega_{p\lambda}(\vartheta)\tilde{\omega}_{q\lambda}(\vartheta)]_0 = \delta_{pq} = \begin{cases} 1 & \text{for } p = q \\ 0 & \text{for } p \neq q \end{cases} \ . \tag{3.33}$$

Those readers who are interested in a connection between $\omega(\vartheta)$ and spherical monogenics [5] will find a brief discussion of this obvious relation at the end of section 6.

4. Radial Algebra

The radial spinor equation (2.19)

$$
-\hbar\gamma_3 \partial_r \eta i_2 + \left(\gamma_0 \frac{E-V}{c} + \gamma_1 \frac{\hbar\kappa}{r}\right)\eta = mc\eta\gamma_0 \quad ,
$$

$$
i\eta = \eta i \quad , \quad \eta = \eta(r) \quad , \quad \partial_r = \frac{\partial}{\partial r} \quad ,
$$

(4.1)

defines a radial algebra $\mathcal{G}(1,2)$ as a subalgebra of the full space-time algebra $\mathcal{G}(1,3)$. Its generic vectors of grade 1 are the timelike vektor γ_0 and the two spacelike vectors γ_3 and γ_1 which span a 3-dimensional Minkowski space. For all grades of $\mathcal{G}(1,2)$ the generic elements are displayed in the following table:

TABLE 1. Generic elements of $\mathcal{G}(1,2) \subset \mathcal{G}(1,3)$

Name	grade	elements		
Scalar	0	$\underline{1}$		
Vector	1	$\gamma_0, \gamma_3, \gamma_1$		
Bivector	2	$\underline{\vec{\sigma}_3 = \gamma_3\gamma_0}$,	$\underline{\vec{\sigma}_1 = \gamma_1\gamma_0}$,	$\underline{i_2 = \gamma_1\gamma_3} = \vec{\sigma}_3\vec{\sigma}_1$
Pseudoscalar Trivector in $\mathcal{G}(1,3)$	3	$\gamma_0\gamma_1\gamma_3 = \gamma_0 i_2 = i_2\gamma_0$		

According to $i\eta = \eta i$ the radial spinor η is composed of all even elements of $\mathcal{G}(1,3)$ and, at the very least, has to be a linear combination of all the underlined even elements of $\mathcal{G}(1,2)$ which span $\mathcal{G}^+(1,2) \simeq \mathcal{G}(2)$, i.e.,

$$
\eta = \eta_0 + i_2\eta_2 + \vec{\sigma}_1(\eta_1 + i_2\eta_3) \quad , \quad \eta_\mu \in I\!R \quad , \quad \mu = 0,1,2,3.
$$

(4.2)

It should be emphasized however that the scalar coefficients may be generalized to include pseudoscalar parts of $\mathcal{G}(1,3)$ as well. In that case the multivector (4.2) would comprise all even products of $\mathcal{G}(1,3)$ and hence represent a general spinor of 8 real components.

In this article I shall employ scalar coefficients η_μ only, such that $\eta \in \mathcal{G}(2)$.
The restriction then induces the following form of the space inversion or parity operation on $\mathcal{G}(2)$

$$
\eta^* = \gamma_0\eta\gamma_0 = -i_2\eta i_2 \quad , \quad i_1\eta^* = i_3\eta i_2 \quad .
$$

(4.3)

Equation (4.1) is forminvariant under a reflection at the $\vec{\sigma}_2\vec{\sigma}_3$ - plane with the normal $\vec{\sigma}_1$

$$\eta \xrightarrow{R} \eta^R = -\vec{\sigma}_1\eta\vec{\sigma}_1 \quad . \tag{4.4}$$

Therefore the general solution $\eta \in \mathcal{G}(2)$ can always be decomposed into an odd part

$$\eta_o = \frac{1}{2}(\eta - \eta^R) = \frac{1}{2}(\eta + \vec{\sigma}_1\eta\vec{\sigma}_1) = \eta_0 + \vec{\sigma}_1\eta_1 \tag{4.5}$$

and an even part

$$\eta_e = \frac{1}{2}(\eta + \eta^R) = \frac{1}{2}(\eta - \vec{\sigma}_1\eta\vec{\sigma}_1) = (\eta_2 + \vec{\sigma}_1\eta_3)i_2 \quad . \tag{4.6}$$

Since the even spinor η_e, up to the phase factor i_2, is of the same form as the odd spinor η_o, a further restriction from $\eta \in \mathcal{G}(2)$ to $\eta_o = \eta_0 + \vec{\sigma}_1\eta_i$ is no loss of generality. With $\eta = \eta_o = \eta_0 + \vec{\sigma}_1\eta_1$ the radial equation (4.1) yields the following set of coupled differential equations for the scalar functions $\eta_o(r)$ and $\eta_1(r)$,

$$
\begin{aligned}
\hbar c\partial_r\eta_1 &= -\frac{\hbar c\kappa}{r}\eta_1 + (E - mc^2 - V)\eta_o \\
\hbar c\partial_r\eta_o &= (V - E - mc^2)\eta_1 + \frac{\hbar c\kappa}{r}\eta_o \quad .
\end{aligned}
\tag{4.7}
$$

After a replacement of m by $-m$, this system is identical with the conventional radial equations [6]. In this sense, *the new radial spinor equation (4.1) may be understood as a natural hypercomplex generalization of the radial equations (4.7) with a well defined geometrical structure*. Both the structural geometric insight and the gain of practicability by avoiding matrices make (4.1) superior to (4.7).

5. Selection of $\beta = 0(\mathrm{mod}\ \pi)$ Solutions

The general form of a spherically separated central field solution of the Dirac equation (2.3) follows from (2.4), (2.8), (2.9) and (2.16),

$$\psi(r_0, r, \vartheta, \varphi) = \Omega(\eta g + i_3\eta f)\ \exp\left\{i_2\left(\lambda\varphi - \frac{Er_0}{\hbar c}\right)\right\} \quad . \tag{5.1}$$

The spinor (5.1) may be brought into the canonical polar form

$$\psi = \varrho^{1/2} \, e^{\frac{i\beta}{2}} \, R \quad , \quad iR = Ri \quad , \quad R\tilde{R} = 1 \quad , \quad \varrho > 0 \quad , \quad \beta \in I\!R \quad , \qquad (5.2)$$

in order to allow the determination of the parameter β of Yvon [7] and Takabayasi [8]. From equation (5.2) one finds

$$\psi\tilde{\psi} = \varrho \, e^{i\beta} = \varrho\cos\beta + i\varrho\sin\beta \quad . \qquad (5.3)$$

One notes in particular that $\beta = 0\,(\mathrm{mod}\ \pi)$ iff $\psi\tilde{\psi} = [\psi\tilde{\psi}]_0 =$ scalar. From (5.1) and the fact that because of (2.17) $f\tilde{f} = |f|^2$ and $g\tilde{g} = |g|^2$ are scalars, one obtains

$$\begin{aligned}
\psi\tilde{\psi} &= \Omega(\eta g + i_3 \eta f)(\tilde{g}\tilde{\eta} - \tilde{f}\tilde{\eta}i_3)\tilde{\Omega} \\
&= \Omega(g\tilde{g}\eta\tilde{\eta} - f\tilde{f}i_3\eta\tilde{\eta}i_3 + i_3\eta f\tilde{g}\tilde{\eta} - \eta g\tilde{f}\tilde{\eta}i_3)\tilde{\Omega}
\end{aligned} \qquad (5.4)$$

and equation (4.2)

$$\begin{aligned}
\eta = a + \vec{\sigma}_1 b \quad , \quad a = \eta_0 + i_2\eta_2 \quad , \quad b = \eta_1 + i_2\eta_3 \quad , \\
\eta_\mu \in I\!R \quad , \quad \mu = 0,1,2,3 \quad ,
\end{aligned} \qquad (5.5)$$

leads to

$$\eta\tilde{\eta} = a\tilde{a} - b\tilde{b} = |a|^2 - |b|^2 . \qquad (5.6)$$

Therefore, equation (5.4) can be rearranged according to

$$\psi\tilde{\psi} = (|f|^2 + |g|^2)(|a|^2 - |b|^2)\Omega\tilde{\Omega} + i\Omega(\vec{\sigma}_3\eta f\tilde{g}\tilde{\eta} - \eta g\tilde{f}\tilde{\eta}\vec{\sigma}_3)\tilde{\Omega} . \qquad (5.7)$$

The decomposition $f\tilde{g} = [f\tilde{g}]_0 - i_2[i_2 f\tilde{g}]_0 = (g\tilde{f})^{\sim}$ implies

$$\begin{aligned}
\vec{\sigma}_3\eta f\tilde{g}\tilde{\eta} - \eta g\tilde{f}\tilde{\eta}\vec{\sigma}_3 &= -(\vec{\sigma}_3\eta i_2\tilde{\eta} + \eta i_2\tilde{\eta}\vec{\sigma}_3)[i_2 f\tilde{g}]_0 \\
&= 2(a\tilde{b} + \tilde{a}b)[i_2 f\tilde{g}]_0
\end{aligned} \qquad (5.8)$$

by making use of (5.5). So, together with (2.8) and (2.9) one finds the result

$$\psi\tilde{\psi}r^2\sin\vartheta = (|f|^2 + |g|^2)(|a|^2 - |b|^2) + 2i(a\tilde{b} + \tilde{a}b)[i_2 f\tilde{g}]_0 \quad , \qquad (5.9)$$

where

$$2a = \eta + \eta^* \quad , \quad 2b = \vec{\sigma}_1(\eta - \eta^*) \quad , \quad |a|^2 - |b|^2 = [\eta\tilde{\eta}]_0 \quad . \tag{5.10}$$

There are only two possibilities for the pseudoscalar in (5.8) to vanish: either $a\tilde{b} = -\tilde{a}b = -(a\tilde{b})^{\sim}$, or, $[i_2 f\tilde{g}]_0 = 0$. The first condition leads to $a\tilde{b} = i_2\varepsilon|a|^2$, $\varepsilon \in \mathbb{R}$, i.e., $b = -i_2\varepsilon a$, which implies that the radial spinor η must be of the very particular form $\eta = (1 + \varepsilon\vec{\sigma}_3)a$. With the exception of e.g. $\varepsilon = \pm 1$, $\kappa = 0$ and $m = 0$ (charged neutrino with no centrifugal repulsion) *the particular spinor* $\eta = (1 + \varepsilon\vec{\sigma}_3)(\eta_0 + i_2\eta_2)$ *can not be a solution of the radial equation (4.1).*

The only possibility to find $\beta = 0(\mathrm{mod}\ \pi)$ solutions for a massive lepton therefore is to fulfill the condition $[i_2 f\tilde{g}]_0 = 0$, or, $f\tilde{g} = \tilde{f}g$. One notes that equation (3.1) or equation (3.2) is sufficient to satisfy this restriction. That either equation (3.1) or equation (3.2) are also necessary, is seen by inspection of (3.9), which then leads to the constraint $[\omega^2]_0 = 0$.

My final conclusion concerning the existence of $\beta = 0(\mathrm{mod}\ \pi)$ solutions of the central field Dirac equation therefore is, that the only ones are those defined by equation (3.1), or , equation (3.2). Hence, we fall back upon those, I have presented at the Antigonish workshop 1990.

6. Parity and Multi-valuedness

The parity operation

$$\psi(r_0, r, \vartheta, \varphi) \rightarrow \psi^P = \gamma_0\psi(r_0, r, \pi - \vartheta, \varphi + \pi)\gamma_0 \tag{6.1}$$

leaves equation (2.3) with potential (2.12) invariant. So, if ψ is a solution of (2.3), then, ψ^P, the higher iterates $(\psi^P)^P$, etc., and their linear combinations are also solutions, provided the potential A is even, i.e., $A = A^P$.

In order to investigate the parity of the various solutions ψ obtained via (5.1), (3.1), (3.2), (3.9), (3.10) and (3.29) to (3.32), let me start with the case $\beta = 0(\mathrm{mod}\ \pi)$, or, equivalently, $\lambda = 0$. Equations (5.1) and (3.1) then yield

$$\psi\pi\sqrt{2} = \Omega\eta\ \exp\left\{i_2\left(\kappa\vartheta - \frac{Er_0}{\hbar c}\right)\right\} \equiv \psi_+ \quad , \tag{6.2}$$

and equation (3.2) with a convenient choice of phase leads to

$$\psi \pi \sqrt{2} = \Omega i_3 \eta i_2 \; \exp\left\{ -i_2 \left(\kappa \vartheta + \frac{E r_0}{\hbar c} \right) \right\} \equiv \psi_- \quad .$$
(6.3)

From equations (2.8) and (2.9) one finds

$$\Omega^P = \Omega i_1 \quad ,$$
(6.4)

and with (4.3)

$$\eta^P = \eta^* = -i_1 i_3 \eta i_2 \quad ,$$
(6.5)

such that finally

$$\psi_+^P = \psi_- \; e^{i_2 \pi \kappa} \quad , \quad \psi_-^P = \psi_+ \; e^{-i_2 \pi (\kappa + 1)} \quad .$$
(6.6)

In the case of the hydrogen atom (see section 7), the constant κ is confined to integer values with the exception of the value zero. Wether this also holds for the valence electron of alkaliatoms remains to be seen! At any rate, for $\kappa \in \mathbb{Z}\backslash\{0\}$, ψ_+ and ψ_- have opposite parities and for $\kappa \in \mathbb{R}$ one obtains

$$(\psi_\pm^P)^P = -\psi_\pm \quad .$$
(6.7)

Turning now to the $\lambda \neq 0$ solutions (5.1), (3.9) and (3.10), two cases must be distinguished according to (3.29) and (3.30), namely,

$$\psi_s = \Omega(\eta_s \tilde{\omega} - i_3 \eta_s i_2 \omega) \; \exp\left\{ i_2 \left(\lambda \varphi - \frac{E r_0}{\hbar c} \right) \right\} ,$$
$$\kappa = -k \quad , \quad k = p + |\lambda| + \frac{1}{2} ,$$
(6.8)

and

$$\psi_g = \Omega(\eta_g \omega - i_3 \eta_g i_2 \tilde{\omega}) \; \exp\left\{ i_2 \left(\lambda \varphi - \frac{E r_0}{\hbar c} \right) \right\} \quad , \quad \kappa = k = p + |\lambda| + \frac{1}{2} ,$$
(6.9)

where

$$\omega = \omega_{p\lambda}(\vartheta)\, e^{\frac{i_2\,\pi(|\lambda|-\lambda)}{4|\lambda|}} \quad , \tag{6.10}$$

and η_s, η_g are solutions of (2.19) and (4.2) for the respective values of $\kappa = -k$ or $\kappa = k$, $k = p + |\lambda| + \frac{1}{2}$,

$$-\hbar\gamma_3\,\partial_r\,\eta_{s\atop g}\,i_2 + \left(\gamma_0\frac{E-V}{c} + \gamma_1\frac{\hbar\kappa}{r}\right)\eta_{s\atop g} = mc\eta_{s\atop g}\gamma_0 \quad , \tag{6.11}$$

$$\kappa = \mp k = \mp\left(p + |\lambda| + \frac{1}{2}\right) \quad .$$

One might be tempted to generate the solutions η_g from η_s by applying the linear mapping

$$\eta \longleftrightarrow l(\eta) = -i_3\eta i_2 \quad , \tag{6.12}$$

since this obviously transforms (up to phase factors) η_s into η_g and vice versa. It should be emphasized however, that for a spinor η satisfying the restriction (4.2), the image $l(\eta)$ no longer lies in the domain (4.2). Consequently the map (6.12) is not disposable to generate such a simple equivalence between η_s and η_g. This aspect is quite important when degeneracies between (6.8) and (6.9) are to be classified.

　　Again, with the help of (6.4), (6.5) and (3.22), the parities of ψ_s and ψ_g are easily found, viz.,

$$\psi_s^P = \psi_s\, e^{i_2\pi(p+\lambda+\frac{1}{2})} \quad , \quad \psi_g^P = \psi_g\, e^{i_2\pi(p+\lambda-\frac{1}{2})} \quad , \quad p \in I\!N_0 \quad . \tag{6.13}$$

As seen from (2.8) and (2.9), the valuedness of Ω for the cycle $\varphi \to \varphi + 2\pi$ is $\Omega(\varphi + 2\pi) = -\Omega(\varphi)$. The valuedness of (6.8) and (6.9) in addition is determined by the phase factor $e^{i_2\lambda\varphi}$. So, one finds

$$\psi_{s\atop g}\ \overset{\varphi\to\varphi+2\pi}{\longrightarrow}\ \psi_{s\atop g}\, e^{i_2\pi(2\lambda+1)} = \left(\psi_{s\atop g}^P\right)^P \quad , \tag{6.14}$$

which proves the conjecture stated in the introduction.

　　At the end of section 3, I promised to briefly establish the relation between spherical harmonics with spin and spherical monogenics. Here it is! Going back from (2.14) to (2.3) and putting $m = 0$, $q = 0$, or, $V = 0$, and $E = 0$, the spinors (6.8) and (6.9)

$$\psi_\bullet = \Omega(\eta_\bullet \tilde{\omega} - i_3 \eta_\bullet i_2 \omega) \, e^{i_2 \lambda \varphi} \tag{6.15}$$

$$\psi_g = \Omega(\eta_g \omega - i_3 \eta_g i_2 \tilde{\omega}) \, e^{i_2 \lambda \varphi} \tag{6.16}$$

are seen to be solutions of

$$\vec{\partial} \psi_{\substack{\bullet \\ g}} = 0 \quad . \tag{6.17}$$

The radial equations (6.11)

$$\partial_r i_2 \eta i_2 + \frac{\kappa}{r} \eta = 0 \quad , \quad \kappa = \mp k = \mp(p + |\lambda| + \frac{1}{2}) \quad , \tag{6.18}$$

according to (4.5), have the solutions

$$\eta = r^\kappa \tag{6.19}$$

and

$$\eta = \vec{\sigma}_1 r^{-\kappa} \quad , \tag{6.20}$$

which generate multi-valued spherical monogenics with the aid of (6.16) and (6.15). In the same way one may construct $\beta = 0 (\mathrm{mod}\ \pi)$ monogenics by making use of (5.1), (3.1), or, (3.2).

7. Hydrogen-like Bound States

In this section, square integrable solutions of type (4.5) are derived for the radial equation

$$- \hbar \gamma_3 \partial_r \eta i_2 + \left(\gamma_0 \frac{E - V}{c} + \gamma_1 \frac{\hbar \kappa}{r} \right) \eta = mc \eta \gamma_0 \quad , \tag{7.1}$$

$$\eta = \eta_0 + \vec{\sigma}_1 \eta_1 \quad , \quad \eta_{0,1} \in I\!\!R \quad ,$$

with the attractive Coulomb potential

$$rV = -\hbar cZ\alpha \quad , \quad \frac{1}{\alpha} = \frac{\hbar c}{q^2} = 137,036 \quad , \tag{7.2}$$

under the conditions, that the charge number Z of the nucleus satisfies

$$0 < Z\alpha < k \quad , \quad 1 \le k = |\kappa| \quad , \tag{7.3}$$

and the energy E is in the range

$$0 < E < mc^2 \quad . \tag{7.4}$$

Introducing a reduced energy ε, an effective fine-structure constant $\bar{\alpha}$

$$E = \varepsilon mc^2 \quad , \quad \bar{\alpha} = Z\alpha \tag{7.5}$$

and instead of $r = |\vec{r}\,|$ the variable

$$y = 2r\frac{mc}{\hbar}\sqrt{1 - \varepsilon^2} \quad , \quad 0 < \varepsilon < 1 \quad , \tag{7.6}$$

the radial equation (7.1) becomes

$$-\gamma_3 \partial_y \eta i_2 + \left[\frac{\varepsilon\gamma_0}{2\sqrt{1 - \varepsilon^2}} + \frac{1}{y}(\bar{\alpha}\gamma_0 + \kappa\gamma_1)\right]\eta = \frac{\eta\gamma_0}{2\sqrt{1 - \varepsilon^2}} \quad . \tag{7.7}$$

Provided condition (7.3) is fulfilled, the vector $\bar{\alpha}\gamma_0 + \kappa\gamma_1$ is spacelike and can be made proportional to γ_1 with the help of Lorentz transformations. The spinors for all these transforms can be traced back to the following definitions

$$\gamma = \sqrt{k^2 - \bar{\alpha}^2} \quad , \quad 1 \le k = |\kappa| \quad , \quad \bar{\alpha} < 1 \quad , \tag{7.8}$$

$$\gamma L^2 = k + \bar{\alpha}\vec{\sigma}_1 \tag{7.9}$$

$$L = \sqrt{\frac{\gamma}{2(\gamma + k)}}(1 + L^2) \quad . \tag{7.10}$$

The spinor L has the properties

$$L^{-1} = \tilde{L} = L^* \tag{7.11}$$

and

$$L\gamma_3 = \gamma_3 L \quad, \quad L\gamma_0 = \gamma_0 \tilde{L} \quad, \quad L\gamma_1 = \gamma_1 \tilde{L} \ . \tag{7.12}$$

So, for $\kappa > 0$ one finds the Lorentz transformation

$$\bar{\alpha}\gamma_0 + \kappa\gamma_1 = \gamma L^2 \gamma_1 = \gamma L\gamma_1 \tilde{L} \quad, \quad \kappa > 0 \tag{7.13}$$

and for $\kappa < 0$

$$\bar{\alpha}\gamma_0 + \kappa\gamma_1 = (\bar{\alpha}\gamma_0 + |\kappa|\gamma_1)^* = \gamma(L^2\gamma_1)^* = -\gamma\tilde{L}^2\gamma_1 = -\gamma\tilde{L}\gamma_1 L \quad, \quad \kappa < 0 \ . \tag{7.14}$$

Making use of (7.14) the radial equation (6.11) in the form (7.7) for the $\kappa < 0$ solution (6.8) (with the suffix s = smaller) may be written

$$\vec{\sigma}_3 \partial_y L\eta_s i_2 + \left(\frac{\varepsilon\tilde{L}^2}{2\sqrt{1-\varepsilon^2}} + \vec{\sigma}_1 \frac{\gamma}{y} \right) L\eta_s = \frac{(L\eta_s)^*}{2\sqrt{1-\varepsilon^2}} \quad, \tag{7.15}$$

and the corresponding one for $\kappa > 0$ (suffix g = greater)

$$\vec{\sigma}_3 \partial_y \tilde{L}\eta_g i_2 + \left(\frac{\varepsilon L^2}{2\sqrt{1-\varepsilon^2}} - \vec{\sigma}_1 \frac{\gamma}{y} \right) \tilde{L}\eta_g = \frac{(\tilde{L}\eta_g)^*}{2\sqrt{1-\varepsilon^2}} \ . \tag{7.16}$$

As seen from (7.8) - (7.10), both equations depend on $|\kappa|$ and not on the sign of κ. Therefore the value

$$|\kappa| = k = p + |\lambda| + \frac{1}{2} \quad, \quad p = 0, 1, 2, \ldots, \in I\!N_0 \tag{7.17}$$

may be taken in both (7.15) and (7.16).

For solving (7.15) and (7.16), I follow the standard method, see e.g. section 28 in [6], and split from η its behaviour at $y = 0$ and $y = \infty$. For $y \to 0_+$ (7.15) and (7.16) behave like (6.18) with κ replaced by γ. So, the regular solutions are

$$\eta_\bullet \xrightarrow[y \to 0_+]{} \vec{\sigma}_1 \tilde{L} y^\gamma \quad , \quad \eta_g \xrightarrow[y \to 0_+]{} L y^\gamma \quad . \tag{7.18}$$

It should be noted that, as distinguished from the incorrect nonrelativistic limit (Schrödinger), the Coulomb potential (7.2) in Dirac's theory acts as a *singular potential*, which modifies the free-particle behaviour to a potential-dependent different form.

At $y \to \infty$ the spinor L cancels out, since its importance is confined to the origin, and equations (7.15) and (7.16) behave asymptotically like

$$2\sqrt{1 - \varepsilon^2} \, \vec{\sigma}_3 \partial_y \eta i_2 + \varepsilon \eta \simeq \eta^* \quad , \quad \eta \simeq \eta_\bullet \simeq \eta_g \quad . \tag{7.19}$$

Looking for square integrable solutions, I write

$$\eta \simeq e^{-\mu y} H \quad , \quad \mu > 0 \quad , \tag{7.20}$$

Where H is a constant spinor of the form (4.5), i.e., with $H_{0,1} \in \mathbb{R}$,

$$H = H_0 + \vec{\sigma}_1 H_1 = \sqrt{H_0^2 - H_1^2} K \quad , \quad K^{-1} = \tilde{K} = K^* \quad . \tag{7.21}$$

Equation (7.19) then imposes on the unimodular spinor K the condition

$$\varepsilon K^2 = 1 + 2\mu \vec{\sigma}_1 \sqrt{1 - \varepsilon^2} = 1 + \vec{\sigma}_1 \sqrt{1 - \varepsilon^2} \quad , \tag{7.22}$$

which because of $K^2 (K^2)^\sim = 1$ can only hold if $4\mu^2 = 1$, or, for $\mu = \frac{1}{2}$ in (7.20) and (7.22). From $1 + K^4 = 2K^2/\varepsilon$ one infers (up to a sign)

$$K = \sqrt{\frac{\varepsilon}{2(1 + \varepsilon)}} (1 + K^2) = \frac{1}{\sqrt{2\varepsilon}} (\sqrt{1 + \varepsilon} + \vec{\sigma}_1 \sqrt{1 - \varepsilon}) \quad . \tag{7.23}$$

Equations (7.18) and (7.20) suggest to put

$$L \eta_\bullet = y^\gamma \, e^{-y/2} \vec{\sigma}_1 (1 + a y \partial_y) \mathcal{L}(y) \quad , \tag{7.24}$$

$$\tilde{L} \eta_g = y^\gamma \, e^{-y/2} (1 + b y \partial_y) \mathcal{M}(y) \quad , \tag{7.25}$$

where a and b are constant spinors and $\mathcal{L}$, $\mathcal{M}$ are *scalar* functions of the variable y. Inserting (7.24) and (7.25) in (7.15) and (7.16) one finds after some straightforward algebra that $\mathcal{L}$ and $\mathcal{M}$ have to solve the following confluent hypergeometric differential equation

$$[y\partial_y^2 + (2\gamma + 1 - y)\partial_y + \nu(\varepsilon)]\mathcal{L}(y) = 0 \quad , \quad \mathcal{M}(y) = \mathcal{L}(y) \quad , \tag{7.26}$$

where

$$\nu(\varepsilon) = \frac{\bar{\alpha}\varepsilon}{\sqrt{1 - \varepsilon^2}} - \gamma \quad , \tag{7.27}$$

provided the spinors a and b are determined by

$$\gamma a(L^2 - K^2) = L^2 \quad , \quad \gamma b(L^2 + \tilde{K}^2) = \tilde{K}^2 \quad . \tag{7.28}$$

With integration constants c_1, c_2, the general solution of (7.26) may be written, see e.g. [4], pp. 252-253,

$$\mathcal{L}(y) = c_{11}\,F_1(-\nu; 2\gamma + 1; y) + c_2 e^y {}_1F_1(\nu + 2\gamma + 1; 2\gamma + 1; y) \quad . \tag{7.29}$$

Square integrable solutions exist if $c_2 = 0$ and only if the Sommerfeld condition

$$\nu(\varepsilon) = \frac{\bar{\alpha}\varepsilon}{\sqrt{1 - \varepsilon^2}} - \gamma = 0, 1, 2, \ldots, \in I\!N_0 \tag{7.30}$$

is fulfilled, which leads to his famous fine-structure formula

$$\varepsilon = \frac{E}{mc^2} = \left[1 + \left(\frac{\bar{\alpha}}{\nu + \gamma}\right)^2\right]^{-1/2} \quad ,$$

$$\gamma = \sqrt{k^2 - \bar{\alpha}^2} \quad , \quad k = p + |\lambda| + \frac{1}{2} \quad , \quad \nu, p \in I\!N_0 \quad . \tag{7.31}$$

For $c_2 = 0$ and $\nu \in I\!N_0$, $\mathcal{L}(y)$ according to (7.29) is proportional to the Laguerre polynomial $L_\nu^{(2\gamma)}(y)$, [4], p.268, and the unnormalized spinors (7.24), (7.25) become

$$\eta_{\bullet} = y^{\gamma}\ e^{-y/2}\vec{\sigma}_1\tilde{L}\left(\gamma+\frac{L^2 y\partial_y}{L^2-K^2}\right)L_{\nu}^{(2\gamma)}(y)\quad,$$

$$\eta_{g} = y^{\gamma}\ e^{-y/2}L\left(\gamma+\frac{\tilde{K}^2 y\partial_y}{L^2+\tilde{K}^2}\right)L_{\nu}^{(2\gamma)}(y)\quad,$$

which up to a scalar factor may as well be written in the form

$$\eta_{\bullet} = y^{\gamma}\ e^{-y/2}\tilde{K}[\gamma(1-K^2\tilde{L}^2)+y\partial_y]L_{\nu}^{(2\gamma)}(y) = \eta_{\bullet}^{\dagger}\quad,$$

$$\eta_{g} = y^{\gamma}\ e^{-y/2}\tilde{K}[\gamma(1+K^2 L^2)+y\partial_y]L_{\nu}^{(2\gamma)}(y) = \eta_{g}^{\dagger}\quad. \tag{7.32}$$

The derivative of the Laguerre polynomial can be eliminated with the help of

$$y\partial_y L_{\nu}^{(2\gamma)}(y) = \nu L_{\nu}^{(2\gamma)}(y)-(\nu+2\gamma)L_{\nu-1}^{(2\gamma)}(y)\quad,\quad \nu\geq 0\quad,\quad L_{-1}^{(2\gamma)}(y)=0\quad, \tag{7.33}$$

which yields

$$\eta_{\bullet} = \eta_{\bullet}^{\dagger} = y^{\gamma}\ e^{-y/2}\tilde{K}[(\nu+\gamma-\gamma K^2\tilde{L}^2)L_{\nu}^{(2\gamma)}(y)-(\nu+2\gamma)L_{\nu-1}^{(2\gamma)}(y)]\quad, \tag{7.34}$$

$$\eta_{g} = \eta_{g}^{\dagger} = y^{\gamma}\ e^{-y/2}\tilde{K}[(\nu+\gamma+\gamma K^2 L^2)L_{\nu}^{(2\gamma)}(y)-(\nu+2\gamma)L_{\nu-1}^{(2\gamma)}(y)]\quad. \tag{7.35}$$

A further simplification is achieved by following Bethe and Salpeter [11] and eliminating ε in favour of the apparent principal quantum-number

$$N = \sqrt{\nu(\nu+2\gamma)+k^2} = \sqrt{n^2+2\nu(\gamma-k)}\quad,$$

$$n = \nu+k\quad,\quad k = p+|\lambda|+\frac{1}{2}\quad, \tag{7.36}$$

whence

$$\varepsilon N = \nu+\gamma\quad,\quad N\sqrt{1-\varepsilon^2} = \bar{\alpha} = Z\alpha\quad,\quad \nu(\nu+2\gamma) = N^2-k^2\quad, \tag{7.37}$$

and

$$(\nu + \gamma)K^2 = N + \vec{\sigma}_1\vec{\alpha} \quad ,$$
$$\nu + \gamma - \gamma K^2 \tilde{L}^2 = (N - k)K^2 \quad , \tag{7.38}$$
$$\nu + \gamma + \gamma K^2 L^2 = (N + k)K^2 \quad .$$

So, (7.34) and (7.35) can be written in the simpler form

$$\eta_s = \eta_s^\dagger = N_s y^\gamma \, e^{-y/2} \, [(N - k)KL_\nu^{(2\gamma)}(y) - (\nu + 2\gamma)\tilde{K}L_{\nu-1}^{(2\gamma)}(y)] \quad , \tag{7.39}$$

$$\eta_g = \eta_g^\dagger = N_g y^\gamma \, e^{-y/2} \, [(N + k)KL_\nu^{(2\gamma)}(y) - (\nu + 2\gamma)\tilde{K}L_{\nu-1}^{(2\gamma)}(y)] \quad . \tag{7.40}$$

The normalization constants N_s and N_g are now determined such that the spinors (6.8) and (6.9) are normalized according to

$$\int d^3r \, [\psi\psi^\dagger]_0 = 1 \quad , \quad \psi = \psi_{s,g} \quad , \tag{7.41}$$

which restricts the total probability in the $I\!\!E^3$ to the value 1, or, the total charge to the value q. Equation (3.33) then implies for the spinor η the normalization condition

$$\int_0^\infty dr \, [\eta\eta^\dagger]_0 = 1 = \frac{a_0 N}{2Z} \int_0^\infty dy \, [\eta^2]_0 \quad , \quad a_0 = \frac{\hbar^2}{mq^2} \quad , \tag{7.42}$$

where a_0 is the *Bohr radius*. The integrals in (7.42) are easily done by making use of

$$\int_0^\infty dy \, y^{2\gamma} \, e^{-y} L_\mu^{(2\gamma)}(y) L_\nu^{(2\gamma)}(y) = \frac{\delta_{\mu\nu}\Gamma(\nu + 2\gamma + 1)}{\Gamma(\nu + 1)} \quad . \tag{7.43}$$

Before displaying the final results in Table 2, it should be noted that $\eta_s = 0$ for $\nu = 0$. Hence, square-integrable nontrivial solutions η_s only exist for $\nu \geq 1$, $\nu \in I\!\!N$. Together with (7.40) they form a twofold degenerated set of solutions of (7.1) for the energy eigenvalues (7.31). Only for $\nu = 0$, equation (7.40) defines nondegenerate solutions, which include the state of lowest energy and Darwin's [10] $1S_{\frac{1}{2}}$ ground state solution. The degeneracy for $\nu \geq 1$ is dynamical, because it originates in the shape of the potential. A variation of the Coulomb potential , e.g., by the selfinteraction removes this degeneracy!

TABLE 2. Normalized one- and two-valued bound state spinors for hydrogen-like atoms

---

Normalization $\qquad\qquad \int d^3r\,[\psi\psi^\dagger]_0 = 1$

$\boxed{\nu = 0}\qquad \gamma = \sqrt{k^2 - Z^2\alpha^2}\ ,\quad N = k = 1,2,3,\ldots \in I\!N\ ,\quad kE = mc^2\gamma$

$$\eta_s = 0\ ,\quad \eta_g = \frac{1}{k}\left(\frac{Z}{a_0\Gamma(2\gamma+1)}\right)^{1/2} y^\gamma\, e^{-y/2}\left(\sqrt{k+\gamma} + \vec{\sigma}_1\sqrt{k-\gamma}\right)\ ,$$

$$ya_0 = 2Zr\ ,\quad a_0 = \frac{\hbar^2}{mq^2}\ .$$

$\boxed{\nu = 1,2,3,\ldots \in I\!N}\qquad \gamma = \sqrt{k^2 - Z^2\alpha^2}\ ,\quad N = \sqrt{\nu(\nu + 2\gamma) + k^2}\ ,$

$$EN = mc^2(\nu + \gamma)\ ,\quad k = 1,2,3,\ldots \in I\!N$$

$$\eta_s = \frac{1}{N}\left[\frac{Z\nu!(\nu+\gamma)(N-k)}{a_0 N\Gamma(\nu+2\gamma+1)}\right]^{1/2} y^\gamma\, e^{-y/2}\left[KL_\nu^{(2\gamma)}(y) - \frac{1}{\nu}(N+k)L_{\nu-1}^{(2\gamma)}(y)\right]$$

$$\eta_g = \frac{1}{N}\left[\frac{Z\nu!(\nu+\gamma)(N+k)}{a_0 N\Gamma(\nu+2\gamma+1)}\right]^{1/2} y^\gamma\, e^{-y/2}\left[KL_\nu^{(2\gamma)}(y) - \frac{1}{\nu}(N-k)L_{\nu-1}^{(2\gamma)}(y)\right]$$

$$y = \frac{2Zr}{a_0 N},\ a_0 = \frac{\hbar^2}{mq^2},\ K\sqrt{2(\nu+\gamma)} = \sqrt{N+\nu+\gamma} + \vec{\sigma}_1\sqrt{N-\nu-\gamma},$$

$$(\nu+1)L_{\nu+1}^{(2\gamma)}(y) = (2\nu+2\gamma+1-y)L_\nu^{(2\gamma)}(y) - (\nu+2\gamma)L_{\nu-1}^{(2\gamma)}(y)\ ,$$

$$\nu \geq 0\ ,\quad L_{-1}^{(2\gamma)}(y) = 0\ ,\quad L_0^{(2\gamma)}(y) = 1\ .$$

$\boxed{\text{two-valued spinors}}$

$$\psi = \frac{\Omega}{\pi\sqrt{2}}\eta_s\,\exp\left\{i_2\left(\mp k\vartheta - \frac{Er_0}{\hbar c}\right)\right\},\quad \psi = \frac{\Omega}{\pi\sqrt{2}}i_3\eta_g\,i_2\,\exp\left\{i_2\left(\pm k\vartheta - \frac{Er_0}{\hbar c}\right)\right\}$$

$\boxed{\text{one-valued spinors} = \text{Darwin}[9]}\qquad k = p + |\lambda| + \frac{1}{2},\ p,m = 0,1,2,\ldots \in I\!N_0,$

$$|\lambda| = m + \frac{1}{2}\ ,\quad \psi_s = \Omega(\eta_s\tilde{\omega} - i_3\eta_s i_2\omega)\,\exp\left\{i_2\left(\lambda\varphi - \frac{Er_0}{\hbar c}\right)\right\},$$

$$\psi_g = \Omega(\eta_g\omega - i_3\eta_g i_2\tilde{\omega})\,\exp\left\{i_2\left(\lambda\varphi - \frac{Er_0}{\hbar c}\right)\right\}$$

$$\omega = \omega_{p\lambda}(\vartheta)\,e^{\frac{i_2\pi(|\lambda|-\lambda)}{4|\lambda|}}\ ,\quad \Omega r\sqrt{\sin\vartheta} = S = e^{-i_3\varphi/2}\,e^{-i_2\vartheta/2}$$

$$\omega_{p\lambda}(\vartheta) = \frac{2^{|\lambda|}\Gamma(|\lambda|)}{4\pi}\left(\frac{p!}{\Gamma(p+2|\lambda|+1)}\right)^{1/2}\sin^{|\lambda|}\vartheta\ e^{i_2\vartheta/2}$$

$$\left[(p+2|\lambda|)C_p^{|\lambda|}(\cos\vartheta) + 2i_2|\lambda|\sin\vartheta\ C_{p-1}^{|\lambda|+1}(\cos\vartheta)\right]\ .$$

Comparing the one-valued solutions with those of Bethe and Salpeter [11], one finds after deriving their matrix form according to equations (2.1)-(2.4) and [1], that they become identical with Darwin's solutions [9] in a nonstandard matrix representation.

8. Anomalous Zeeman Effect, Magnetic Moments

Richard Gurtler [12], to my knowledge, has given the first variational formulation of the real space-time version of Dirac's theory. His action functional, a mapping from a sufficiently well behaved set of real spinor fields ψ into the scalars $I \in \mathbb{R}$,

$$
I(\psi) = \int_{ct_1}^{ct_2} dr_0 \int_{v \subseteq \mathbb{E}^3} d^3 r \, L(x) \quad , \quad x = (r_0 + \vec{r}\,)\gamma_0 \quad , \tag{8.1}
$$

corresponding to equation (2.3) then is defined in terms of the Lagrange density

$$
L = \left[(\hbar \partial \psi i \vec{n} - mc\psi\gamma_0 - \frac{q}{c} A\psi)\gamma_0 \tilde{\psi} \right]_0 \quad , \quad \vec{n} = \vec{\sigma}_2 \quad . \tag{8.2}
$$

If for time-independent potentials $A = A(\vec{r}\,)$ this set of spinor fields is restricted to square-integrable stationary ones satisfying

$$
\hbar \partial_t \psi i \vec{n} = E\psi = \hbar c \partial_0 \psi i \vec{n} \quad , \tag{8.3}
$$

equation (8.1) gives rise to the hamiltonian action

$$
\mathcal{A} = \int_{\mathbb{E}^3} d^3 r \left[(E\psi - \mathcal{H}(\psi))\psi^\dagger \right]_0 \quad , \quad \psi^\dagger = \gamma_0 \tilde{\psi} \gamma_0 = \tilde{\psi}^* \quad , \tag{8.4}
$$

with the linear and spinor-valued hamiltonian operator

$$
\mathcal{H}(\psi) = -\hbar c \vec{\partial} \psi i \vec{n} + q\gamma_0 A\psi + mc^2 \psi^* \quad , \tag{8.5}
$$

which includes non-commutative pre-and postmultiplication by multivectors. It should be noted, that the scalar product of the two spinors ψ and φ, defined by $[\varphi\psi^\dagger]_0 = [\psi\varphi^\dagger]_0$, is just the natural one in $\mathbb{R}^8$ to which the space of Dirac-Hestenes spinors $\mathcal{G}(3) \cong \mathcal{G}^+(1,3)$ is isomorphic. The euclidean magnitude $|\psi| = \sqrt{[\psi\psi^\dagger]_0}$ of a spinor ψ fulfills all laws of a Minkowski-Banach metric. So, on the linear space of (Lebesgue) square-integrable spinor fields (on $I\!E^3$), the bilinear functional (into $I\!R$ and <u>not</u> into $\mathcal{C}$ as in conventional quantum mechanics)

$$< \varphi, \psi > = \int d^3r \, [\varphi\psi^\dagger]_0 \quad , \quad < \psi, \psi > \geq 0 \quad , \tag{8.6}$$

leads to the norm $\sqrt{< \psi, \psi >}$ (and vice versa) and endows this linear space with the structure of a complete normed space, i.e., of a Banach space. By application of Stokes's theorem, one notes that (8.5) is symmetric, viz.

$$< \mathcal{H}(\psi), \psi > = < \psi, \mathcal{H}(\psi) > \quad , \tag{8.7}$$

whence equation (8.4) alternatively may be written in the form

$$\mathcal{A} = < E\psi - \mathcal{H}(\psi), \psi > = < \psi, E\psi - \mathcal{H}(\psi) > = E < \psi, \psi > - < \mathcal{H}(\psi), \psi > \quad . \tag{8.8}$$

In the case of the Zeeman effect the potential A is composed of a central field and the vector potential $\vec{A}$ of a constant magnetic field $\vec{B} = \vec{\partial} \times \vec{A} = B\vec{\sigma}_3$,

$$A = \frac{1}{q}V(r)\gamma_0 + \vec{A}\gamma_0 \tag{8.9}$$

$$\vec{A} = \frac{1}{2}\vec{B} \times \vec{r} = \frac{1}{2}B\vec{\sigma}_3 \times \vec{r} = \Omega\vec{A}'\Omega^{-1} \quad ,$$

$$\vec{A}' = \frac{1}{2}Br\vec{\sigma}_2 \sin \vartheta \quad , \quad \Omega r\sqrt{\sin \vartheta} = S = e^{-i_3\varphi/2} \, e^{-i_2\vartheta/2} \quad . \tag{8.10}$$

The hamiltonian (8.5) may be decomposed according to (8.9)

$$\mathcal{H}(\psi) = \mathcal{H}_0(\psi) + \mathcal{H}_1(\psi) \tag{8.11}$$

into the unperturbed part

$$\mathcal{H}_0(\psi) = -\hbar c\vec{\partial}\psi i_2 + V\psi + mc^2\psi^* \tag{8.12}$$

and the perturbation due to the magnetic field

$$\mathcal{H}_1(\psi) = -q\vec{A}\psi \quad , \tag{8.13}$$

which entails a corresponding splitting of the action (8.8)

$$\mathcal{A} = \mathcal{A}_0 + \mathcal{A}_1 \quad ,$$
$$\mathcal{A}_0 = E < \psi, \psi > \; - \; < \mathcal{H}_0(\psi), \psi >$$
$$= E \int d^3r \, [\psi \psi^\dagger]_0 - \int d^3r \, [\mathcal{H}_0(\psi)\psi^\dagger]_0 \tag{8.14}$$

$$\mathcal{A}_1 = < \mathcal{H}_1(\psi), \psi > \; = q \int d^3r \, [\vec{A}\psi\psi^\dagger]_0 \quad . \tag{8.15}$$

In order to establish a simple relation between $\mathcal{A}_1$ and the magnetic dipole moment of Dirac's current $j = \psi\gamma_0\tilde{\psi}$, one may rearrange the term $[\vec{A}\psi\psi^\dagger]_0$ for $\vec{A} = \frac{1}{2}\vec{B} \times \vec{r}$, $\partial\vec{B} = 0$ as follows:

$$[\vec{A}\psi\psi^\dagger]_0 = [\gamma_0 \vec{A}\psi\gamma_0 \tilde{\psi}]_0 = -[\vec{A}\gamma_0 \psi\gamma_0 \tilde{\psi}]_0 = -(\vec{A}\gamma_0) \cdot j \quad ,$$
$$j = \psi\gamma_0\tilde{\psi} = (j_0 + \vec{j})\gamma_0 \quad ,$$

Now because of $(\vec{A}\gamma_0) \cdot \gamma_0 = 0$ one finds

$$-(\vec{A}\gamma_0) \cdot j = -(\vec{A}\gamma_0) \cdot (\vec{j}\gamma_0) = \frac{1}{2}(\vec{A}\vec{j} + \vec{j}\vec{A}) \equiv \vec{A} \cdot \vec{j}$$
$$= \frac{1}{2}(\vec{B} \times \vec{r}) \cdot \vec{j} = \frac{1}{2}\vec{B} \cdot (\vec{r} \times \vec{j}) = [\vec{A}\psi\psi^\dagger]_0 \quad . \tag{8.16}$$

So, defining the magnetic dipole moment $\vec{\mu}$ of the Dirac charge current $cq\vec{j}$ according to

$$\vec{\mu} = \frac{q}{2} \int d^3r \, \vec{r} \times \vec{j} \quad , \quad \vec{j} = (\psi\gamma_0\tilde{\psi}) \wedge \gamma_0 = [\psi\psi^\dagger]_1 \quad , \tag{8.17}$$
suffix 1 in $[\psi\psi^\dagger]_1$ means: grade 1 part in $\mathcal{G}(3) = \mathcal{G}^+(1,3)$,

equation (8.16) implies that (8.15) may be written in the wellknown simple form

$$\mathcal{A}_1 = \vec{B} \cdot \vec{\mu} \quad . \tag{8.18}$$

It is obvious, that if the special magnetic field $\vec{B}$ is replaced by a time dependent general electromagnetic field $F = \vec{E} + i\vec{B}$, the interaction term in (8.2) may be brought into a form, which generalizes (8.18) in terms of the full electromagnetic polarization bivector of the Dirac current. Applications of this modified action [13] in quantum electrodynamics will be published elsewhere.

The application of the principle of least action on the Zeeman effect now is conventional and straightforward. Assume that a finite number of unperturbed solutions ψ_l is known,

$$\mathcal{H}_0(\psi_l) = E_l \psi_l \quad , \quad l = 1, \ldots, l_{max} \quad , \tag{8.19}$$

and extremize (minimize) the action (8.14) for the trial spinor

$$\psi = \sum_l c_l \psi_l \tag{8.20}$$

with respect to the scalar, linear variation parameters c_l. (A generalization to nonlinear spinor-valued variations is evident.) In this way, one finds

$$\mathcal{A} \equiv \mathcal{A}(c) = \sum_{l_1 l_2} \left[(E - E_{l_1}) \Delta_{l_1 l_2} + \vec{B} \cdot \vec{\mu}_{l_1 l_2} \right] c_{l_1} c_{l_2} \geq 0 \quad , \tag{8.21}$$

where the overlap matrix $\Delta_{l_1 l_2}$ is

$$\Delta_{l_1 l_2} = \; < \psi_{l_1}, \psi_{l_2} > \; = \int d^3 r \, [\psi_{l_1} \psi_{l_2}^\dagger]_0 = \Delta_{l_2 l_1} \quad , \tag{8.22}$$

and the $\vec{\mu}_{l_1 l_2}$ are the magnetic dipole moments of the current matrix

$$\vec{j}_{l_1 l_2} = \vec{j}_{l_2 l_1} = [\psi_{l_1} \psi_{l_2}^\dagger]_1 \quad , \quad \vec{\mu}_{l_1 l_2} = \frac{q}{2} \int d^3 r \, \vec{r} \times \vec{j}_{l_1 l_2} = \vec{\mu}_{l_2 l_1} \quad , \tag{8.23}$$

suffix 1 means: grade 1 part in $\mathcal{G}(3)$.

The characteristic system of Euler-Lagrange equations then is

$$\frac{1}{2} \frac{\partial \mathcal{A}(c)}{\partial c_l} = \sum_{l_1} \left[(E - E_l) \Delta_{l l_1} + \vec{B} \cdot \vec{\mu}_{l l_1} \right] c_{l_1} = 0 \quad , \quad l = 1, 2, \ldots, l_{max} \quad , \tag{8.24}$$

and the energy eigenvalues E are obtained by solving the characteristic equation

$$\det \left[(E - E_l)\Delta_{ll_1} + \vec{B} \cdot \vec{\mu}_{ll_1} \right] = 0 \quad . \tag{8.25}$$

Equations (8.24) and (8.25) are now applied to a few states in TABLE 2. As representative two- and one-valued spinors I select the following solutions for the lowest energy $E_0 = mc^2\gamma$, $Z = 1$, $\gamma = \sqrt{1 - \alpha^2}$ with $\nu = 0$, $k = 1$, $\lambda = 0$, or, $p = 0$, $|\lambda| = \frac{1}{2}$, $N = 1$:

two-valued solutions

$$\psi_a \pi\sqrt{2} = \Omega\eta \, \exp\left\{ i_2 \left(\vartheta - \frac{E_0 r_0}{\hbar c} \right) \right\},$$

$$\psi_b \pi\sqrt{2} = \Omega i_3\eta \, \exp\left\{ -i_2 \left(\vartheta + \frac{E_0 r_0}{\hbar c} \right) \right\} \quad, \tag{8.26}$$

$$\Omega r\sqrt{\sin\vartheta} = S = e^{-i_3\varphi/2} \, e^{-i_2\vartheta/2} \quad, \tag{8.27}$$

one-valued solutions (Darwin)

$$r\sqrt{8\pi}\psi_{D\pm} = S\left(\eta \, e^{i_2\vartheta/2} \mp i_3\eta i_2 \, e^{-i_2\vartheta/2} \right) \exp\left\{ i_2 \left(\pm\frac{\varphi}{2} - \frac{E_0 r_0}{\hbar c} \right) \right\} \quad . \tag{8.28}$$

Apart from $\int_0^\infty dr \, r[\vec{\sigma}_1 \eta^2]_0 < \infty$, and the normalization condition

$$\int_0^\infty dr \, [\eta^2]_0 = 1 \quad, \tag{8.29}$$

the spinor η may be left arbitrary in principle. In the case of the hydrogen atom η is given by $\eta = \eta_g$, i.e.,

$$\eta[a_0\Gamma(2\gamma + 1)]^{1/2} = y^\gamma \, e^{-y/2} \left(\sqrt{1+\gamma} + \vec{\sigma}_1\sqrt{1-\gamma} \right) \quad,$$

$$a_0 y = 2r \quad, \quad a_0 = \frac{\hbar^2}{mq^2} \quad. \tag{8.30}$$

As a consequence of the normalization (8.29) the overlap matrix for the set ψ_a, ψ_b and the set ψ_{D+}, ψ_{D-} is the unitmatrix respectively

$$\Delta_{aa} = 1 = \Delta_{bb} \quad , \quad \Delta_{ab} = 0 = \Delta_{ba} \quad ;$$
$$\Delta_{D++} = 1 = \Delta_{D--} \quad , \quad \Delta_{D+-} = 0 = \Delta_{D-+} \quad .$$
(8.31)

Evaluating the current matrix one finds that the currents $\vec{j}_{\substack{aa \\ bb}}$ are poloidal

$$2\pi^2 r^2 \sin \vartheta \vec{j}_{\substack{aa \\ bb}} = \pm [\vec{\sigma}_1 \eta^2]_0 S \vec{\sigma}_1 \tilde{S} \quad ,$$
(8.32)

and the transition current $\vec{j}_{ab} = \vec{j}_{ba}$ is equatorial

$$2\pi^2 r^2 \sin \vartheta \, \vec{j}_{ab} = -[\vec{\sigma}_1 \eta^2]_0 \cos(2\vartheta) S \vec{\sigma}_2 \tilde{S} \quad ,$$
(8.33)

which implies that

$$\vec{\mu}_{\substack{aa \\ bb}} = \frac{q}{2} \int d^3 r \, \vec{r} \times \vec{j}_{\substack{aa \\ bb}} = \vec{0} \quad ,$$
(8.34)

and

$$\vec{\mu}_{ab} = \vec{\mu}_{ba} = \frac{q}{2} \int d^3 r \, \vec{r} \times \vec{j}_{ab} = \frac{q \vec{\sigma}_3}{3\pi} \int_0^\infty dr \, r[\vec{\sigma}_1 \eta^2]_0 \equiv -\mu \vec{\sigma}_3 \quad .$$
(8.35)

The result (8.34) may be interpreted that *a hydrogen atom in one of the two-valued states (8.26) has a vanishing magnetic dipole moment as long as there is no external magnetic field (or an internal selffield) which magnetizes it.* If the presence of the external Zeeman field $\vec{B} = B\vec{\sigma}_3$, $B \geq 0$, is taken into account by the trial spinor $\psi = c_a \psi_a + c_b \psi_b$, equation (8.24) leads to the characteristic system

$$(E - E_0)c_a - \mu B c_b = 0 \quad , \quad -\mu B c_a + (E - E_0)c_b = 0 \quad .$$
(8.36)

From the characteristic equation (8.25) one obtains for an electron ($\mu > 0$)

$$E = E_\pm = E_0 \pm |\mu B| = E_0 \pm \mu B \quad , \quad -q = |q| \quad , \quad \mu > 0 \quad , \quad B > 0 \quad ,$$
(8.37)

and the corresponding normalized solutions of (8.36), (8.26) are

$$\psi_\pm \sqrt{2} = \psi_a \pm \psi_b \quad , \tag{8.38}$$

which yield the following magnetic dipole moments and energies

$$\vec{\mu}_{\pm\pm} = \pm\vec{\mu}_{ab} = \mp\mu\vec{\sigma}_3 \quad ,$$
$$E_\pm = E_0 \pm \mu B = E_0 - \vec{B}\cdot\vec{\mu}_{\pm\pm} \quad , \quad \vec{\mu}_{\pm\mp} = \vec{0} \quad . \tag{8.39}$$

In the same way one finds for the one-valued Darwin solutions (8.28) the results

$$\vec{\mu}_{D\pm\pm} = \mp\pi\mu\vec{\sigma}_3 \quad , \quad \vec{\mu}_{D\pm\mp} = \frac{\pi}{4}\mu\vec{\sigma}_1 \quad , \quad \mu = -\frac{q}{3\pi}\int\limits_0^\infty dr\, r[\vec{\sigma}_1\eta^2]_0 \quad . \tag{8.40}$$

Therefore in the case of the Zeeman field $\vec{B} = B\vec{\sigma}_3$, $B \geq 0$, the characteristic matrix is diagonal and hence one obtains the following energies and corresponding normalized spinors

$$E_{D\pm} = E_0 \pm \pi\mu B = E_0 - \vec{B}\cdot\vec{\mu}_{D\pm\pm} \quad , \quad \psi = \psi_{D\pm} \quad . \tag{8.41}$$

For a hydrogen electron $(-q = |q|)$ the quantity μ in (8.40) is easily calculated with the help of (8.30). One derives the result

$$\mu = \frac{|q|}{3\pi}\int\limits_0^\infty dr\, r[\vec{\sigma}_1\eta^2]_0 = \frac{|q|a_0^2}{12\pi}\int\limits_0^\infty dy\, y[\vec{\sigma}_1\eta^2]_0$$
$$= \frac{|q|a_0\,\alpha}{6\pi}(2\gamma+1) = \frac{2\gamma+1}{3\pi}\mu_B \quad , \tag{8.42}$$

where

$$\mu_B = \frac{|q|\hbar}{2mc} \tag{8.43}$$

is the magneton of Bohr. Equations (8.39)–(8.43) confirm the announcement made in the abstract $\mu\pi = \mu_{Darwin}$, which means that the magnetic dipole moment for the one-valued Darwin solutions (8.28) is a factor of π times larger than for the two-valued realizations (8.38) of this state of lowest energy. Also, it should be emphasized that although ψ_a and ψ_b according to (8.26) and (8.27) are $\beta = 0(\mathrm{mod}\ \pi)$ solutions, the two-valued spinors (8.38) no longer are of this type. This fact indicates a sensitive dependence of the Yvon-Takabayasi parameter β on the magnetic properties of a lepton.

9. Summary

By means of a few selected examples it has been shown that Hestenes's space-time formulation of Dirac's theory provides an efficient geometric alternative to the traditional relativistic quantum theory.

REFERENCES

1. Hestenes, D. (1975) 'Observables, Operators, and Complex Numbers in the Dirac Theory', J. Math. Phys. 16, 556-572.
2. Weyl, H. (1930) The Theory of Groups and Quantum Mechanics, Dover, U.S.A..
3. Bopp, F. and Haag, R. (1950) 'Über die Möglichkeit von Spinmodellen', Z. Naturforschg. 5a, 644-653.
4. Erdélyi, A. (1953) Bateman Manuscript Project, Higher Transcendental Functions, Volume I, pp. 178-179, McGraw-Hill, New York.
5. Delanghe, R. and Sommen, F. (1986) 'Spingroups and spherical monogenics' , in J.S.R. Chisholm and A.K.Commen (eds.), Clifford Algebras and Their Applications in Mathematical Physics, D. Reidel Publishing Company, Dordrecht, pp. 115-132.
6. Rose, M.E. (1961) Relativistic Electron Theory, John Wilney, New York, p. 159, equation 5.5..
7. Yvon, J. (1940) 'Équations de Dirac-Madelung',J. Phys. et le Radium 1, 18-24.
8. Takabayasi, T. (1957) 'Relativistic Hydrodynamics of the Dirac Matter', Prog. Theor. Phys. Suppl. 4, 1-80.
9. Hestenes, D. (1967) 'Real Spinor Fields', J. Math. Phys. 8, 798-808, p. 805, equations (5.7)-(5.11).
10. Darwin, C. G. (1928) 'The Wave Equations of the Electron', Proc. Roy. Soc. Lond., A118, 654-680.
11. Bethe, H. A. and Salpeter, E. E. (1957) Quantum Mechanics of One- and Two-Electron Atoms, Springer, Berlin, pp. 68-70.
12. Gurtler, R.W. (1972) 'Local observables in the Pauli and Dirac formulations of quantum theory', Dissertation, Arizona State University, Appendix A2.
13. Babiker, M. and Loudon, R. (1983) 'Derivation of the Power-Zienau-Woolley Hamiltonian in Quantum Electrodynamics by Gauge Transformation', Proc. Roy. Soc. Lond. A385, 439-460.

THE ROLE OF THE DUALITY ROTATION IN THE DIRAC THEORY. COMPARISON BETWEEN THE DARWIN AND THE KRÜGER SOLUTIONS FOR THE CENTRAL POTENTIAL PROBLEM.

Roger BOUDET
Université de Provence
Pl. Hugo
13331 Marseille
France

ABSTRACT. The Dirac particle is mostly to be represented by a spacelike plane $P(x)$, the "spin plane" considered at each point x of the Minkowski spacetime M. The infinitesimal motion of this plane expresses, after multiplication by the constant $\hbar c/2$, the local energy of the particle.

The spin plane is subjected to a transformation of an euclidean nature, but particular to the geometry of M and situated outside our usual understanding of the geometry of the euclidean spaces, the duality rotation by an "angle" β. Such a transformation can act on planes but leaves invariant the straight lines. It plays a fundamental role in the passage from the equation of the particle to the one of the antiparticle: this passage is achieved by the change $\beta \to \beta + \pi$ which reverses the orientation of planes without changing the one of straight lines. It brings arguments against the physical interest of the PT transform.

The solutions of the Dirac equation for a central potential problem are studied from the point of view of the behaviour of the energy—momentum tensor and of the angle β. All would be clear if one would have, everywhere, $\beta = 0$ for the electron and $\beta = \pi$ for the positron! But it is not the case in the Darwin solutions. A comparison between these solutions and the ones recently established by H. Krüger, for which $\beta = 0$ or π everywhere, is carried out.

1. Introduction

This paper is mostly concerned with the role played in the Dirac theory by the "angle" β. The "mysterious" angle of Yvon–Takabayasi (the qualification "mysterious" is due to L. de Broglie), recognized in the Dirac theory in 1940 (or even before) has found in 1967 its geometrical meaning with the use of the multivector Clifford algebra $C(\mathcal{M})$, or Space Time Algebra (STA), associated with the Minkowski spacetime $\mathcal{M} = \mathbb{R}^{1,3}$.

In this algebra, introduced for the first time in Quantum Mechanics, in a convenient way by D. Hestenes [2], β is an "angle" of a "rotation", the duality rotation, which acts on the antisymmetric tensors of rank two (bivectors) of $\mathcal{M}$, but leaves the vectors of $\mathcal{M}$ invariant. Its meaning is beyond the field of properties which can be reached by means of the orthogonal group $O(\mathcal{M})$ of $\mathcal{M}$.

83

D. Hestenes and A. Weingartshofer (eds.), The Electron, 83–104.

The role of β in Quantum Mechanics is both mathematically clear and physically obscure.

It is clear that the change $\beta \to \beta + \pi$ plays an essential role in the passage from the equation of a particle to the one of its associated antiparticle. But the angle β has never been measured in an experiment.

However, one might deduce, as an indirect result, from the Darwin solutions of the Dirac equation for the central potential problem (see the article of P. Quilichini or the Gurtler thesis [3]) that this angle could have a value distinct from zero for an electron, or π for a positron, and depending on the point x of M where the wave function ψ is considered.

But the solutions, recently established by H. Krüger ("New solutions of the Dirac Equation", in the present "Proceedings"), in which $\beta = 0$ or $\beta = \pi$ everywhere give to the question a new aspect. Should not the values $\beta = 0$ or $\beta = \pi$ be considered as eigenstates of the ψ function, which assign to a Dirac particle the quality of particle or antiparticle?

This is a crucial question with implications for the theory of all elementary particles.

In part I, we recall some results previously established about the geometrical interpretations of the local energy of a Dirac particle, and the role played by the duality rotation, both in this interpretation and in the passage from the electron equation to the positron one ([4] − [9]). Furthermore, using a pure geometrical construction, the Dirac equation is presented here as a simple generalization of the equality $E = \pm mc^2$, relative to the particle or antiparticle at rest, which takes into account its spin.

The aim of this part is, in particular, to avoid a widespread misunderstanding: the use of STA is more than a change of formalism (with respect to the usual matrices and operators representation of Quantum Mechanics); it also brings an important clarification and simplification in Quantum Theory.

STA is to be considered as the fundamental structure of the geometry of spacetime, and the euclidean geometric properties that it allows one to describe cannot be obtained by the usual methods of the matric representations on complex spaces.

Hence, to the extent that it could not be possible that there exists disagreement between the laws of physics and the geometry of spacetime, STA is to be considered as a fundamental structure for Quantum Mechanics.

In part II, using the form given by P. Quilichini [3] and myself [6] to the Darwin solutions, we achieve a comparison between these solutions and Krüger's.

The reading of the first part needs, for its complete comprehension, some knowledge about STA (see Ref. [2] or the article of D. Hestenes in the present "Proceedings").

Readers non acquainted with STA may read the second part, only by using the usual matricial formalism. The sympols γ_μ and $\sigma_k = \gamma_k \gamma_0$ are to be thought as

representing vectors of the euclidean spaces $M = \mathbb{R}^{1,3}$ and $E^3 = \mathbb{R}^{3,0}$ (the latter being to be considered as a space of bivectors inside the former). They are employed in part II only with the help of operating rules which are the same as those used for the Dirac and Pauli matrices.

2. The Dirac Particle and the Duality Rotation.

2.1. THE MOMENTUM–ENERGY TENSOR OF A DIRAC PARTICLE AND THE DUALITY ROTATION. A GEOMETRICAL INTERPRETATION OF THE QUANTAL ENERGY.

2.1.1. *Kinematical définition*. If I had to propose a definition of the energy associated with a Dirac particle, I would say first that such a definition implies a) the mathematical knowledge of the Minkowski spacetime $\mathcal{M} = \mathbb{R}^{1,3}$, b) the physical notions (which are not pecular to Quantum Mechanics) of the local energy and momentum–energy of a system.

Then I would continue to suggest: "We will define a Dirac particle as being an oriented plane P(x) of $\mathcal{M}$, called the "spin plane", passing through a point x of $\mathcal{M}$. P(x) is orthogonal, at each point $x \in \mathcal{M}$, to an unit time–like vector v(x) of $\mathcal{M}$, called the "velocity unit vector" of the particle.

Furthermore, we will suppose that the plane P(x) satisfies the following properties that we will call "*Some of the Geometrical Principles of Quantum Mechanics.*"

First Principle (Principle of Inertia).
"If the energy associated with the particle would be null, v would be a constant vector, and furthermore the plane P(x) would keep the same direction when x varies, without having any movement of rotation on itself."

Second Principle.
"The kinetic part (i.e. the part which does not imply a potential energy) of the momentum–energy tensor T associated with the particle is equal to the product T = kL of a scalar constant k by a linear mapping

$$L: \; n \in \mathcal{M} \longrightarrow L(n) \in \mathcal{M}$$

which expresses the infinitesimal motion in $\mathcal{M}$ of the plane P(x) when the point x varies.

The constant k is equal to $\hbar c/2$, where $h = 2\pi\hbar$ is the so–called Planck constant, and c the velocity of the light".

How is the tensor L to be defined? We have both to express the proper motion of P(x) on itself, and the motion in $\mathcal{M}$ of the direction of the plane P(x).

Suppose that P(x) is defined by a couple (n_1, n_2) of two orthonormal (space–like) vectors, each one orthogonal to v.

Let us complete the set $\mathcal{S}(x) = \{v, n_1, n_2\}$ by a fourth unit (space– like) vector s in such a way that the moving frame $\mathcal{R}(x) = \{v, n_1, n_2, s\}$, or "proper frame of the particle" is orthonormal. We suppose furthermore that $\mathcal{R}(x)$ has the same orientation as a fixed frame of $\mathcal{M}$, $B = \{\gamma_0, \gamma_1, \gamma_2, \gamma_3\}$ on which are defined the components x^μ of x:

$$x = x^\mu \gamma_\mu = x_\mu \gamma^\mu \quad (\gamma_\mu, \gamma^\mu \in \mathcal{M}, \; \gamma_\mu \cdot \gamma^\nu = g^\nu_\mu)$$

The infinitesimal motion of P(x) may be described by the four following mappings (App. I):

$$n \in \mathcal{M} \longrightarrow L_{\mu}(n) = \Omega_{\mu} \cdot (i(n \wedge s)) \in \mathbb{R} \tag{1}$$

in which
— the symbol $X \cdot Y$ means the inner (or "contracted") product, associated with the euclidean structure of $\mathcal{M}$, of two tensors of $\mathcal{M}$;
— Ω_{μ} is a bivector of $\mathcal{M}$ which defines the infinitesimal rotation of the frame $\mathcal{R}(x)$ when x moves in the direction of the vector γ_{μ} (App. I,II);

— the symbol $i(a \wedge b)$ represents the bivector which is the "dual" of the bivector $a \wedge b$ $(a,b \in \mathcal{M})$.
For example one has

$$i(v \wedge s) = n_1 \wedge n_2, \qquad i(n_1 \wedge n_2) = s \wedge v. \tag{2}$$

We will define

$$L(n) = L_{\mu}(n) \, \gamma^{\mu} \in \mathcal{M} \tag{3}$$

in such a way that the spacetime vector $L(n)$ is invariant in all change of frame B.
    The spacetime vector ω:

$$\omega = L(v) \tag{4}$$

is such that the scalar $N \cdot \omega = N^{\mu} \omega_{\mu}$ expresses the infinitesimal proper rotation of the plane $P(x)$ on itself when x moves in the direction of a vector N.
    If one uses the formalism of STA (see Ref [2]) $i(n \wedge s)$ may be interpreted as the Clifford product of the unit pseudoscalar of $\mathcal{M}$

$$i = \gamma_0 \wedge \gamma_1 \wedge \gamma_2 \wedge \gamma_3 \tag{5}$$

by the bivector $n \wedge s$.
    However, we do emphasize that the use of STA is not a necessity for the construction of the tensor L, and that the traditional tensor algebra is sufficient.

Note. Relation with the Zitterbewegung.
    What we have called "the infinitesimal proper rotation of the plane $P(x)$", represented by the spacetime vector ω, corresponds to what is called "Zitterbewegung" in the articles of D. Hestenes and H. Krüger, of the present Proceedings.

The duality rotation. The definition of the tensor L is to be completed by taking into account a transformation, the duality rotation, which concerns the planes of $\mathcal{M}$, and consequently may affect the plane $P(x)$ (App. II).
    We have said that the duality rotation leaves all the vectors of $\mathcal{M}$ invariant and acts on the bivectors. In particular, by using such a mapping, it is possible to change the orientation of planes without changing the one of straight lines.
    In this way, such a transformation is beyond the traditional understanding of the geometry of the euclidean spaces because:

a) one uses to consider mostly the transformations which acts on vectors,
b) or the ones acting on multivectors which can be only deduced from the ortho-gonal group.

Nevertheless, the duality rotation is a pure euclidean transformation, and such a mapping is a trivial fact for mathematicians who are acquainted with the modern approach of the study of the euclidean spaces [10].

For the purpose of taking into account the geometry of planes in spacetime, we consider that the bivector $\sigma = n_1 \wedge n_2$, associated with the direction of $P(x)$, may be submitted to a duality rotation by an angle $\beta(x)$:

$$\sigma = n_1 \wedge n_2 \longrightarrow \sigma' = e^{i\beta}\sigma = \cos \beta \; n_1 \wedge n_2 + \sin \beta \; s \wedge v \tag{6}$$

Then $L_\mu(n)$ becomes (App. II)

$$L_\mu(n) = \Omega_\mu \cdot (i(n \wedge s)) + (n \cdot s)\partial_\mu \beta. \tag{7}$$

2.1.2. *Introduction of the Planck constant.* One can notice that, up to now, we have said nothing about physics. We have used only the euclidean properties of space-time.

Physics is introduced now by means of the relation

$$T(n) = \frac{\hbar c}{2} L(n) \tag{8}$$

as defining the momentum–energy tensor of the particle. One deduces from T the momentum–energy spacetime vector p (reduced to its kinetic part),

$$p = T(v) = \frac{\hbar c}{2} \omega, \quad p_\mu = \frac{\hbar c}{2} (\partial_\mu n_1) \cdot n_2. \tag{9}$$

We are far away from the obscure "Rule of Correspondence" $p_\mu \rightarrow -i\hbar\partial_\mu$!

We use the expression "obscure", because in such a relation, the links which could exist between the symbol "i" and the constant $\hbar$ on one side, and the euclidean structure of spacetime on the other, are in a total obscurity.

Certainly, such an obscurity has been the source of some aberrations one can find [11] in the Theory of Quantum Fields.

If one takes into account the effect of an electromagnetic potential $A \in \mathcal{M}$ on the particle; furthermore if one multiplies T by the density $\rho(x)$ associated, at each point x, with the particle, one obtains the tensor of density of momentum energy (Tetrode tensor), in the form we have established in [6]:

$$\rho T(n) = \rho \left[\frac{\hbar c}{2} (\Omega_\mu \cdot (i(n \wedge s)) + (n \cdot s)\partial_\mu \beta) - q(n \cdot v)A_\mu \right] \gamma^\mu. \tag{10}$$

where q is the charge of the particle.

We recall that, as a consequence of the Dirac equation, the divergence of the Tetrode tensor is equal to the density of the Lorentz force.

The momentum–energy vector p becomes

$$p = T(v) = \frac{\hbar c}{2} \omega - qA.$$ (11)

The electromagnetic gauge invariance of p is obtained by replacing the couple (n_1, n_2) by an other couple (n_1', n_2'), defining a same direction and orientation for $P'(x)$ as for $P(x)$, and A_μ by

$$A'_\mu = A_\mu + \frac{\hbar c}{2q} \partial_\mu \chi$$ (12)

where $\chi(x)$ is the angle of n_1' and n_1.

Note. Topological implications due to the geometrical nature of the momentum—energy vector.

In a survey devoted to monopoles [12], F. Gliozzi has introduced the infinitesimal proper rotation of a space—like plane $\pi(x)$ for expressing the electromagnetic gauge potential. Some of his remarks, concerning the topological defects of such a rotation, might be applied to the momentum—energy p of the Dirac particle.

The question of the physical incidence of such defects is all the more important as what we call a photon is nothing else but a discontinuity of the vector $\hbar c \omega / 2$ associated with the proper rotations of the spin plane in the Dirac theory.

2.1.3. *Introduction of the mass of the particle* The definition of the energy of a Dirac particle proposed above is to be related to the definition $E = mc^2$ of the energy of a material particle whose mass is m. So we will add to the two first principles, the following one.

"Third principle"

"We will say that the Dirac particle is at rest if v, the direction of the plane $P(x)$, and the proper rotation on itself of $P(x)$ are constant.

For a Dirac particle at rest one has $T(v) \cdot v = mc^2$; for a Dirac antiparticle at rest one has $T(v) \cdot v = - mc^2$."

In this particular case, one has $\omega = \omega_0 v$, $\omega_0 = (dn_1/d\tau) \cdot n_2$ where τ/c is the proper time of the particle and one can write $T(v) = p = (\hbar c/2)\omega_0 v$. Furthermore if one links the quality of particle or antiparticle to the duality rotation, the two above relations may be unified in a single one:

$$\frac{\hbar c}{2} \omega = mc^2 e^{i\beta_0} v,$$ (13)

in which $\beta_0 = 0$ for a particle and $\beta = \pi$ for an anti—particle.

But the duality rotation has been related to a transformation of the spin plane.

Multiplying eq. (13) from the right by the bivector σ which represents the direction of $P(x)$, it is convenient to write eq. (13) in the form

$$\frac{\hbar c}{2} \omega\sigma = mc^2 (e^{i\beta_0} \sigma)v.$$ (14)

Eq. (14) is exactly the Dirac equation (see eq. (17) and the relation (0), App. II) in the particular case where the potential is null; v, σ, the density ρ, are constant, and $\beta = 0$ or $\beta = \pi$ everywhere.

So the Dirac equation appears as an immediate generalization of the relations $p = (\hbar c/2)\omega_0 v$, $p = \pm mc^2 v$, which defines the momentum–energy of a material particle (anti–particle) at rest.

Note. The Louis de Broglie corpuscle.

We recall the L. de Broglie has used a comparison with a clock to describe the stationary state associated, as the foundation of Quantum Mechanics, with his corpuscle.

Here, one can imagine that the dial–plate of de Broglie's clock is the spin plane $P(x)$, and the vector n_1 is the hour–hand of the clock.

The duality rotation by an angle π may be seen as a returning of the graduations of the dial–plate of the clock, in the sense inverse of the one which corresponds to the increasing proper time of the corpuscle.

Thus, the notions of spin and anti–particle, the germ of Dirac equation were already included in the de Broglie clock–corpuscle!

2.2. THE PASSAGE FROM THE EQUATION OF THE ELECTRON TO THE ONE OF THE POSITRON, AND THE DUALITY ROTATION. A GEOMETRICAL INTERPRETATION OF THE DIRAC EQUATION.

2.2.1. *The invariant form of the Dirac equation.* In Ref. [2], D. Hestenes has written the Dirac equation for an electron in the form

$$\hbar c\, \gamma^\mu \partial_\mu \psi = -(mc^2 \psi \gamma_0 + eA\psi)\gamma_1 \gamma_2, \qquad A = A^\mu \gamma_\mu, \qquad (15)$$

in which e is the charge of the electron and

$$\psi = \sqrt{\rho}\, e^{i\beta/2}\, R. \qquad (16)$$

R corresponds to a Lorentz rotation such that

$$v = R\gamma_0 R^{-1}, \qquad \sigma = n_1 \wedge n_2 = R\gamma_1 \gamma_2 R^{-1}.$$

In Ref. [5], we have put this equation in a form independent of the frame B:
Multiplying eq. (15) from the right by

$$\psi^{-1} = \frac{1}{\sqrt{\rho}}\, e^{-i\beta/2}\, R^{-1}$$

one finds immediately the intrinsic equation

$$-\frac{\hbar c}{2}\, \eta = mc^2 (e^{i\beta}\sigma)v + eA\sigma, \qquad \sigma = n_1 \wedge n_2. \qquad (17)$$

Here

$$\eta = \gamma^{\mu}\eta_{\mu}, \quad \eta_{\mu} = 2(\partial_{\mu}\psi)\psi^{-1} = \Omega_{\mu} + i\partial_{\mu}\beta + \partial_{\mu}(\ell n\rho) \qquad (18)$$

is the invariant infinitesimal operator associated with the mapping $X \to \psi X \tilde{\psi}$ (App. II).

All the terms of eq. (17) have an intrinsic geometrical meaning. The presence of the angle β in the mass term, and of its gradient in the infinitesimal operator of the left hand side of the equation, is to be related to what we said about the construction of the momentum–energy tensor, and the generalization of the equality $p = \pm mc^2v$ verified by an (anti–)particle at rest.

The scalar part of the equation which may be deduced from eq. (17) after multiplication from the right by σv gives the relation $Tr(T) = mc^2\cos\beta$ (see Ref. [7]). This equality cancels the Lagrangian density which is usually associated with the Dirac particle (see Ref. [1–d], and eq. (8–2) in Krüger's article of these "Proceedings").

2.2.2. *The fundamental role of the duality rotation.* The intrinsic form (17) of the Dirac equation allows us to discuss the passage from the electron equation to the positron one.

There are in actual two possible orientations of the spin bivector σ, the "spin up" and the "spin down" ones. The equation corresponding to one of these orientations is transformed into the other by changing σ into $-\sigma$ in eq. (17) (or $\gamma_1\gamma_2$ into $\gamma_2\gamma_1$ in eq. (15)).

Suppose we want to pass from the equation of an electron to the corresponding one of a positron. In any way, the sign of the charge e is to be changed.

It is convenient, for the conservation of the spin in the creation or an nihilation of a pair, to associate a "spin down" positron with a "spin up" electron, and vice–versa. Thus, the sign in front of σ is to be changed.

As a result, we note that the sign of the charge term of the right hand side of eq. (17) (or eq. (15)) is unchanged. Thus the sign of the mass term is to be unchanged.

Because of the change of sign of σ, we have to change the sign of one of the other factors of the mass term.

Because mc^2 is positive, we have the choice between the changes $v \to -v$ or $\beta \to \beta + \pi$ ([1–d], [8]).

Let us notice that the change $v \to -v$ conforms to the PT transform. The P and T transformations have separately no sense in the consideration of eq. (17) because they need the choice of a particular frame of $\mathcal{M}$ and because eq. (17) is invariant. But PT changes the sense of the vectors of $\mathcal{M}$ independently of all frames and may be applied to the vectors associated with the particle in eq. (17).

This implies that one has $-v$ for the positron, and its physical meaning is that this particle runs from the future to the past (Stückeberg, Feynmann)!

The other possible change $\beta \to \beta + \pi$ (Takabayasi) corresponds simply to a duality rotation by an angle π.

If we do not accept that the positrons which are observed come from the future, we are obliged to consider the change of the orientation of the planes of $\mathcal{M}$, due to a duality rotation by an angle π, as the key of the explanation one can give to the difference between the world of the particles and the one of the antiparticles.

But such a point of view is an infirmation of the physical interest of the PT transform.

Historical note

All the results established in the article [2] of Hestenes and in my paper [5] may be considered as already included in the works achieved by "Louis de Broglie's School", concerning the "Relativistic Hydrodynamics Theory of the Dirac Matter".
The angle β was introduced in [1–a]. But furthermore, for example, a systematic use of the multivectors of $\mathcal{M}$ appears in [1–b]; the form (16) given to the Dirac wave function is included (in the matricial formalism) in [1–c]; the notion of "proper frame" in [1–e]; the one of "spin plane" in [1–f]. The invariant equation (17) is strictly equivalent to the set of invariant equations of [1–d]. And the point of view of Hestenes by which only real quantities are to be considered is the basis of this "hydrodynamics".

(Note. The identification of the β parameter of Hestenes with the angle of Yvon–Takabayasi was made by G. Casanova [1–e]. The relation between the equation (17) and the ones of Takabayasi was pointed out by O. Costa de Beauregard [private communication, 1968]).

But with respect to three points, the Hestenes approach has supplied an indispustable progress:
– The use of STA. For example, without employing the Clifford product, it is impossible to express the Dirac equation in an invariant form by a single equation as eq. [17].
– The elimination of a useless "spinor unity" of the traditional formalism which is the source of a lof of complications and ambiguities in relativistic Quantum Mechanics (see No. 3.1).
– But, beyond these conveniences, the fact that STA contains a specific property of spacetime, the duality rotation, allows one to understand the geometrical links between a particle and its antiparticle.

3. **Comparison Between the Darwin and the Krüger Solutions of the Coulomb Problem.**

3.1. QUATERNIONIC SOLUTIONS OF THE DIRAC EQUATION

The readers non acquainted with the formalism of STA have simply to consider the symbols γ_μ, σ_k, i, as representing the Dirac, Pauli and γ_5 matrices, that they are used to employ, instead of vectors, time–like bivectors and pseudoscalar, respectively, of $\mathcal{M} = \mathbb{R}^{1,3}$.

We recall the relations one can associate with a particular galilean frame B of $\mathcal{M}$,

$$\tfrac{1}{2}\left(\gamma_\mu\gamma_\nu + \gamma_\nu\gamma_\mu\right) = g_{\mu\nu}, \ \gamma^0 = \gamma_0, \ \gamma^k = -\gamma_k, \ (k = 1,2,3) \ (19)$$

and, as consequences of (19),

$$\tfrac{1}{2}(\sigma_j\sigma_k + \sigma_k\sigma_j) = \delta_{jk}, \quad \sigma_j = \gamma_k\gamma_0, \quad (j,k = 1,2,3), \tag{20}$$

$$i = \gamma_0\gamma_1\gamma_2\gamma_3 = \sigma_1\sigma_2\sigma_3, \quad i^2 = -1. \tag{21}$$

The set

$$\mathcal{X} = \{\alpha + i\mathbf{a}: \mathbf{a} = \sum_k \alpha_k\sigma_k, \ \alpha, \alpha_k \in \mathbb{R}\} \tag{22}$$

(which depends on the choice of γ_0) is isomorphic to the field of the Hamiltonian quaternions.

The ring $\mathcal{X} + i\mathcal{X}$ of the biquaternions is isomorphic (independenly of γ_0) to the direct sum of the vectorial spaces of the scalars, bivectors and pseudoscalars of $\mathcal{M}$, endowed with the Clifford product.

We recall the Hestenes presentation of the Dirac equation (see Ref. [2]).

The Dirac spinor wave function Ψ is put in the form $\Psi = \psi u$, in which u is a "spinor unity" and $\psi \in \mathcal{X} + i\mathcal{X}$. (As a consequence ψ may be interpreted as in eq. (16)).

The elimination of u leads to eq. (15). If the potential A is such that

$$\frac{e}{\hbar c} A = V(r)\gamma_0, \quad V(r) \in \mathbb{R} \tag{23}$$

(the source of the potential is then at rest in the frame B) the Dirac– Hestenes equation may be written

$$\gamma^\mu\partial_\mu\psi = (\mathcal{E}_0\psi\gamma_0 + V\gamma_0\psi)i\sigma_3, \quad \psi \in \mathcal{X} + i\mathcal{X}, \tag{24}$$

in which $i\sigma_3 = \gamma_2\gamma_1$, $\mathcal{E}_0 = mc^2/(\hbar c)$.

Denoting $x = x^0\gamma_0 + \vec{r}$, $\mathbf{r} = \vec{r}\gamma_0 = rn$, $n^2 = 1$, we look for solutions in the form

$$\psi = \phi(r)e^{i\sigma_3\mathcal{E}x_0}, \quad \mathcal{E} \in \mathbb{R}, \phi(r) \in \mathcal{X} + i\mathcal{X}, \tag{25}$$

in which $\mathcal{E} = E/(\hbar c)$, E being a constant energy.

Because $\gamma_0 i\sigma_3 = i\sigma_3\gamma_0$ one can write

$$\mathcal{E}\gamma^0\phi i\sigma_3 + \gamma^k\partial_k\phi = (\mathcal{E}_0\phi\gamma_0 + V\gamma_0\phi)i\sigma_3.$$

Multiplying from the left by γ_0 and because $\gamma_0\gamma^0 = 1$, $\gamma_0\gamma^k = -\gamma^k\gamma_0 = \gamma_k\gamma_0 = \sigma_k$, one obtains

$$\nabla\phi = (\mathcal{E}_0\gamma_0\phi\gamma_0 + (-\mathcal{E}+V)\phi)i\sigma_3, \qquad \nabla = \sum_k \sigma_k\partial_k \tag{26}$$

3.2.   THE DARWIN SOLUTIONS

One writes

$$\phi = \phi_1 + i\phi_2, \qquad \phi_1, \phi_2 \in \mathcal{X}. \tag{27}$$

Because $i\nabla = \nabla i$, $i\gamma_0 = -\gamma_0 i$, $\gamma_0^2 = 1$, one obtains the equation

$$\nabla\phi_1 + i\nabla\phi_2 = (\mathcal{E}_0(\phi_1-i\phi_2) + (-\mathcal{E}+V)(\phi_1+i\phi_2))i\sigma_3. \tag{28}$$

from which one deduces the system

$$\nabla\phi_1 = (\mathcal{E}_0+\mathcal{E}-V)\phi_2\sigma_3, \quad \nabla\phi_2 = (\mathcal{E}_0-\mathcal{E}+V)\phi_1\sigma_3. \tag{29}$$

Multiplying from the left $(27)_1$ by $\mathbf{n}$, one may solve this system by writing

$$\phi_1 = g(r)S(\mathbf{n}), \quad \phi_2 = f(r)\mathbf{n}S(\mathbf{n})\sigma_3, \quad \mathbf{n} = \mathbf{r}/r, \tag{30}$$

in which $g(r), f(r) \in \mathbb{R}$, and $S(\mathbf{n}) \in \mathcal{X}$ is such that (App. III)

$$i(\mathbf{r}\mathbf{x}\nabla)S = (1+\kappa)S, \qquad \kappa \in \mathbb{Z}^*, \; S \in \mathcal{X}. \tag{31}$$

Indeed: one has $\mathbf{n}\nabla(gS) = \frac{dg}{dr}S + g\mathbf{n}\nabla S$, $\nabla(f\mathbf{n}S) = \frac{df}{dz}S + f\nabla(\mathbf{n}S)$, $\mathbf{n}\nabla S = i(\mathbf{r}\mathbf{x}\nabla)S/r$, $\nabla\mathbf{n} = 2/r$, $\nabla(\mathbf{n}S) = (\nabla\mathbf{n})S - \mathbf{n}\nabla S = (1-\kappa)S/r$.

The functions g,f must satisfy the system

$$\frac{dg}{dr} + \frac{1+\kappa}{r}g = (\mathcal{E}_0+\mathcal{E}-V)f, \quad \frac{df}{dr} + \frac{1-\kappa}{r}f = (\mathcal{E}_0-\mathcal{E}+V)g. \tag{32}$$

with the condition of normalization $\int_0^\infty (g^2+f^2)r^2dr = 1$.

The normalized solutions of eq. (29) are (App. III)

$$S = N\sigma_3 e^{i\sigma_3 m\varphi}, \quad N = L\sigma_3 + Mu, \tag{33}$$
$$\text{with } u = \cos\varphi\sigma_1 + \sin\varphi\sigma_2, \quad \mathbf{n} = \cos\theta\sigma_3 + \sin\theta u,$$

and $P_\ell^m(x) = (-1)^m \left[\frac{(\ell-m)!\,(2\ell+1)}{(\ell+m)!\,2}\right]^{\frac{1}{2}} \dfrac{(1-x^2)^{\frac{m}{2}} [(x^2-1)^\ell]^{(\ell+m)}}{2^\ell \ell!}$,

$\ell \in \mathbb{N}$, $m \in \mathbb{Z}$, $-\ell \le m \le \ell$, or $P_\ell^m(x) = 0$ if $|m| > \ell$, and the alternative

if $\kappa = \ell$, $\ L = -\left[\dfrac{\ell-m}{2\pi(2\ell+1)}\right]^{\frac{1}{2}} P_\ell^m(\cos\theta)$, $\ M = \left[\dfrac{\ell+m+1}{2\pi(2\ell+1)}\right]^{\frac{1}{2}} P_\ell^{m+1}(\cos\theta)$,

$$(34)$$

if $\kappa = -(\ell+1)$, $\ L = \left[\dfrac{\ell+m+1}{2\pi(2\ell+1)}\right]^{\frac{1}{2}} P_\ell^m(\cos\theta)$, $\ M = \left[\dfrac{\ell-m}{2\pi(2\ell+1)}\right]^{\frac{1}{2}} P_\ell^{m+1}(\cos\theta)$.

$$(35)$$

Thus, one obtains [6]

$$\psi = (gN\sigma_3 + \text{fi}nN)e^{i\sigma_3(m\varphi + \frac{E}{\hbar c}x^0)}.$$
$$(36)$$

For example,

Levels $nS_{\frac{1}{2}}$ ($\ell = 0$, $\kappa = -1$, $m = 0$): $\psi = \dfrac{1}{\sqrt{4\pi}}(g + \text{fi}n\sigma_3)e^{i\sigma_3 \frac{E}{\hbar c}x^0}$,

Levels $nP_{\frac{1}{2}}$ ($\ell = 1$, $\kappa = 1$, $m = 0$): $\psi = -\dfrac{1}{\sqrt{4\pi}}(gn\sigma_3 + \text{fi})e^{i\sigma_3 \frac{E}{\hbar c}x^0}$.

One passes from the "spin up" solutions to the "spin down" ones by changing i into −i.

3.3. THE KRÜGER SOLUTIONS

The Krüger solutions are based on the following transformation of the operator ∇:

$$\nabla\phi = \Omega\nabla_0(\Omega^{-1}\phi), \quad \Omega = \frac{e^{-i\sigma_3 \frac{\varphi}{2}}\,e^{-i\sigma_2 \frac{\theta}{2}}}{r(\sin\theta)^{\frac{1}{2}}},$$
$$(37)$$

$$\nabla_0 = \sigma_3\partial_r + \frac{1}{r}(\sigma_1\partial_\theta + \frac{1}{\sin\theta}\sigma_2\partial_\varphi).$$
$$(38)$$

One deduces from eq. (24)

$$\nabla_0\phi_0 = (\mathcal{E}_0\gamma_0\phi_0\gamma_0 + (-\mathcal{E}+V)\phi_0)i\sigma_3, \quad \phi_0 = \Omega^{-1}\phi.$$
$$(39)$$

Eq. (39) has exactly the same form as eq. (26) except that the operator ∇_0 uses the fixed frame $(\sigma_3, \sigma_1, \sigma_2)$. This allows one to write solutions in the form (see the article of H. Krüger in the present "Proceedings" and App. IV)

$$\phi = \Omega \phi_0 = \frac{1}{\sqrt{2\pi}} \, \Omega \mathrm{ir}(g(r)\sigma_1 + f(r)) \, e^{i\sigma_2 \kappa \theta} \, e^{-i\sigma_1 \frac{\pi}{4}}, \quad \kappa \in \mathbb{Z}^*. \tag{40}$$

in which the functions g,f satisfy the system (32).

3.4.   INVARIANTS ASSOCIATED WITH THE WAVE FUNCTION.

We calculate $\rho e^{i\beta} = \psi\tilde{\psi}$, the Dirac current $j = \psi\gamma_0\tilde{\psi}$.

We recall (see Ref. [6]) that if one denotes

$$N^2 = L^2 + M^2, \quad \mathrm{tg}\, \nu = \frac{f}{g}, \quad \mathrm{tg}\, \tau = \frac{M}{L} \tag{41}$$

one obtains for the Darwin solutions

$$\mathrm{tg}\, \beta = \mathrm{tg}2\nu \, \cos(\theta - 2\tau); \quad j = N^2[(g^2 + f^2)\gamma_0 + 2fg \, \sin(\theta - 2\tau) \, \vec{v}] \tag{42}$$

in which $\vec{u} = \cos\varphi \gamma_1 + \sin\varphi \gamma_2$, $\vec{v} = \frac{d\vec{u}}{d\varphi}$.

After some calculations, one can obtain the spacetime vector (see also Ref. [3-a]).

$$p + eA = E\gamma_0 + \frac{\hbar c}{r \, \sin\theta} \left| \frac{B^2}{z^2} \right| \vec{v} \quad \begin{cases} B = -\cos\nu \, \sin\tau + i \, \sin\nu \, \sin(\tau - \theta) \\ z^2 = \cos 2\nu + i \, \sin 2\nu \, \cos(2\tau - \theta) \end{cases}$$

$$\tag{43}$$

whose time component is the energy E associated with the solution.

The calculation of the angle β gives, for example, for the fundamental level,

$$\mathrm{tg}\beta = \mathrm{tg}\alpha \, \cos\theta, \quad \alpha = e^2/(\hbar c)$$

The angle β is non null except for $\theta = \pi/2$.

One obtains for the Krüger solutions

$$\beta = 0, \quad \rho = \frac{g^2 - f^2}{2\pi^2 \, \sin\theta}, \quad j = \frac{(g^2 + f^2)\gamma_0 + 2fg\vec{w}}{2\pi^2 \, \sin\theta}, \tag{44}$$

$$p + eA = E\gamma_0 + \frac{\hbar c}{r}\left[\kappa + \frac{g^2+f^2}{2(g^2-f^2)}\right]\vec{w}, \tag{45}$$

where $\vec{n} = \cos\Theta\gamma_3 + \sin\Theta\vec{u}$, $\vec{w} = \frac{\partial\vec{n}}{\partial\Theta}$.

The transition Dirac current between two levels.

If ψ_1, ψ_2 correspond to the two levels, one has

$$\left.\begin{array}{l}j = \psi_1\gamma_0\tilde{\psi}_2 + \psi_2\gamma_0\tilde{\psi}_1 = j^I\cos(\omega_{12}x^0) + j^{II}\sin(\omega_{12}x^0),\quad \omega_{12} = \frac{E_1-E_2}{\hbar c}\\[2mm] j^I = \phi_1\gamma_0\tilde{\phi}_2 + \phi_2\gamma_0\tilde{\phi}_1,\quad j^{II} = \phi_1 i\gamma_3\tilde{\phi}_2 - \phi_2 i\gamma_3\tilde{\phi}_1\end{array}\right\} \tag{46}$$

For the Darwins solutions one obtains (if $m_1 = m_2$)

$$j^I = 2[(g_1g_2+f_1f_2)(L_1L_2+M_1M_2)\gamma_0 + (g_1f_2+g_2f_1)\,(\sin\Theta(L_1L_2-M_1M_1)$$

$$-\cos\Theta(M_2L_1+L_2M_1))\vec{v}]$$

$$j^{II} = 2[(g_1f_2-g_2f_1)(L_1L_2+M_1M_2)\vec{n} + (g_2f_1+g_1f_2)(M_1L_2-L_1M_2)\vec{w}] \tag{47}$$

For the Krüger solutions one has

$$\left.\begin{array}{l}j = \frac{1}{\pi^2\sin\Theta}\,[((g_1g_2+f_1f_2)\gamma_0 + (g_1f_2+g_2f_1)\vec{w})\cos\xi + (g_2f_1-g_1f_2)\vec{n}\sin\xi]\\[2mm] \xi = \kappa_{12}\Theta - \omega_{12}x^0,\quad \kappa_{12} = \kappa_1-\kappa_2\,.\end{array}\right\} \tag{48}$$

3.5. COMPARISON BETWEEN THE DARWIN AND THE KRÜGER SOLUTIONS.

Contrary to Darwin's, the Krüger solutions ψ and the associated invariants, ρ, β are not defined for $\Theta = 0$ and $\Theta = \pi$. These singularities are to be added to the ones which affect the functions g, f for r = 0, when $|\kappa| = 1$.

However the integration of ρ and j over any space volume is finite.

A remark is to be made about the transverse component (orthogonal to the vector $\vec{n}$) of the Dirac transition current between two levels. In both Darwin and Krüger solutions, this component is not defined for $\Theta = 0$ or π. But these defects disappear after integration over a space volume.

One can notice that, in both cases, the spacetime vector $p+eA = \hbar c\omega/2$ is not defined on the poles axis ($\Theta = 0$ or π). To the extend that a) what we call the energy of a real photon emitted in the passage from a level to another, is nothing else but a discontinuity of the time component of the vector ω, and b) no integration over a space volume is to be made for expressing this energy, this property may confer to the poles axis a particular significance (perhaps related to the topo-logical defects one can associate to the vector ω; see No. 2.1.2).

Note that in the Krüger solutions, the vector ω is endowed with an inter-esting property which does not exist for the Darwin solution. The fundamental level $1S\frac{1}{2}$ is the only level for which ω is a gradient.

4. Conclusion

Certainly, from a mathematical point of view, the Krüger solutions are as convenient as the Darwin ones. But do they represent a physical reality?

The criterion cannot be the exactitude of the energy level values in the bare problem, where only the exterior potential is considered since the values are the same in both solutions. The verification must concern phenomena in which the radiation of the electron is implied as the Lamb shift or the spontaneous emission. In our opinion, it would be sufficient that the Krüger solutions are in good agree-ment with the experimental results of spontaneous emission, for having a strong presumption that the agreement with the Lamb shift is good. (See in Ref. [14] the Barut complex formula whose spontaneous emission is the imaginary part and the Lamb shift the real part).

If it would appear that the Darwin and the Krüger solutions equally account for the phenomenas we are used to associate with an electron or a positron in a central potential, we could abandon the former as non representative of a pure state of particle or antiparticle, because of the non equality to zero or π, everywhere, of β. The mystery of the angle β would be dissipated!

But if the Krüger solutions were not in accordance with these phenomenas, we would be in the situation where one should have to consider exact solutions of the Dirac equations, which are in actual fact much simpler than Darwin's, but which would not have a physical meaning!

In any way these solutions impose a reconsideration of the traditional opi-nions about the Dirac theory of the Coulomb problem.

Appendix I

Let us define, in agreement with the orthonormality relations

$$v^2 = 1, \quad n_k^2 = -1, \quad v \cdot n_k = 0, \quad n_1 \cdot n_2 = 0 \quad (k = 1,2) \tag{a}$$

the three scalars

$$\omega_\mu^{12} = (\partial_\mu n_1) \cdot n_2 \ (= -n_1 \cdot (\partial_\mu n_2)), \quad \omega_\mu^{20} = (\partial_\mu n_2) \cdot v, \quad \omega_\mu^{01} = (\partial_\mu v) \cdot n_1. \tag{b}$$

They represent, at the point x, the infinitesimal proper rotation of the ortho-normal set of vectors $S(x) = \{v, n_1, n_2\}$ when the point x moves in the x^μ direction.

This definition is exactly the same as the ones of the components of the instantaneous rotation vector, associated with the movement of a three–dimensional rigid body. But here, we are in a four dimensional space and it is convenient to consider the set $S(x)$ as a sub–set of the orthonormal moving frame $\mathcal{R}(x) = \{v, n_1, n_2, s\}$.

Now, we define a linear mapping $n \in \mathcal{M} \longrightarrow L_\mu(n) \in \mathbb{R}$, such that

$$L_\mu(v) = \omega_\mu^{12}, \; L_\mu(n_1) = \omega_\mu^{20}, \; L_\mu(n_2) = \omega_\mu^{01}, \; L_\mu(s) = 0 \qquad (c)$$

A simple expression of this mapping may be given by the use of STA.

Recalling that the associative Clifford product of vectors $(a_1, \ldots, a_p) \longrightarrow a_1 \ldots a_p$ is equal to their Grassmann product $a_1 \wedge \ldots \wedge a_p$ when these vectors are orthogonal, one can write

$$i = \gamma_0 \wedge \gamma_1 \wedge \gamma_2 \wedge \gamma_3 = v \wedge n_1 \wedge n_2 \wedge s = v n_1 n_2 s, \qquad (d)$$

because the frames B and $\mathcal{R}(x)$ are both orthonormal and have the same orientation.

The mapping $X \longrightarrow iX$ allows one to define the "dual" (in the tensor sense) of a multivector X.

For example, one has

$$i(v \wedge s) = ivs = -v^2 n_1 n_2 s^2 = n_1 n_2 = n_1 \wedge n_2,$$

because $v n_1 n_2 sv = v n_1 n_2 (-vs) = \ldots = (-1)^3 v^2 n_1 n_2 s.$

Because the Clifford square i^2 of the pseudo–scalar i is equal to -1, one has

$$i(n_1 \wedge n_2) = -v \wedge s = s \wedge v.$$

If Ω_μ is the bivector which expresses the infinitesimal rotation of the moving frame $\mathcal{R}(x)$ when x moves in the x^μ direction (App. II):

$$\Omega_\mu: \; \partial_\mu N = \Omega_\mu \cdot N, \quad \forall N \in \mathcal{R}(x), \qquad (e)$$

one can write

$$L_\mu(n) = \Omega_\mu \cdot (i(n \wedge s)). \qquad (f)$$

Indeed, replacing n by v, n_1, n_2, s respectively and using the following relation of the tensor algebra $X \cdot (a \wedge b) = (X \cdot a) \cdot b$, $(a, b \in \mathcal{M})$, one can immediately verify (c).

Note that one can write

$$L_\mu(n) = [\Omega_\mu v n_1 n_2 n]_{(0)} \qquad (g)$$

where $[X]_{(0)}$ means the scalar part of X.

Appendix II.

The duality rotation in M belongs to the following type of mapping

$$X \in C(E) \longrightarrow Y = \phi X \tilde{\phi} \in C(E), \quad \phi \in C(E). \tag{a}$$

$C(E)$ is the Clifford algebra associated with some euclidean space $E = \mathbb{R}^{q,n-q}$ and $\phi \to \tilde{\phi}$ means the operation of reversion, or principal antiautomorphism in $C(E)$. One recalls that

$$\tilde{\lambda} = \lambda, \quad \tilde{a} = a, \quad (a_1 ... a_p)^{\sim} = a_p ... a_1 \quad (\lambda \in \mathbb{R}, \; a, a_k \in E).$$

The operation of reversion has a particular importance in $C(E)$. For example the mappings

$$x \in E \to y = \pm U \, x \, \tilde{U}, \quad U = u_1 ... u_p, \quad u_k \in E, \quad |u_k^2| = 1 \tag{b}$$

generate the orthogonal group $O(E)$.

If $\phi = f(t)$, where t is a scalar parameter and f some differential function, such that ϕ is invertible for all t, one can associate with the mapping (a) the following element of $C(E)$ [8]

$$\xi = 2 \frac{d\phi}{dt} \phi^{-1}. \tag{c}$$

We will call ξ the infinitesimal operator at t of the mapping (a). ξ allows one to calculate dY/dt by means of the relation

$$\frac{dY}{dt} = \frac{1}{2} (\xi Y + Y \tilde{\xi}). \tag{d}$$

We recall [2] that if $E = M$, and if i means the unit pseudo–scalar of M (or antisymmetric tensor of rank 4 associated with the orthonormal frame), we have

$$i^2 = -1, \quad \tilde{i} = i, \quad ai = -ia, \quad ia\tilde{i} = a, \quad iab\tilde{i} = -ab, \quad \forall a,b \in M \tag{e}$$

Thus, the mapping or "duality rotation"

$$X \to Y = dX\tilde{d}, \quad d = e^{i\beta/2}, \quad \beta \in \mathbb{R} \tag{f}$$

is such that $Y = X$ if $X \in M$, and $Y = \exp(i\beta)X$ if $X = a \wedge b = (ab-ba)/2$ is a bivector.

In this way, the duality rotation "rotates" the bivectors by an "angle" β, but leaves the vectors invariant.

Consider the mapping

$$X \to Y = gX\tilde{g}, \quad g = e^{i\beta/2}R \tag{g}$$

in which R is the sum of a scalar, a bivector and a pseudo scalar, and such that $\tilde{R} = R^{-1}$. We recall that the mapping

$$\gamma_\nu \in \mathcal{M} \longrightarrow e_\nu = R\gamma_\nu\tilde{R} \in \mathcal{M} \tag{h}$$

transforms the orthonormal frame $\{\gamma_\nu\}$ into an orthonormal frame $\{e_\nu\}$ by a proper Lorentz rotation (see Ref. [2]).

If g depends on x, the infinitesimal operator, at x, when x moves in the x^μ direction, associated with the mapping (h) is

$$\xi_\mu = 2(\partial_\mu g)g^{-1} = \Omega_\mu + i\partial_\mu\beta, \tag{i}$$

$\Omega_\mu = 2(\partial_\mu R)R^{-1}$ is the bivector (see Ref. [4]) which corresponds to the infinitesimal rotation of the frame $\{e_\nu\}$. Indeed one has, using (d)

$$\partial_\mu N = \tfrac{1}{2}(\Omega_\mu N + N\tilde{\Omega}_\mu) = \tfrac{1}{2}(\Omega_\mu N - N\Omega_\mu) = \Omega_\mu \cdot N, \forall N \in \{e_\nu\} \tag{j}$$

where the fundamental relations $aX = a \cdot X + a \wedge X$, $Xa = X \cdot a + X \wedge a$, $a \in \mathcal{M}$ have been applied.

Now, if we denote $e_0 = v$, $e_1 = n_1$, $e_2 = n_2$, $e_3 = s$ and if we replace in the right hand side of eq. (g), App. I, the infinitesimal operator Ω_μ by ξ_μ as given in (i), one obtains

$$L_\mu(n) = [\xi_\mu v n_1 n_2 n]_{(0)} = \Omega_\mu \cdot (i(n \wedge s)) + (n \cdot s)\partial_\mu\beta. \tag{k}$$

The infinitesimal operator associated with the mapping

$$X \to \psi X \tilde{\psi}, \quad \psi = \sqrt{\rho}\, e^{i\frac{\beta}{2}}R, \tag{l}$$

is

$$2(\partial_\mu\psi)\psi^{-1} = \xi_\mu + \partial_\mu(\ln \rho). \tag{m}$$

We recall the following relation (see Ref. [4], [5])

$$\gamma^\mu \Omega_\mu = -(\gamma^\mu \partial_\mu \sigma + \omega + \bar{\omega} i)\,\sigma \tag{o}$$

in which $\bar{\omega} = ((\partial_\mu v)\cdot s)\gamma^\mu$.

Appendix III

Denoting $\mathbf{w} = \dfrac{\partial \mathbf{n}}{\partial \theta}$, $\mathbf{v} = \dfrac{d\mathbf{u}}{d\varphi}$ one can write

$$\nabla = \mathbf{n}\partial_r + \frac{1}{2}\left(\mathbf{w}\partial_\theta + \frac{\mathbf{v}}{\sin\theta}\partial_\varphi\right), \quad \nabla\mathbf{n} = \frac{1}{r}\left(\mathbf{w}^2 + \frac{\sin\theta}{\sin\theta}\mathbf{v}^2\right) = \frac{2}{r} \tag{a}$$

One has to solve the equation

$$\mathbf{rn}\nabla S = \mathbf{n}\left(\mathbf{w}\partial_\theta + \frac{\mathbf{v}}{\sin\theta}\partial_\varphi\right)S = \lambda S, \quad \lambda \in \mathbb{R} \tag{b}$$

where $S = (L(\theta)\sigma_3 + M(\theta)\mathbf{u})\,\sigma_3 e^{i\sigma_3 m\varphi}$.

Because $i = \sigma_3 \mathbf{u}\mathbf{v} = \mathbf{n}\mathbf{w}\mathbf{v}$, $\mathbf{n}\mathbf{w}\sigma_3 = \mathbf{n}\mathbf{w}(\mathbf{v}^2)\sigma_3 = i\mathbf{v}\sigma_3 = -\mathbf{u}$, $\mathbf{n}\mathbf{w}\mathbf{u} = i\mathbf{v}\mathbf{u} = \sigma_3$, $\mathbf{n}\mathbf{v}\sigma_3 i\sigma_3 = \mathbf{w}$, $\mathbf{n}\mathbf{v}\mathbf{u}i\sigma_3 = \mathbf{n}$, $\mathbf{n}\mathbf{v}^2 = \mathbf{n}$, one obtains

$$-\frac{\partial L}{d\theta}\mathbf{u} + \frac{dM}{d\theta}\sigma_3 + \frac{m}{\sin\theta}L\mathbf{w} + \frac{m+1}{\sin\theta}M\mathbf{n} = \lambda(L\sigma_3 + M\mathbf{u}) \tag{c}$$

from which one deduces the system

$$\frac{dM}{d\theta} + (1+m)\cot g\,\theta M - mL = \lambda L, \quad -\frac{dL}{d\theta} + m\cot g\,\theta L + (1+m)M = \lambda M \tag{d}$$

whose solutions are, if $(\lambda-1)\lambda = \ell(\ell+1)$, $\ell \in \mathbb{N}$,

$$L = q\,p_\ell^m(\cos\theta), \quad M = p_\ell^{m+1}(\cos\theta), \quad p_\ell^m(x) = \frac{(1-x^2)^{\frac{m}{2}}}{2^\ell \ell!}\left[(x^2-1)^\ell\right]^{(\ell)}, \quad m\in\mathbb{Z}. \tag{e}$$

The number q may be chosen in such a way that the compatibility of the system (d) is achieved. Using the relation (see Gel'fand "Representations of the rotation and Lorentz group" Pergamon Press, 1963) in which $x = \cos\theta$:

$$-\frac{dM}{d\theta} = \sin\theta \frac{d}{dx}\left[p_\ell^{m+1}(x)\right] = (1+m)\cot g\,\theta\,p_\ell^{m+1}(x) - (\ell+m+1)(\ell-m)p_\ell^m(x) \tag{f}$$

one obtains

$$\frac{dM}{d\theta} + (1+m)\cot g\,\theta\,M = (\ell+m+1)(\ell-m)\frac{L}{q} = (\lambda+m)L \qquad (g)$$

and thus $q(\lambda+m) = (\ell+m+1)(\ell-m)$ which implies $q = \ell-m$ if $\lambda = \ell+1$, $q = -(\ell+m+1)$ if $\lambda = -\ell$.

The case where $q = 0$ may be obtained by taking $\ell = m = 0$, $\lambda = 1$. But the correspondent solution $N = u/\sin\theta$ is not acceptable because the relation (h) is then impossible. So, if one denotes $\lambda = 1 + \kappa$, one must have $\kappa \in \mathbb{Z}^*$. Note that the case $\lambda = -\ell = 0$ corresponds to the levels $nS_{\frac{1}{2}}$.

The condition of normalization of the current

$$\iiint (\bar{\psi}\gamma_0\tilde{\psi}) \cdot \gamma_0 \, dx^1 dx^2 dx^3 = 1 \qquad (h)$$

impose $\int_0^{2\pi}\int_0^{\pi} N^2 \sin\theta \, d\varphi \, d\theta = 1$. This is achieved by taking L, M as in (34) or (35).

<u>Note</u>. For the general solution of eq. (b), see Ref. [13].

Note that the solution of the central potential problem described here (Ref. [3], [6]) is very close to the one written by A. Sommerfeld [15].

Appendix IV

Multiplying from the right eq. (39) by $\sqrt{2}\pi\,\exp(-i\sigma_2\kappa\theta)\,\exp(i\sigma_1\frac{\pi}{4})$, and because

$$e^{-i\sigma_1\frac{\pi}{4}}\sigma_3 e^{i\sigma_1\frac{\pi}{4}} = \sigma_3 i\sigma_1 = -\sigma_2,\ \gamma_0 i\gamma_0 = -i,\ \gamma_0 i\sigma_1\gamma_0 = i\sigma_1,$$

one obtains

$$i[\sigma_3((rg' + g)\sigma_1 + rf' + f) + \kappa(g\sigma_1 + f)\sigma_1 i\sigma_2]$$

$$= ir[-\mathcal{E}_0(g\sigma_1 - f) + (\mathcal{E} - V)(g\sigma_1 + f)]\,i\sigma_2.$$

Because $\sigma_3\sigma_1 = i\sigma_2$, $\sigma_1 i\sigma_2 = -\sigma_3$, one has

$$(g' + \frac{1+\kappa}{r}g)i\sigma_2 + (f' + \frac{1-\kappa}{r}f)\sigma_3 = (\mathcal{E}_0 + \mathcal{E} - V)fi\sigma_2 + (\mathcal{E}_0 - \mathcal{E} + V)g\sigma_3$$

i.e. the system (32).

REFERENCES

[1]   —a— Yvon, J. (1940) 'Equations de Dirac—Madelung', J. Phys. et le Radium' VIII, 18.

—b— Costa de Beauregard, O. (1943) Contribution à l'étude de la Théorie de l'électron", Ed. Gauthiers—Villars, Paris .

—c— Jakobi, G. and Lochak, G. (1956) 'Introduction des paramètres relativistes de Cayley—Klein dans la représentation hydrodynamique de l'équation de Dirac', C.R. Acad. Sc. Paris 243, 234.

—d— Takabayasi, T. (1957) 'Relativistic hydrodynamics of the Dirac matter', Suppl. Prog. Theor. Phys. 4, 1.

—e— Halbwachs, F. (1960) Théorie relativiste de fluides à spin, Ed. Gauthiers—Villars, Paris.

—f— Halbwachs, F. Souriau, J. M. and Vigier, J.R. (1961) 'Le groupe d'invariance associé aux rotateurs relativistes et la théorie bilocale', J. Phys. et le Radium, 22, 293.

—e— Casanova, G. (1968) 'Sur l'angle de Takabayasi', C.R. Acad. Sc. Paris, 266 B, 1551.

[2]   Hestenes, D. (1967) 'Real spinor fields', J. Math. Phys., 8, 798.

[3]   —a— Quilichini, P. (1971) 'Calcul de l'angle de Takabayasi dans le cas de l'atome d'hydrogène', C.R. Acad. Sc. Paris 273 B, 829.

—b— Gurtler, R. (1972), Thesis, Arizona State University.

[4]   Boudet, R. and Quilichini, P. (1969) 'Sur les champs de multivecteurs unitaires et les champs de rotations', C.R. Acad. Sc. Paris 268 A, 725.

[5]   Boudet, R. (1971) 'Sur une forme intrinsèque de l'équation de Dirac et son interprétation géometrique', C.R. Acad. Sc. Paris 272 A, 767.

[6]   Boudet,R. (1974) 'Sur le tenseur de Tetrode et l'angle de Takabayasi. Cas du potential central', C.R. Acad. Sc. Paris 278 A, 1063.

[7]   Boudet, R. (1985) 'Conservation laws in the Dirac theory', J. Math. Phys., 26, 718.

[8]   Boudet, R. (1988) 'La géométrie des particules du groupe SU(2)', Annales Fond. L. de Broglie (Paris), 13, 105.

[9]   Boudet, R. 'The role of Planck's constant in Dirac and Maxwell theories' ("Jounées Relativistes" Tours 1989), Ann. de Physiques, Paris 14, No. 6 suppl. 1, 27.

[10]   Micali, A. (1986) 'Groupes de Clifford et groupes des spineurs' in Clifford Algebras and Their Applications in Mathematical Physics, 67—78, Ed. Chrisholm J. and Common, K., Reidel Publ. Co., Dordrecht, Holland, The Netherland.

[11]   Boudet, R. (1990) 'The role of Planck's constant in the Lamb shift standard formulas' in Quantum Mechanics and Quantum Optics, Ed. Barut, A.O. Plenum Press.

[12]   Gliozzi, F. (1978) 'String—like topological excitations of the electromagnetic field', Nucl. Phys., B 141, 379.

[13]   Boudet, R. (1975) 'Sur les fonctions propres des opérateurs différentiels invariants des espaces euclidiens, et les fonctions spéciales', C.R. Acad. Sc. Paris, 280 A, 1365.

[14] —a— Barut, A.O. and Kraus, J. (1983), 'Nonperturbative Quantum Electrodynamics: The Lamb Shift', Found. of Phys., 13, 189.

−b− Barut, A.O. and van Huele, J.F. (1985) 'Quantum electrodynamics based on self—energy: Lamb shift and spontaneous emission without field quantization', Phys. Rev. A, <u>32</u> 3187.

[15]   Sommerfeld, A. (1960), Atombau und Spektrallinien, Ed. Friedr. Vieweg and Sohn, Braunschweig.

BRIEF HISTORY AND RECENT DEVELOPMENTS IN ELECTRON THEORY AND QUANTUMELECTRODYNAMICS

A. O. Barut
Department of Physics
University of Colorado
Boulder, CO 80309, USA

ABSTRACT. Major steps in the hundred years history of the electron concerning its selfenergy due to its own electromagnetic field are outlined and the present status and revival of the selfenergy and radiative problems of the electron are discussed.

1. HISTORY OF THE ELECTRON

1.1. Concept and Discovery

The first and the foremost of all elementary particles, the electron, has not been discovered suddenly, as the muon, for example. It was "in the air" for a long time. 1990 is a good date to remember the hundredth anniversary of its conception. The evidence came from at least three widely different phenomena.

Its study in the cathode rays can be said to begin with the work of the mathematician J.Plücker in 1858. Plücker is also a pioneer in differential geometry in introducing the rays as coordinates rather than the points (Plücker coordinates) which anticipates the space of all light rays or light cones. Wilhelm Hittorf (1869) talks about the straightline trajectories of the negatively charged "glowrays". Other important names in the early studies of cathode rays are W. Crookes (1879), H.Hertz (1881), P.Lenard, J.Perrin, W.Wien and J.J.Thomson.

Secondly, from Faraday's equivalence law for electrolysis, Helmholtz in 1881 concluded that each ion must carry a multiple of an elementary charge. Independently, the irish physicist G.Johnstone Stoney in 1881 also talks of fundamental units of both positive and negative electric charges. Already in 1874 Stoney had the idea of an "atom" of electricity. And it was Stoney who introduced the name "electron" a bit later in 1894.

The third phenomenon in which the electron appears independently was its identification in radioctivity by Jean Becquerel (1 March 1896).

The first precise determination of the ratio e/m were made by Peter Zeeman (31 October 1896), E.Wiechert (7 Jan. 1897) and J.J.Thomson (30 April 1897). Zeemann also gave an explanation of what is now called the Zeeman effect on the basis of

105

D. Hestenes and A. Weingartshofer (eds.), The Electron, 105–148.
© 1991 *Kluwer Academic Publishers. Printed in the Netherlands.*

the electron-hypothesis. These early experimental developments we may call the first period of electron's history.

The second phase, up to the discovery of the wave properties of the electron, is the history of the electron as a relativistic particle, the electron according to Lorentz, and the models by Abraham and Lorentz. The relativistic equations of motion derived by Poincaré and Einstein in 1904 and 1905 have been verified e.g the increase of the effective mass with velocity.

The third period of electron's history opens with the unexpected wave and spin properties of the electron and leads to the picture according to Schrödinger and Dirac,the electron described by a wave equation, and ends with the QED picture of the electron. It is remarkable that such a seemingly simple object as an electron has produced so many surprises. In view of the appearance of heavy leptons, like muon and tau which are very much like the electron, I am tempted to conjecture a fourth period in which we may be again surprised by a nonperturbative internal structure of the electron. The perturbative treatment of electron interactions by Feynman graphs has somewhat diminished the preeminence of the electron; it is just one of the many "elementary" particles. But these results have not solved the structure problem. In the last sentence of his famous review article on Quantum Theory of Radiation[1] Fermi writes: "In conclusion, we may therefore say that practically all the problems in radiation theory which do not involve the structure of the electron have their satisfactory explanation; while the problems connected with the internal properties of the electron are still very far from their solution".

Because the structure of the electron is the main topic of these lectures I would like also to review briefly the history of the selfenergy of the electron.

1.2. History of Selfenergy

It is remarkable that the force law between two charged particles in motion including the magnetic force was written down as early as 1875 by R. Clausius, and this in "relativistic" form namely, what would correspond to an interaction Lagrangian of the form

$$L = \frac{1}{2}m_1 v_1^2 + \frac{1}{2}m_2 v_2^2 - \frac{e_1 e_2}{4\pi\varepsilon_0} \left(1 - \frac{1}{c^2} \boldsymbol{v}_1 \cdot \boldsymbol{v}_2\right) \frac{1}{|r_1 - r_2|}$$

This is not surprising because the laws of relativity were essentially deduced later from electrodynamics. J.J.Thomson in 1881 confirmed this law of force (up to a factor) from Maxwell equations. In 1892 H.A.Lorentz began to combine the action of the particles with that of the electromagnetic field culminating in his *Theory of the Electrons* as described in his famous 1904 Encyclopedie article and in his book. This is essentially the present "classical" electron theory, the first comprehensive selfconsistent treatment of charged particles and their electromagnetic field in mutual interaction. We shall see that the basic tenets of this theory remains unchanged today, only the way we describe matter has undergone several changes over the years. Because the selfenergy of the electron for a point particle is infinite at the position of the particles, Lorentz and Abraham tried to model the electron as an extended charge distribution in order to understand whether the mass of the electron is wholly of electromagnetic origin, that is an electromagnetic origin of the concept of mass and an electromagnetic view of all matter. The advent of the relativity principle, independently formulated by Poincaré and Einstein, put a temporary end to these endeavours by elevating the concept of the rest mass to the level of an invariant, like the velocity of light. Also the covariant laws of mechanics were promoted to the same footing as the Maxwell's equations; in fact every classical theory could be relativized. Even if we accept an unexplained invariant rest mass the problems of the infinite selfenergy and the structure of the electron remained. Dirac in 1938 returned to the classical electron theory and extracted the covariant form of the selfenergy contribution to the motion of the electron. This is the radiation reaction force in addition to the external force and the resultant equation is now known as the Lorentz-Dirac equation. But an infinite renormalization was necessary by putting part of the selfenergy into the rest mass. Dirac returned to the classical electron theory in 1962 by modelling it as an extensible charged shell held together by a surface tension. The parameter of the surface tension can be eliminated in terms of the mass and charge so that this model has also only two parameters, rest mass and charge, just like a point particle. Unfortunately,to my knowledge, the relativistic motion of such a shell in space-time, its radiation reaction and renormalization have not been studied at all.

The reason might be that meanwhile we have developed first a wave mechanics for the electron, then a relativistic Dirac wave equation incorporating the new spin property of the electron and finally a quantumelectrodynamics so that classical models seemed to be obsolete. However, as exemplified by the above quote from Fermi the selfenergy problem remained.At first it might be thought that the description of the electron by an extended wave function instead of a point charge might alleviate the selfenergy infinity. Schrödinger himself in 1926 tried to include additional radiation reaction or selfenergy terms to his famous equation, but obtained, due to an incomplete treatment, wrong results. In the meantime the statistical interpretation of the wave function became the dominant paradigm; the self energy terms were dropped, an independent quantized radiation field was introduced and the self energy infinities were absorbed by renormalization into mass and charge. So the expected finite, closed and selfconsistent electrodynamics did not materialize.

1.3. The Importance of the Structure and Selfenergy of the Electron

It is with this background that I wish to reexamine the problem of the selfenergy of the electron. As Lorentz had already prophesied, "in speculating on the structure of these minute particles we must not forget that there may be many possibilities not dreamt of at present". It is not only a problem of having a mathematically sound finite theory of the electron, but a question of curiosity which may have farreaching consequences : what is really an electron? What is the structure which gives its spin? Why must there be a positron? What is mass? Why and how does the electron manifests wave properties? And what is the interaction between two electrons or between an electron and a positron at short distances? The Dirac equation gives a mathematical description of these questions, except the last one, but not a clear intuitive picture.

These questions, in fact the history of the electron,show that although there has been tremendous progress in details and applications of the electron theory, the progress in the foundations of physics is very slow. We are normally busy with our calculations and with building models. But at occasions like this, which I welcome, we may reflect on these larger issues , fundamental ideas and unifying concepts. I think some of the most fundamental and soluble problems of basic physics at the present time are: (1) a clear understanding of the wave and particle properties of electron, photon and other particles within simple logic, without paradoxes, as objective material properties; (2) the completion of quantumelectrodynamics; a selfconsistent intutive rendering of the interactions of matter and electromagnetic field without infinities ; (3) different forms of matter : how many really fundamental particles do we have, the electron-muon puzzle, for example; and (4) how many distinct fundamental interactions do we have to introduce, and how do we unify the four seemingly different forces ? Into a large mathematical framework containing all these, or showing that they are just different manifestations of a single interaction? For example, the chemical force and alpha decay, although extremely week, have been shown to be different manifestations of the electromagnetic force. It may seem incredulous, but I think we may have a much better understanding of these four basic problems on the basis of electron's structure and specially on the basis of the strong selfenergy effects between electron and positron at short distances. This would be a fullfilment of Einstein's statement "You know, it would be sufficient to really understand the electron". The framework for this program,I think,exists, but we must justify it with precise mathematical calculations.

2. SOME RECENT RESULTS FROM ELECTRON THEORY

2.1 Selfenergy of the Point Electron: The Lorentz-Dirac Equation

Because the basic postulate of both the classical electrodynamics and the selffield quantumelectrodynamics , the action principle, is essentially the same, we begin with the longstanding apparent problem of causality violation and runaway solutions of the Lorentz-Dirac equation. The classical action for spinless charged point particles in terms of invariant time parameters is given by

$$A = \int d\tau p^\mu \dot{x}_\mu - \int \left[e A_\mu j^\mu - \frac{1}{4} F_{\mu\nu} F^{\mu\nu} \right] dx$$

where the current is

$$j_\mu(x) = \int d\tau \delta(x - x(t)) \dot{x}(\tau)$$

The action leads to Maxwell's equations and to the particle equations

$$F'^\nu_{\mu\nu} = -j_\mu$$

$$m\ddot{x}_\mu = e F_{\mu\nu} x^\nu$$

Each particle sees the external field of all other particles plus its own selffield which is a necessary consequence of the selfconsistent action principle. The selffield is obtained from the Lienard-Wiechert potential

$$A_\mu(x) = e \int d\tau \dot{x}_\mu(t) \delta(x - x(t)) D(x - x(t))$$

A regularization or renormalization is necessary, because the selffield at the position of the point particle is not defined, which can be done in different ways. The result is the covariant equation

$$m\ddot{x}_\mu = eF_{\mu\nu}\dot{x}^\nu + \frac{2}{3}e^2\left(\dddot{x}_\mu + (\ddot{x})^2\dot{x}_\mu\right)$$

which contains all the radiative processes for point particles in a closed nonperturbative manner which, I think, is one of the most important feature of this equation. However, the complete consistency of this equation has been questioned. This comes about if one tries to solve this equation in regions where the external field vanishes. The nonlinear term by itself then leads to preacceleration and to runaway solutions, the exponential increase of x with time. I have recently pointed out that renormalization of the theory has two aspects : one is putting an infinite inertial term $\lim (1/u)\ddot{x}^\mu$, into the mass term, the other to make sure that when the external field vanishes the charge must move like a free particle with an experimental mass. In renormalizing a theory we must know beforehand to what we are renormalizing. This means that in regions where the external field is zero the correct solution must coincide with that for a free particle. The socalled causality violation and runaway solutions all have been shown in the examples where the external field is switched on and off at finite times. On the basis of explicit solutions it has been shown that with the above physical requirement no preacceleration or runaway solutions arise[2]. Furthermore, with the radiated power and the finite change of mass correctly included the Lorentz Dirac equation conserves energy[3] removing some other doubts expressed in the literature. This does not mean that we should be completely happy with the Lorentz Dirac equation. First of all it does not contain spin which however will be added in the next Section. But the infinity in the mass renormalization is still with us. But then the electron has wave and spin properties which are not yet in the point particle model.

2.2. Classical Relativistic Spinning Electron

The spin properties of the electron are very well described by the quantum Dirac equation. For many practical applications of the Dirac equation there is no need to make a model of spin. But eventually it becomes important to understand the physical mechanism underlying the spin or magnetic moment degree of freedom of the electron.

Dirac[4] has found (by chance as he says) his equation without quantizing of an existing classical model. Ever since there was no lack of effort to find an intuitive model of this remarkable relativistic spinning particle. A model of spin may help to discuss possible excited states of the electron, the existence of antiparticles, heavy leptons, and perhaps shed some light on Pauli exclusion principle, and short distance extrapolation of electrodynamics.

If a spinning particle is not quite a point particle, nor a solid three dimensional top, what can it be? What is the structure which can appear under probing with electromagnetic fields as a point charge, yet as far as spin and wave properties are concerned exhibits a size of the order of the Compton wave length? I want to describe a model in which a point charge performs as its natural motion a helix which gives an effective structure and size scale to the particle and accounts for the spin as the intrinsic angular momentum of the helix, and the frequency of the helical motion determines an internal clock and attributes a mass to the particle. Furthermore the sense of the orientation of the helix is related to the particle-antiparticle duality. And all this already in a purely classical framework.

The Action

The phase space consists of the usual conjugate pair of variables $(x_\mu(\tau), p_\mu(\tau))$ plus another conjugate pair of internal variables $(\bar{z}(\tau), z(\tau))$. Here τ is an invariant time parameter, $\bar{z}$ and z are 4-component classical c-number spinors, thus in C_4. We could have used the real and imaginary parts of z, but for a symplectic formulation the spinor form is much more economical and elegant. The notation is such that

$$\bar{z} = z^+ \gamma \cdot n \quad , \quad \gamma \cdot n \equiv \gamma^\mu n_\mu$$

where n^μ is the normal to a space-like surface $\sum$. The action is the integral of Cartan's symplectic 1-form

$$\omega = pdq - \mathcal{H} d\tau$$

where $\mathcal{H}$ is the "Hamiltonian" with respect to τ or the mass operator. In our case we have explicitly

$$\omega = pdx + i\lambda \bar{z} \gamma \cdot nz - \mathcal{H} d\tau$$

with

$$\mathcal{H} = \bar{z} \gamma^\mu z \left(p_\mu - e A_\mu \right)$$

so that the action including that of the coupled electromagnetic field A is[5]

$$\int w = \int \left(i\lambda \bar{z} \gamma \cdot nz + p\dot{x} - \pi_\mu \bar{z} \gamma^\mu z \right) d\tau - \frac{1}{4} \int dx F_{\mu\nu} F^{\mu\nu}$$

where we have introduced the kinetic momenta

$$\pi_\mu - p_\mu - e A_\mu$$

Properties of the Particle and Solutions

1. There are only two fundamental constants in the theory: the coupling constant e, the charge, and the constant λ of dimension of action ($\hbar$) multiplying z for dimensional reasons. The mass m will enter as the value of an integral of motion in the solutions.
2. The system is integrable. Two integrals of motion are

$$\mathcal{H} = \pi_\mu \bar{z} \gamma^\mu z \quad , \quad \dot{\mathcal{H}} = 0$$
$$\mathcal{N} = \bar{z} \gamma \cdot nz \quad , \quad \dot{\mathcal{N}} = 0$$

We can choose $\mathcal{N} = 1$ (normalization), and $\mathcal{H} = m$.

3. The equations of motion are

$$\dot{z}\gamma \cdot n = \frac{i}{\lambda}\bar{z}\pi \quad , \quad \pi = \gamma^\mu\left(p_\mu - eA_\mu\right) = \gamma^\mu\pi_\mu$$

$$\gamma \cdot n\dot{z} = -\frac{i}{\lambda}\pi z$$

$$\dot{\pi}_\mu = eF_{\mu\nu}\dot{x}^\nu$$

$$\dot{x}_\mu = \bar{z}\gamma_\mu z$$

4. The velocities $\dot{x}_\mu$ are thus constrained by the internal variables (similar to the rolling condition of a rigid body on a surface) by $\dot{x}_\mu = \bar{z}\gamma_\mu z$. In particular

$$\dot{x}_0 = \bar{z}\gamma_0 z = \frac{dx_o}{d\tau}$$

which for $n = (1000)$ is equal to $\mathcal{N} = 1$. Hence τ has the meaning of proper time in the frame determined by n. It is also possible to rewrite the action so that τ has the meaning of the proper time of the center of mass.

5. Solutions for a free particle. We give only the solution for $\dot{x}_\mu$.[6]

$$\dot{x}_\mu(\tau) = \dot{x}_\mu(0) + \ddot{x}_\mu(0)\frac{\sin 2p\tau}{2p} + 2\frac{\sin^2 p\tau}{p^2}V_\mu$$

where

$$V_\mu = \delta_{\mu 0}\left(mp_0 - m^2\right) - p_\mu(m - p_0)$$

Thus

$$\dot{x}_0(\tau) = \dot{x}_0(0) + \ddot{x}_0(0)\frac{\sin^2 p\tau}{2p} + 2\sin^2 p\tau$$

$$\vec{\dot{x}}(\tau) = \vec{\dot{x}}(0) + \vec{\ddot{x}}(0)\frac{\sin^2 p\tau}{2p} + \frac{\vec{p}}{p_0 + m}\sin^2 p\tau$$

Compare this with the solution when the invariant parameter is taken to be the proper time of the center of mass:

$$\dot{x}_0 = \frac{p_0}{m} + \left(\dot{x}_0(0) - \frac{p_0}{m}\right)\cos 2m\tau + \frac{\ddot{x}_0(0)}{2m}\sin 2m\tau$$

$$\vec{\dot{x}} = \frac{\vec{p}}{m} + \left(\vec{\dot{x}}(0) - \frac{\vec{p}}{m}\right)\cos 2m\tau + \frac{\vec{\ddot{x}}(0)}{2m}\sin 2m\tau.$$

6. We see in either of the above forms the helical motion of the velocities, hence after integration, of the coordinates x_μ . This is the natural motion of the particle and the helix does not radiate. The frequency of the helical; motion is $2m$ in the proper time of the center of mass.

7. Instead of $\bar{z}$ and z we can introduce the more physical velocity and spin variables. Defining $S_{\mu\nu} = \frac{1}{4}\bar{z}[\gamma_\mu\gamma_\nu]z$ we have the dynamical system

$$\dot{x}_\mu = v_\mu$$
$$\dot{v}_\mu = 4S_{\mu\varrho}\pi^\varrho$$
$$\dot{\pi}_\mu = eF_{\mu\varrho}v^\varrho$$
$$\dot{S}_{\mu\nu} = \pi_\mu u_\nu - \pi_\nu u_\mu$$

These equations are identical in form to the Heisenberg equations of the Dirac equation.

8. The system is symplectic given by the Poisson brackets

$$\{f,g\} = \left(\frac{\partial f}{\partial x^\alpha}\frac{\partial g}{\partial p_\alpha} - \frac{\partial g}{\partial x^\alpha}\frac{\partial f}{\partial p_\alpha}\right) - i\left(\frac{\partial f}{\partial z}\frac{\partial g}{\partial \bar{z}} - \frac{\partial g}{\partial z}\frac{\partial f}{\partial \bar{z}}\right)$$
$$\{z, i\bar{z}\} = 1 \quad , \quad \{x^\mu, p_\nu\} = \delta^\mu_\nu$$

Consequently the equations of motion have the Poisson bracket as well as the Hamiltonian forms

$$\dot{x}_\mu = \{x_\mu, \mathcal{H}\} = \partial\mathcal{H}/\partial p^\mu$$
$$i\dot{\bar{z}} = \{i\bar{z}, \mathcal{H}\} = -i\,\partial\mathcal{H}/\partial z$$

similarly for all the other dynamical variables.

9. Quantization can be performed in three different forms.

(i) Canonical quantization: This is the replacement of Poisson brackets by the commutators and of the dynamical variables by the Heisenberg operators. And as we mentioned above the resultant equations of motion coincide with those of the Dirac theory. Thus we have a correct classical model of the relativistic spinning electron.

(ii) Path Integral quantization. There is a longstanding problem of how to obtain discrete quantum spin values from continuous classical spin variables by path integration. This problem has been solved with our classical spin variables $\bar{z}$ and z. In fact it is possible to formulate precisely the whole of QED perturbation theory directly from classical particle trajectories by path integration[7]. The Dirac propagator, by the way, has the physical interpretation that its matrix element corresponds to a path integral not only with the endpoints fixed, but also with the initial and final spin components fixed at α and β:

$$K_{\alpha\beta} = \left(\frac{1}{\gamma\cdot p - m}\right)_{\alpha\beta} = \int_{x_{a,\alpha}}^{x_{b,\beta}} \mathcal{D}(x)\mathcal{D}(p)\mathcal{D}(z)\mathcal{D}(\bar{z})e^{1/\hbar\int d\tau\omega}$$

(iii) Schrödinger quantization. Here we start from a function $\phi(\bar{z}, x; \tau)$ in the configuration space and represent the conjugate momenta by first order differential operators. The function ϕ satisfies

$$i\frac{\partial}{\partial\tau}\phi(\bar{z}, x; \tau) = \mathcal{H}\phi(\bar{z}, x; \tau)$$

where $\mathcal{H}$ is now the differential operator

$$\mathcal{H} = \bar{z}\gamma^\mu \frac{\partial}{\partial z}\left(i\frac{\partial}{\partial x^\mu} - eA_\mu\right)$$

We expand ϕ on both sides of this equation in powers of $\bar{z}_\alpha$ as

$$\phi(\bar{z}, x, \tau) = \phi(x, \tau) + \bar{z}^\alpha \psi_\alpha(x, \tau) + \bar{z}^\alpha \bar{z}^\beta \psi_\alpha(x, \tau) + \ldots$$

Acting with the operator $\mathcal{H}$ and comparing the coefficients of $\bar{z}_\alpha$ the first equation is the Dirac equation

$$i\frac{\partial}{\partial \tau}\psi_\alpha = \gamma^\mu \left(i\partial_\mu - eA_\mu\right)\psi_\alpha$$

10. **Excited States of Zitterbewegung.** The Schrödinger quantization gives in addition to Dirac equation a set of higher spin equations[8]. The next one is

$$i\frac{\partial}{\partial \tau}\psi_{\alpha\beta} = (\gamma^\mu \otimes I + I \otimes \gamma^\mu)(i\partial_\mu - eA_\mu)\psi_{\alpha\beta} \equiv \beta^\mu D_\mu \psi$$

where the matrices β satisfy the Kemmer algebra

$$\beta^\mu \beta^\lambda \beta^\nu + \beta^\nu \beta^\lambda \beta^\mu = \beta^\mu \delta^{\lambda\nu} + \beta^\nu \delta^{\lambda\mu}$$

the irreducible parts of which are ,as is well known, the 5 and 10—dimensional matrices representing spin 0 and spin 1 particles. The general β-matrices are of the form

$$\beta^\mu(\gamma) = \gamma^\mu \otimes I \otimes I \ldots + I \otimes \gamma^\mu \otimes I \ldots + \ldots + I \otimes I \otimes \ldots \otimes \gamma^\mu$$

and satisfy the commutation relations

$$[(\beta(\gamma), \beta(\gamma')] = \beta\left([\gamma, \gamma']\right)$$

where γ, γ' are the elements of a Lie algebra.
There are also supersymmetric structure given by the maps, for example,

$$\psi_\alpha \to \psi_{\alpha\beta}$$
$$\gamma^\mu \to \gamma^\mu \otimes I + I \otimes \gamma^\mu$$
$$(A \to A \otimes A)$$

11. **Lorentz-Dirac equation with radiation reaction and spin.** Now that we have included the correct spin terms into our classical equation we can evaluate self-energies and radiation reaction. The generalized Lorentz-Dirac equation so derived reads[9]

$$\dot{\pi}_\mu = eF_{\mu\nu}^{\text{ext}}u^\nu + e^2 \tilde{g}_{\mu\nu}\left(\frac{2}{3}\frac{\ddot{v}^\nu}{v^2} - \frac{9}{4}\frac{(v\cdot\dot{v})\dot{v}^\nu}{v^4}\right)$$

where

$$\tilde{g}_{\mu\nu} = \tilde{g}_{\mu\nu} - \frac{1}{v^2} v_\mu v_\nu$$

We have a new term coming from spin with coefficient $-\frac{9}{4}$ in addition to the standard Larmore term. In the spinless limit $v^2 \to 1, v \cdot \dot{v} \to 0, \dot{\pi}_\mu \to m\ddot{x}_\mu$, we obtain back the Lorentz Dirac equation.

12. Generalization to curved spaces is obtained by adding to the momenta a spin connection $\Gamma_\mu : \pi_\mu = p_\mu - eA_\mu + i\bar{z}\Gamma_\mu z$. The action in terms of π_μ is the same as before. The equations of motion are $(\lambda = 1)$[10].

$$\dot{\bar{z}}\gamma \cdot n = i\left(\bar{z}\dot{\pi} + i\bar{z}\Gamma_\mu v^\nu\right)$$

$$\gamma \cdot n\dot{z} = -i\left(\pi z + iv^\nu\Gamma_\mu z\right)$$

$$\dot{x}_\mu = \bar{z}\gamma_\mu z = v_\mu$$

$$\dot{\pi}_\mu - \Gamma^\alpha_{\mu\nu}\pi_\alpha v^\nu = eF_{\mu\nu}v^\nu - \frac{1}{2}R_{\alpha\beta\mu\nu}S^{\alpha\beta}v^\nu$$

Papapetrou equations for spin follow.

13. Generalizations to spinning strings and membranes can be obtained by introducing world sheet variables and functions $x^\nu(\sigma^\alpha, \tau)$ and $\bar{z}(\sigma^\alpha, \tau)$.[11]

14. Internal Algebras SO(5) and SO(6). If one separates center of mass and relative coordinates and momenta one can exhibit a remarkable algebraic structure of the zitterbewegung which shows also the interesting geometry of the internal phase space. Again this structure is the same for the classical and quantum case. The brackets of the dynamical variables close to a SO(6) algebra. We only indicate here how the brackets of the relative conjugate coordinates is related to spin[12]:

$$\{\mathcal{Q}^\mu, \mathcal{Q}^\nu\} = \frac{1}{m^2}S^{\mu\nu} \;, \quad \{\mathcal{Q}^\mu, \mathcal{P}^\nu\} = -\tilde{g}^{\mu\nu}\frac{\mathcal{H}}{m} \;, \quad \{\mathcal{P}^\mu, \mathcal{P}^\nu\} = 4m^2 S^{\mu\nu}$$

15. Two and many-body equations with spin and radiation reaction. In analogy to recently established covariant many-body equations in quantumelectrodynamics we can also derive many -body equations for classical spinning particles. The method consists of defining composite spinors from the tensor product of the spinors for each particles, $Z = z_1 \otimes z_2$ rewrite the action in terms of these composite spinors and derive their equations. For the two body problem one obtains in particular the following covariant Hamiltonian[13]:

$$H = Z\vec{\beta}Z \cdot \vec{P} + \bar{Z}\left(\vec{\alpha}_1 - \vec{\alpha}_2\right)Z \cdot \vec{p} - e_1 e_2 \frac{\bar{Z}(1 - \alpha_1 \cdot \alpha_2)Z}{r} + M$$

where $\vec{\beta} = \vec{\gamma} \cdot 1 + 1 \cdot \vec{\gamma}$ and $\vec{\alpha}$'s are the Dirac matrices.

2.3 Modelling Mass and Wave-Particle Duality

In the previous Section we have presented an intuitive model for spin, but the mass entered the theory as a constant of integratioin for which we do not have a more physical representation. Moreover after quantization we have the Heisenberg equations of motion for the dynamical variables, but how do we picture the operator-valued Heisenberg equations? We picture the classical equations of motion by trajectories, but not so for the Heisenberg equations. In this Section we shall present a complementary classical wave approach to the internal structure of the particle which gives us more insight into the mass and wave properties of the quantum particle. At the end then we can attempt a synthesis between the wave and particle approaches to pinpoint what a "quantum" particle is.

The wave approach is based on the explicit construction of localized oscillating nonspreading wavelets which move like relativistic particles. This is best explained in the simplest case of ordinary scalar linear wave equation

$$\Box \phi = 0$$

We look for a localized solution initially at a point x_0 of the form

$$\phi(\vec{x}, t) = F(\vec{x} - \vec{x}_0)e^{-i\Omega t}$$

The wave form oscillates with an internal frequency Ω; it is not static and this is a crucial point. The function F then satisfies the Helmholtz equation

$$\Delta F + (\Omega/c)^2 F = 0$$

whose solutions in spherical coordinates are

$$F(\vec{x} - \vec{x}_0) = \sum_{\ell m} C_{\ell m} \frac{1}{\sqrt{r}} J_{\ell+1/2}\left(\frac{\Omega}{c}r\right) P_\ell^m(\cos\theta)e^{im\varphi}, \quad r = \sqrt{(\vec{x} - \vec{x}_o)^2}$$

As an example, the simplest spherically symmetric solution is

$$F = C_{00} \frac{1}{\sqrt{r}} J_{1/2}\left(\frac{\Omega}{c}r\right)$$

We can associate a size c/Ω to the large central region of the wavelet, although it has a small but infinitely long oscillating tail (Fig.1). The moving solution is obtained by a Lorentz transformation with parameters

$$\vec{\beta} = \frac{v}{c}, \quad \gamma^2 = (1 - \beta^2)^{-1}$$

and has the form[13]

$$\phi(\vec{x}, t) = F\left(r_\perp(\vec{x} - \vec{x}_0,)t\right) e^{i(\vec{k}\cdot\vec{x} - \omega t)}$$

where
$$r_\perp^2 = (r_\mu n^\mu) - (r_\nu n^\mu) \quad , \quad n^\mu = \gamma(1, \vec{\beta}) \quad , \quad r^\mu = (t, \vec{x} - \vec{x}_0)$$

or

$$r_\perp = \left[(\vec{x} - \vec{x}_0)^2 c^2 \beta^2 \gamma^2 t^2 - 2\gamma^2 ct \left(\vec{\beta} \cdot (\vec{x} - \vec{x}_0) \right) + \gamma^2 \left(\vec{\beta} \cdot (\vec{x} - \vec{x}_0) \right)^2 \right]^{1/2}$$

is the space-like distance on the surface $\sum$ perpendicular to n^ν. We see now the de Broglie phase with $\vec{k} = \frac{\Omega}{c}\gamma\vec{\beta}$ and $\omega = \gamma\Omega$ and the dispersion relation

$$(\omega/c)^2 - \vec{k}^2 = (\Omega/c)^2$$

The group velocity of the wave lump is v, whereas the phase velocity is $u = \frac{\omega}{k}$ satisfying $uv = c^2$.

We have here the remarkable result that the dispersion relation for the localized solution of the massless wave equation $\Box\varphi = 0$ exactly coincides with that of the plane wave solution of the massive Klein-Gordon equation $\left(\Box + \frac{m^2 c^2}{\hbar^2}\right)\phi = 0$. Historically, when wave properties of the electron were discovered one has added by hand a mass term to the wave equation; it would have been equally possible to consider the localized solutions of the massless equation of the above type. We thus obtain an identification between the internal frequency Ω and the mass m:

$$\frac{mc}{\hbar} = \frac{\Omega}{c} \quad \text{or} \quad \Omega = \frac{mc^2}{\hbar}$$

and it is here that, for dimensional reasons, the Planck's constant enters. Thus the concept of mass is now related to the internal oscillations of a wave lump, the frequency of an internal clock, Ω and without such internal oscillations we could neither construct localized solutions, nor make a Lorentz transformation. We shall see again this connection between mass and frequency when we calculate the total field energy contained in the lump. It is given, in the rest frame of the lump, by

$$\mathcal{E} = N_E \int \left\{ \left| \frac{1}{c^2} \frac{\partial \phi}{\partial t} \right|^2 + (\nabla \phi)^2 \right\} d_x^3$$

where N_E is a normalization constant, for dimensional reasons since a free field has no scale. In order to obtain a finite energy we take a superposition of different frequencies Ω in the neighborhood of some basic frequency Ω with a distribution function $f(\Omega)$. Using the orthogonality properties of the Bessel functions we then obtain a finite energy which, normalizing the charge of the complex scalar field, becomes proportional to the frequency Ω in the rest frame. In the moving frame we obtain

$$\mathcal{E} \sim \omega \quad , \quad \vec{P} \sim \vec{k}$$

and the relativistic relation
$$\mathcal{E}^2 = c^2 \vec{P}^2 + m^2 c^4$$

In the nonrelativistic limit the wave equation goes over into the Schrödinger equation and our solution into

$$\psi = F(\vec{x} - \vec{x}_0 - \vec{v}t)e^{i(m\vec{v}\cdot\vec{x} - \frac{\hbar^2 k^2}{2m}t)}e^{-i\Omega t}$$

One can verify directly that this localized solution satisfies the Schrödinger equation with F given as before and with the dispersion relation

$$\omega - \frac{\hbar k^2}{wm} = \Omega$$

It is the Galilei-boosted rest frame solution $\psi = F(x - x_0)e^{-i\Omega t}$ The phase differs from the usual Schrödinger phase by the constant rapid mass oscillations $e^{-i\Omega t}$ which drops out in the calculation of phase differences in interference experiments. The energy and momentum calculated in the same manner as above are given by

$$\mathcal{E} = \hbar\omega \quad , \quad \vec{\mathcal{P}} = \hbar\vec{k}$$

In the presence of a potential $V(x)$ in the Schrödinger equation one can generalize the above localized solutions into the form[14]

$$\psi = F(\vec{x} - \vec{x}_0 - g(t)) \Psi(\vec{x}, t)e^{i\hbar(x,t)}$$

where F again is the localization function and Ψ is the usual solution of the Schrodinger equation in the given potential. In this form we obtain a new deterministic interpretation of a single quantum particle, as well as the statistical interpretation of in repeated experiments as follow. In the limit $\Omega \to \infty$ the localization function F approaches $\delta(x - x_0 - g(t))$ hence we get a classical trajectory of a point particle. On the other hand if we average over the parameters of the solution, for example over x_0 and v_0, F approaches unity and we are left with Ψ. Thus Ψ represents the typical or averaged behaviour of the particle in repeated experiments, and this is the standard Born statistical interpretation of quantum theory applicable only to repeated experiments. But now inbetween these two limits we get something new, namely a description of a single particle which has both particle-like and wave-like behaviour. Further details about these interpretation questions are given elsewhere[15].

Similar localized solutions have been obtained for the electromagnetic field[16], the spinor Dirac field[17] and for the linearized gravity[18]. With these highly localized solutions of wave equations we have modelled the mass in terms of the internal oscillations. We also modelled the wave-particle duality because the wave lumps move like relativistic particles and have the correct wave properties in phase and in dispersion relations. The frequency of the internal oscillations is the same as that of the helical motion of the spinning particle model of Section II. In fact if we imagine a point charge at a distance $\hbar/mc$ from the center of our wavelet solution, it would perform a helical motion during the time evolution of the solution. We may thus view the localized solution as the "quantum" picture of the helical solution, or as the picture of the Heisenberg equations of motion.

References to Chapter 2

1. E.Fermi, *Rev. Mod. Physics*, 4, 87 (1932)

2. A.O.Barut, *Physics Lett. A*, **145**, 387 (1990)

3. A.O.Barut, *Physics Lett. A*, **131**, 11 (1988)

4. P.A.M.Dirac, *Proc. Roy. Soc. (London)*, **A117,610** and **A118**, 351 (21928)

5. A.O.Barut and N.Zanghi, *Phys. Rev. Lett*, **52**, 2009 (1984)

6. A.O.Barut, C.Onem and N.Unal, *J. Phys. A*, . **23**,1113 (1990)

7. A.O.Barut and I.H.Duru, *Physics Reports*, **172**, 1 (1989)

8. A.O.Barut, *Phys. Letters B*, **237**, 436 (1990)

9. A.O.Barut and N.Unal, *Phys. Rev. A*, **40**, 5404 (1989)

10. A.O.Barut and M. Pavcic, *Class. Quant. Gravity*, *4*, **L41** and **L131** (1987); **5**, 707 (1988)

11. A.O.Barut and M.Pavcic, *Lett. Math. Phys*,. **16**, 333 (1988)

12. A.O.Barut and W.D.Thacker, *Phys. Rev.*, **31**, 1386 (1985)

13. A.O.Barut, *Phys. Lett. A*, **143**, 349 (1990)

14. A.O.Barut, *Found. Phys.*, **20**, 1233 (1990)

15. A.O.Barut, Quantum Theory of Single Events, in *Symposium on the Foundations of Modern Physics*, 1990, World Scientific

16. A.O.Barut and A.Grant, *Found. Phys. Lett.*, **3**, 303 (1990)

17. A.O.Barut, E.Okan and G.Akdeniz, to be published

18. A.O.Barut and Y. Sabouti, to be published

3. SELF-FIELD QUANTUMELECTRODYNAMICS

3.1 Introduction

There are many approaches to radiative processes, or more generally, to electromagnetic interactions of charged particles. We should welcome this multitude because different ways of looking at the same physical phenomena can only bring clarity and hopefully enlightenment. I list those different formulations which are definite and more or less complete:

(i) Second quantized quantum field theory, or the perturbative QED[1].
(ii) The S matrix theory of electromagnetic interactions, either from unitarity, analyticity and successive pole approximation[2], or from regularization of the product of distributions[3]. Both of these lead to the renormalized perturbation theory with particles on the mass shell.
(iii) Path integral method. Either path integrals of Maxwell–Dirac fields[4], or path integrals directly from the classical particle trajectories[5].
(iv) Source theory[6].
(v) Selffield quantumelectrodynamics.

Of these only the selffield approach is in the long tradition of classical radiation theory and classical electrodynamics and is the subject of these lectures.

It is often stated that a large number of radiative phenomena conclusively show that the electromagnetic field, and further the electron's field, is quantized as a system of infinitely many oscillators with their zero point energies. The radiative phenomena are listed in Table I. We shall show that all these processes can also be understood and calculated in the selffield approach which does not quantize the fields. The quantum properties of the electromagnetic field are reduced here to the quantum properties of the source. One avoids thereby some of the difficulties of the quantized fields, such as the infinite zero point energy and other infinities of the perturbative QED.

TABLE I: RADIATIVE PROCESSES

Spontaneous emission

Lamb shift

Anomalous magnetic moment

Vacuum polarization

Casimir effect between parallel plates

Casimir Polder potentials

Planck-distribution law for blackbody radiation

Unruh effect

QED in cavities

e^+– e^- system:

      positronium spectrum

      positronium annihilation

      pair production and annihilation

      e^+– e^- scattering

Relativistic many body problem with retardation

Electron - photon system:

      photoelectric effect

      Compton effect

      Bremsstrahlung

This lecture tells the story of the developments of selffield QED and it is good to begin from the beginning, namely the classical electrodynamics.

3.2 Classical Electrodynamics

The selfconsistent treatment of coupled matter and electromagnetic field goes back to H.A. Lorentz[7]. The electromagnetic field has as its source all the charged particles which in turn move in this total electromagnetic field. We have thus the Maxwell's equations coupled to the equations for matter:

$$1) \qquad F_{\mu\nu}{}^{,\nu} = -j_\mu$$

$$\left.\begin{array}{l} \\ 2) \text{ Equation of motion of matter in the field } F \end{array}\right\} \qquad (1)$$

These equations, both, can be derived from a single action principle. It has the general from

$$W = \int [\text{Kinetic energy of matter } -j_\mu A^\mu - 1/4 \ F_{\mu\nu} F^{\mu\nu}] \tag{2}$$

The last term is the action density of the field and the middle term represents the interaction of the matter current with the field.

We shall keep this general framework throughout also for quantum electrodynamics. The only change will be in the specific form of the current or how we describe the matter, the electron.

Classical electrodynamics *per se* is usually associated with the current of point particles moving along wordlines. But we can have more general extended sources of currents, as we shall see. For a number of point particles the current is given by

$$j_\mu(x) = \sum_i e_i \int ds_i \dot{x}_{i\mu}(s_i) \delta\left(x - x_i(s_i)\right) \tag{3}$$

Hence the fundamental equations are

$$F_{\mu\nu}{}^{,\nu} = -j_\mu = -\sum_i e_i \int ds_i \dot{x}_i(s_i) \delta\left(x - x_i(s_i)\right) \tag{4}$$

Here s_i are invariant time-parameters on the worldlines of the particles, and dots represent differentiation with respect to these times.

The equations of motions of the worldlines are

$$m_i \ddot{x}_{i\mu} = e_i F_{\mu\nu} \dot{x}_i^\nu \quad , i = 1, 2, 3, \ldots \tag{5}$$

It is essential for the selfconsistency of our system that the field F entering the last equation is the field produced by all the particles including the particle i, namely the selffield. Hence we divide F into two parts

$$m_i \ddot{x}_{i\mu} = e_i F_{\mu\nu}^{(\text{other particles})} \dot{x}_i^\nu + e_i F_{\mu\nu}^{\text{self}} \dot{x}_i^\nu \tag{6}$$

The selffield can be obtained from the Lienard-Wiechert potential

$$A_\mu(x) = \int dx_\mu(s) D\left(x - x(s)\right) = e \int ds \dot{x}_\mu(s) D\left(x - x(s)\right) \tag{7}$$

but is formally infinite at the position of the particle. It must be treated properly, for example, by analytic continuation onto the world line[8]. This leads to the final Lorentz-Dirac equation for each particle (in natural units $c = \hbar = 1$)

$$m\ddot{x}_\mu = eF_{\mu\nu}^{\text{ext}} \dot{x}^\nu + \frac{2}{3} \ e^2 \left(\dddot{x}_\mu + (\ddot{x})^2 \dot{x}_\mu\right) \tag{8}$$

This is the basic nonperturbative equation of classical electrodynamics. Here m is now the renormalized mass. Furthermore we must find solutions of this equation which have the property that whenever the external force is zero the electron moves like a free particle, $m\ddot{x}_\mu = 0$, that is the second term must vanish together with the external field. This is part of the renormalization program. The important feature of this equation is that all radiative effects are now expressed in a closed, we repeat, nonperturbative way. The price we pay for this is that the equation is not only nonlinear but also contains the third derivatives. The selffield approach to quantumelectrodynamics has the goal of finding the analogous nonlinear, nonperturbative equation in the case of quantum currents. It is clear that radiative effects like the Lamb shift, anomalous magnetic moment, spontaneous emission, etc. have their counterparts also in classical electrodynamics.

As a second example of a classical current we consider the classical model of the Dirac electron which describes a spinning and charged relativistic point particle. In this model the worldline of the point particle is a helix, called zitterbewegung, and the orbital angular momentum of the helix in the rest frame of the center of mass accounts for the spin and the magnetic moment of the particle. The generalization of the Lorentz-Dirac equation for this case has recently been given[9]:

$$\dot{\pi}_\mu = eF^{\text{ext}}_{\mu\nu}v^\nu + e^2\left(g_{\mu\nu} - \frac{v_\mu v_\nu}{v^2}\right)\left[\frac{2}{3}\frac{\ddot{v}^\nu}{v^2} - \frac{9}{4}\frac{(v\cdot\dot{v})\dot{v}^\nu}{v^4}\right] \tag{9}$$

where

$$\pi_\mu = p_\mu - eA_\mu, \;\; v = \dot{x} \text{ and } v^2 \neq 1 \text{ due to spin.}$$

There are other classical models of the electron. A remarkable one is due to Lees[10] amd Dirac[11] in which a charged shell is held stable with a surface tension. In the equilibrium position the surface tension can be expressed in terms of the mass of the electron so that this model has exactly again two parameters, mass and charge, like the point worldline. The Lorentz-Dirac equation for this model to my knowledge has not been worked out yet.

3.3 Schrödinger and Dirac Currents Quantumelectrodynamics

Quantumelectrodynamics has the same two basic equations (1). Only the form of the current j is different. According to Schrödinger and Dirac the electron is described not by a worldline but by a field $\psi(x,t)$ and the basic coupled equations (1) become

$$F^{,\nu}_{\mu\nu} = -j_\mu \;\;\; , \;\; F_{\mu\nu} = A_{\nu,\mu} - A_{\mu,\nu}$$

and

$$(\gamma^\mu i\partial_\mu - m)\psi(x) = e\gamma^\mu\psi(x)A_\mu(x) \tag{10}$$

for the relativistic Dirac case, and

$$i\frac{\partial\psi}{\partial t} = \left(-\frac{1}{2m}\left[(\vec{p} - e\vec{A})^2\right] + eA_0\right)\psi \tag{11}$$

for the nonrelativistic Schrödinger case. The currents for a number of electrons is

$$j_\mu(x) = \sum_i e_i \bar\psi_i(x) \overset{(i)}{\gamma}_\mu \psi_i(x) \tag{12}$$

with a similar expression for the Schrödinger current.

Again the field A_μ is the sum of an external and a selffield parts:

$$A_\mu = A_\mu^{\text{ext}} + A_\mu^{\text{self}} \tag{13}$$

With the choice of gauge $A^\mu_{,\mu} = 0$ the Maxwell equations become

$$\Box A_\mu = j_\mu(x) = \sum_i e_i \bar\psi_i(x) \overset{(i)}{\gamma}_\mu \psi_\mu(x) \tag{14}$$

so that the selffield can be expressed in terms of the current as

$$A_\mu(x) = \int dy D(x-y) j_\mu(x) \tag{15}$$

where $D(x-y)$ is the appropriate Green's function corresponding to initial and boundary conditions. Equation (15) is our generalized Lienard-Wiechert potential. Thus the light emitted by a source depends on the nature and preparation of the current, and also on the nature of the environment determining the Green's function. Furthermore the whole light cone where ψ is different from zero contributes to the field at the field point and not just a single intersection of the worldline with the light cone, as in the case of a point particle.

Thus the selffield can be eliminated from the coupled Maxwell-Dirac equations. Inserting A_μ into the equation of motion we obtain

$$\left\{ \gamma^\mu \left(i\partial_\mu - e_k A_\mu^{\text{ext}} \right) - m_k \right\} \psi_k(x) = e_k \gamma^\mu \psi_k(x) \int dy D(x-y) \sum_i e_i \bar\psi_i(y) \gamma_\mu \psi_i(y) \tag{16}$$

Here A^{ext} is a fixed external field whose sources are far away and not dynamically relevant. In the next Section we shall treat two or many body systems in which we shall eliminate completely all the fields in favor of the currents. Eq. (16) is a nonlinear integral equation for ψ analogous to the nonlinear equation of the classical electrodynamics. The corresponding equation for the Schrödinger case is ($\hbar = 1$)

$$i\frac{\partial \psi}{\partial t} = \left[-\frac{1}{2m} \left(\vec p - e\vec A^{\text{ext}} - e\vec A^{\text{self}} \right)^2 + e \left(A_0^{\text{ext}} + A_0^{\text{self}} \right) \right] \psi \tag{17}$$

where the selfpotentials are

$$A_0^{\text{self}} = \int dy D(x-y) \sum_k e_k \psi_k^*(y) \psi_k(y), \ \vec A^{\text{self}}(x) = \int dy D(x-y) \sum_k \psi_k^*(y) \frac{\nabla}{i} \psi_k(y) \tag{18}$$

In writing these equations we have assumed that the ψ-current is an actual material charge current, and not just a probability current. Thus we are inevitably led to contemplate the interpretation and foundations of quantum theory. The foundations of quantumelectrodynamics and that of quantum theory must be the same, for quantum mechanics was invented to understand the interactions between light and matter. Not surprisingly, it was Schrödinger who first formulated the selfconsistent coupled Maxwell and matter field equations, i.e., the program of Lorentz, for the new wave mechanics and insisted that for the selfconsistency of the theory the self field of the electron must be included as a nonlinear term. Schrödinger however calculated only the static part of the selfenergy and obtained unacceptable large selfenergies. Subsequently quantum electrodynamics went into a different direction. The selffield was dropped completely. Instead, one introduced a separate quantized radiation field with its own new degrees of freedom and coupled this to the quantized matter field. In the selffield approach the electromagnetic field has no separate degrees of freedom, they are determined by the source's degrees of freedom, but then we must include the full nonlinear selffield term. We shall come to this duality between the two approaches and to the questions of interpretation of quantum theory after the developments of the selffield QED.

3.4 Radiative Processes in an External (Coulomb) Field

The basis of selffield quantumelectrodynamics is conceptually very simple and is completely expressed by the single equation (16). All QED processes in an external field listed in Table I should be derived from this single equation. To perform actual calculations it is much simpler and more direct to work with the action rather than with the equations of motion. The action W can, up to an overall δ-function, be related to the energies of the system for bound state problems, and to the scattering amplitude for scattering problems.

The action for the system (10) is

$$W = \int dx \left[\bar{\psi}(x)(\gamma^\mu i\partial_\mu - m)\psi(x) - e\bar{\psi}(x)\gamma^\mu \psi(x)A_\mu(x) - \frac{1}{4}F_{\mu\nu}F^{\mu\nu} \right] \qquad (19)$$

Here we shall express $A_\mu(x)$ in terms of ψ using (15). For bound state problems the action of the electromagnetic field can be reexpressed by a partial integration, using (10), as

$$-\frac{1}{4}\int dx F_{\mu\nu}F^{\mu\nu} = +\frac{1}{2}\int dx j_\mu(x)A^\mu(x) \qquad (20)$$

Putting all together we have the action underlying our nonlinear equation (16), namely

$$W = \int dx \left[\bar{\psi}(x)\left(\gamma^\mu \left(i\partial_\mu - eA_\mu^{\text{ext}} \right) - m \right)\psi(x) \right.$$
$$\left. -\frac{e^2}{2}\int dy\bar{\psi}(x)\gamma^\mu(x)\psi(x)D(x-y)\bar{\psi}(y)\gamma_\mu\psi(y) \right] \qquad (21)$$

We shall consider now the single electron problem in an external field.

We expand the classical field ψ into a Fourier series

$$\psi(x) = \sum_n \psi_n(\vec{x}) e^{-iE_n t} \tag{22}$$

and shall try to determine the expansion coefficients $\psi_n(\vec{x})$ and the spectrum E_n—discrete and continous. This expansion is quite different than the one used in standard QED and quantumoptics, namely the Coulomb series expansion, for example, in the Coulomb field,

$$\psi(x) = \sum_n c_n(t) \psi_n^c(\vec{x})$$

Here one derives equations for the time-dependent coefficients $c_n(t)$. The idea behind is that the system has definite levels and the perturbation will cause transitions between these levels. In our formulation, due to selfenergy, there are no definite (discrete) levels as exact eigenstates of the system to begin with, but the equations will determine the spectrum. In fact it will turn out that only the ground state of the system will be a stable eigenstate followed by a continuum with spectral concentrations around the unperturbed spectrum.

If we insert the Fourier expansion into the action we obtain

$$W = \sum_{nm} \oint \int dx \left\{ \bar{\psi}_n(\vec{x}) e^{iE_n x^0} \left[\gamma^\mu \left(i\partial_\mu - eA_\mu^{\text{ext}} \right) - m \right] \psi_m(\vec{x}) e^{-iE_m x^0} - \right.$$

$$\left. - \frac{e^2}{2} \int dy \bar{\psi}_n(\vec{x}) \gamma^\mu \psi_m(\vec{x}) e^{i(E_n - E_m)x^0} D(x-y) \bar{\psi}_r(\vec{y}) \gamma_\mu \psi_s(\vec{y}) e^{i(E_r - E_s)y^0} \right\} \tag{23}$$

Time integrations can be performed using

$$D(x-y) = -\frac{1}{(2\pi)^4} \int dk \frac{e^{-ik(x-y)}}{k^2} \tag{24}$$

and we can write the interaction part of the action entirely in terms of the Fourier components of the current

$$W_{\text{int}} = +\frac{e^2}{2} \sum_{nmrs} \delta(E_n - E_m + E_r - E_s) \int \bar{\psi}_n(\vec{x}) \gamma^\mu \psi_m(\vec{x})$$

$$\times \frac{e^{i\vec{k}\cdot(\vec{x}-\vec{y})}}{(E_n - E_m)^2 - \vec{k}^2} \bar{\psi}_r(\vec{y}) \gamma_\mu \psi_s(\vec{y}) d\vec{x} d\vec{y} d\vec{k} \tag{25}$$

For the exact solutions of our equations the action W will vanish identically. We will now solve the system iteratively.

To lowest order of iteration we take the field to be given by the solutions of the external field problem without the selfenergy terms, and the energies to be shifted by a small amount:

$$\psi_n(x) = \psi_n^{\text{ext}}(x)$$
$$E_n = E_n^{\text{ext}} + \Delta E_n \tag{26}$$

The first term in (22) therefore gives simply, using the orthonormality of ψ_n's,

$$W_0 = \int d\vec{x} \sum_{nm} \bar{\psi}_n \left(\gamma^0 E_n^{\text{ext}} - \vec{\gamma} \cdot \vec{p} - m - e A_\mu^{\text{ext}} \right) \psi_m \delta(E_n - E_m)$$
$$\Rightarrow \sum_{nm} \Delta E_n \delta(E_n - E_m) \delta nm$$

In the second term we separate the terms according to $E_n = E_m$, $E_r = E_s$ and according to $E_n = E_s$, $E_m = E_r$, the two ways of satisfying the overall δ-function. And since $W = 0$ to this order of iteration we can solve for ΔE_n. The action and the total energy of the system are related by a δ-function. Cancelling this δ-function and also the sum over n to obtain the energy shift of a fixed level n, we obtain

$$\Delta E_n = \frac{e^2}{2} \int d\vec{x} \bar{\psi}_n(\vec{x}) \gamma_\mu \psi_n(\vec{x}) P \int \frac{d\vec{k}}{(2\pi)^3} \int d\vec{y} \frac{e^{i\vec{k}\cdot(\vec{x}-\vec{y})}}{k^2} \cdot \oint_s \bar{\psi}_s(\vec{y}) \gamma^\mu \psi_s(\vec{y})$$
$$- \frac{e^2}{2} \oint_s \int d\vec{x} d\vec{y} \bar{\psi}_n(\vec{x}) \gamma_\mu \psi_s(\vec{x}) \int \frac{d\vec{k}}{(2\pi)^3} e^{i\vec{k}\cdot(\vec{x}-\vec{y})} \bar{\psi}_s(\vec{y}) \gamma^\mu \psi_n(\vec{y}) \cdot$$
$$\cdot \left[\frac{1}{E_s - E_n - k} - \frac{1}{E_s - E_n + k} \right] \tag{27}$$
$$- \frac{e^2}{2} \oint_{\substack{s \\ (s<n)}} \int d\vec{x} d\vec{y} \bar{\psi}_n(\vec{x}) \gamma_\mu \psi_s(\vec{x}) \int \frac{d\vec{k}}{(2\pi)^3} e^{i\vec{k}\cdot(\vec{x}-\vec{y})} \bar{\psi}_s(\vec{y}) \gamma^\mu \psi_n(\vec{y}) \cdot$$
$$\cdot \frac{i\pi}{2k} \delta(E_s - E_n - k)$$

This can be written in the form

$$\Delta E_n = -\frac{e^2}{2} \oint_m \frac{d\vec{k}}{(2\pi)^3} \frac{j_{nn}^\mu(\vec{k}) j_\mu^{mm}(-\vec{k})}{\vec{k}^2} - \frac{e^2}{2} \oint_{m<n} \frac{d\vec{k}}{(2\pi)^3} j_{nm}^\mu(\vec{k}) j_\mu^{mn}(-\vec{k}) \frac{i\pi}{2} \delta(E_m - E_n -$$
$$- \frac{e^2}{2} \oint_m \frac{d\vec{k}}{(2\pi)^3} j_{nm}^\mu(\vec{k}) j_\mu^{mn}(-\vec{k}) \frac{1}{2k} \left[\frac{1}{E_m - E_n - k} - \frac{1}{E_m - E_n + k} \right] \tag{28}$$

Thus the energy shifts are entirely expressed in terms of the integrals over the Fourier spectra of currents of all states. The first term corresponds to vacuum polarization, the second to spontaneous emission, and the third term to the Lamb shift proper. In arriving at these results we have used the causal Green's function and separated the integrals into a principal and a imaginary part according to the formula

$$\frac{1}{x} = P\frac{1}{x} \pm i\pi\delta(x)$$

(29)

All the main QED effects are obtained here from a single expression. In fact one can also read off the anomalous magnetic moment $(g-2)$ from this expression as we shall show in Section VI.

The evaluation of these expressions is a rather laborious technical problem. We have to use relativistic Coulomb wave functions for both the discrete and continous spectrum and integrate the products of such functions and sum over the whole spectrum. We shall indicate some of these calculations and give results in Section VIII. The most important feature of the present formulation is that there are no infrared nor ultraviolet divergences.

The spontaneous emission term in Eq. (27) has been exactly evaluated[12]. We have now complete relativistic spontaneous decay rates for all hydrogenic states[13]. Table II shows some of these results.

TABLE II. Decay rates (s^{-1}) in hydrogen and muonium

Transition	Hydrogen	Muonium
$2S_{1/2} \rightarrow 1S_{1/2}$	2.4964×10^{-6}	2.3997×10^{-6}
$2S_{1/2} \rightarrow 1P_{1/2}$	5.194×10^{-10}	5.172×10^{-10}
$2P_{1/2} \rightarrow 1S_{1/2}$	2.0883×10^{8}	2.0794×10^{8}
$2P_{3/2} \rightarrow 1S_{1/2}$	4.1766×10^{8}	4.1587×10^{8}
$2P \rightarrow 1S_{1/2}$	6.2649×10^{8}	6.2382×10^{8}

The vacuum polarization term has also been evaluated analytically[14] to lowest order term in $\alpha(Z\alpha)^4$. This is the most divergent term in perturbative QED and vanishes in the nonrelativistic limit.

The Lamb shift term which correctly reduces to the standard expressions in the dipole approximation has also been shown to be finite and will be evaluated in closed form[15].

In all these calcuations, since we are using Coulomb wave fuctions instead of the plane waves, the individual integrals are all finite. The summation over all the discrete and continous levels are done by means of the relativistic Coulomb Green's functions.

3.5 Quantumeletrodynamics of the Relativistic Two-Body System

One of the most important and perhaps unexpected features of the selffield formulation of quantumelectrodynamics turned out to be a nonperturbative treatment of two and many body systems in closed from. It is well known that bound state problems cannot be treated in perturbative QED starting from first principles. Instead

one begins from a Schrödinger or Dirac-like equation obtained from some approximation to the Bethe-Salpeter relations and then calculates the perturbation diagrams
to the bound state solutions of these equations. What one really needs is a genuine
two-body relativistic equation which includes all the radiative terms as well as all the
recoil corrections at once. We shall now discuss the principles of this theory.

In nonrelativistic quantum theory the many body problem is formulated in configuration space by a wave equation with pair potentials $v_{ij}(x_i - x_j)$ of the form

$$\left(\frac{p_1^2}{2m_1} + \frac{p_2^2}{2m_2} + \ldots V_{12} + V_{13} + V_{23} + \ldots \right) \psi(x_1, \ldots, x_n; t) = i\hbar \frac{\partial \psi}{\partial t}$$

This a priori not obvious. We may also think that each particle has its own field $\psi(x)$
and satisfy a wave equation with a potential coming from the charge distribution of the
other particles. For two particles, for example, we would have the coupled Hartre-type
equations

$$i\hbar \frac{\partial \psi_1(\vec{x}_1, t)}{\partial t} = \left(-\frac{\hbar^2}{2m_1} \Delta + \int \frac{\psi_2^*(x_2, t)\psi_2(x_2, t)}{|\vec{x}_2 - \vec{x}_1|} d\vec{x}_2 \right) \psi_1(\vec{x}_1, t)$$

$$i\hbar \frac{\partial \psi_2(\vec{x}_2, t)}{\partial t} = \left(-\frac{\hbar^2}{2m_2} \Delta + \int \frac{\psi_1^*(\vec{x}_1, t)\psi_1(\vec{x}_1, t)}{|\vec{x}_1 - \vec{x}_2|} d\vec{x}_1 \right) \psi_2(\vec{x}_2, t)$$

These two formulations are closely related but not identical. We shall see that they
correspond to two different types of variational principles and actually describe two
different types of physical situations. Quantum theory has a separate new postulate for
two or more particles, namely that the state space is the tensor product of one particle
state spaces. This leads immediately to the first formulation in configuration space.
Such combined systems are called in the axiomatic of quantum theory "nonseparated"
systems with all the nonlocal properties of quantum theory. But this postulate does
not apply universally. There are other systems, namely the "separated" systems, which
are described by the second type of equations. For example, for the system hydrogen
molecule the two protons are separated, whereas the two electrons are nonseparated.
The superposition principle holds for the nonseparated systems only. We shall now
see how all this comes about from two different basic variational princples in the
relativistic case (the nonrelativistic case is similar).

Consider a number of matter fields $\psi_1(x), \psi_2(x) \ldots$ The action of these fields interacting via the electromagnetic field is

$$W = \int dx \left\{ \sum_k \bar{\psi}_k \left(\gamma^\mu i \partial_\mu - m_k \right) \psi_k - j_\mu(x) A^\mu(x) - \frac{1}{4} F_{\mu\nu} F^{\mu\nu} \right\} \tag{30}$$

where the current j^μ is the sum of Dirac currents for each field

$$j^\mu(x) = \sum_k e_k \bar{\psi}_k(x) \gamma^\mu \psi_\mu(x) \tag{31}$$

Again in the gauge $A^\mu,_\mu = 0$ we obtain the equations for the electromagnetic field as

$$\Box A_\mu = j_\mu = \sum_k j_\mu^{(k)} \tag{32}$$

with the solution

$$A_\mu(x) = \int dy D(x-y) j_\mu(x). \tag{33}$$

If we insert this into the action both in the $j^\mu \cdot A_\mu$ term as well as in the term $-(1/4) F_{\mu\nu} F^{\mu\nu}$, and using the identity (20), we obtain

$$W = \int dx \sum_k \bar{\psi}_k \left(\gamma^\mu i \partial_\mu - m_k\right) \psi_k - \sum_{k,\ell} \frac{1}{2} \int dx dy j_\mu^{(k)}(x) D(x-y) j_{(\ell)}^\mu(y) \tag{34}$$

The interaction action is a sum of current-current interactions containing both the mutual interaction terms, e.g.

$$-\frac{e_1 e_2}{2} \int dx dy \bar{\psi}_1(x) \gamma^\mu \psi_1(x) D(x-y) \bar{\psi}_2(y) \gamma_\mu \psi_2(y) - (1 \leftrightarrow 2)$$

and the self interaction terms like

$$-\frac{e_1^2}{2} \int dx dy \bar{\psi}_1(x) \gamma^\mu \psi_1(x) D(x-y) \bar{\psi}_1(y) \gamma_\mu \psi_1(y)$$

If we vary this action with respect to each field ψ_k separately we obtain coupled nonlinear equations. For example for two particles

$$
\begin{aligned}
\left(\gamma^\mu i \partial_\mu - m_1\right) \psi_1 &= \frac{e_1 e_2}{2} \gamma^\mu \psi_1 \int dy D(x-y) \bar{\psi}_2(y) \gamma_\mu \psi_2(y) \\
&+ \frac{e_1^2}{2} \gamma^\mu \psi_1 \int dy D(x-y) \bar{\psi}_1(y) \gamma_\mu \psi_1(y) \\
\left(\gamma^\mu i \partial_\mu - m_2\right) \psi_2 &= \frac{e_1 e_2}{2} \gamma^\mu \psi_2 \int dy D(x-y) \bar{\psi}_1(y) \gamma_\mu \psi_1(y) \\
&+ \frac{e_2^2}{2} \gamma^\mu \psi_2 \int dy D(x-y) \bar{\psi}_2(y) \gamma_\mu \psi_2(y)
\end{aligned}
\tag{35}
$$

Next let us define a composite field Φ by

$$\Phi(x_1, x_2) = \psi_1(x_1) \psi_2(x_2) \tag{36}$$

This is a 16-component spinor field. We can rewrite our action (34) entirely in terms of the composite field. This is straightforward in the mutual interaction terms. In the kinetic energy and selfinteraction terms we multiply suitable by normalization factors. For example for the first kinetic energy term we get

$$\int dx_1 \bar\psi_1(x_1)(\gamma^\mu i\partial_\mu - m_1)\psi_1(x_1) \cdot \int d\sigma_2 \bar\psi_2(x_2)\gamma \cdot n\psi_2(x_2)$$

where $d\sigma_2 n^\mu = d\sigma_2^\mu$ is a 3-dimensional volume element perpendicular to the normal n^μ. Similarly for the other kinetic energy term. The selfenergy terms need two such normalization factors. The resultant action in terms of the composite field is then

$$
\begin{aligned}
W = \Bigg[&\int dx_1 d\sigma_2 \bar\Phi(x_1 x_2)(\gamma^\mu \pi_{1\mu} - m_1) \otimes \gamma \cdot n\Phi(x_1 x_2) \\
&+ \int dx_2 d\sigma_1 \bar\Phi(x_2 x_1)\gamma \cdot n \otimes (\gamma^\mu \pi_{2\mu} - m_2)\Phi(x_2 x_1) \\
&- e_1 e_2 \int dx_1 dx_2 \bar\Phi(x_1 x_2)\gamma^\mu \otimes \gamma_\mu D(x_1 - x_2)\Phi(x_1 x_2) \Bigg]
\end{aligned}
\tag{37}
$$

The generalized canonical momenta $\pi_{i\mu}$ are given further below. Here and through the rest of the paper we shall write spin matrices in the form of tensor products $\otimes$, the first factor always referring to the spin space of particle 1, the second to particle 2. We shall give the selfenergy terms explicitly below.

Now our second variational principle is that the action be stationary not with respect to the variations of the individual fields but with respect to the total composite field only. This is a weaker condition than before and leads to an equation for Φ in configuration space. For bound state problems only the symmetric Green's function contributes and it contains a $\delta(x^2)$-function which we decompose relative to the space-like surface with normal n^μ as follows

$$\delta(r^2) = \frac{\delta[(r \cdot n) - r_\perp] \pm \delta[(r \cdot n) + r_\perp]}{2r_\perp}, \quad r_\perp = [(r \cdot n)^2 - r^2]^{1/2} \tag{38}$$

where $r_\perp$ is a relativistic three dimensional distance which for $n = (1000)$ reduces to the ordinary distance r. All the integrals in the action (37) are 7-dimensional. For covariance purposes it is necessary to have the vector n^μ. It tells us how to choose the time axis. The vector n is also present, in principle, in the one-body Dirac equation but we usally do not write it when discussing the solutions, but automatically choose it to be $n = (1000)$, i.e., the rest frame. The final form of our two-body equation is then

$$\left\{ (\gamma^\mu \pi_1^\mu - m_1) \otimes \gamma \cdot n + \gamma \cdot n \otimes (\gamma^\mu \pi_2^\mu - m_2) - e_1 e_2 \frac{\gamma^\mu \otimes \gamma_\mu}{r_\perp} \right\} \Phi(x_1 x_2) = 0 \tag{39}$$

where now the selfpotentials are inside the generalized momenta

$$\pi_i^\mu = p_i^\mu - e_i A_i^{\mu\,\text{self}} - e_i A_i^{\mu\,\text{ext}} \tag{40}$$

with

$$A_{\mu,1}^{\text{self}}(x) = \frac{e_1}{2} \int dz d\sigma_u D(x-z) \bar{\Phi}(z,u) \gamma_\mu \otimes \gamma \cdot n \Phi(z,u)$$

$$A_{\mu,2}^{\text{self}}(x) = \frac{e_2}{2} \int d\sigma_z du D(x-u) \bar{\Phi}(z,u) \gamma \cdot n \otimes \gamma_\mu \Phi(z,u)$$

(41)

We note that the last term in (39) can also be put into the potential A_μ, one half for each particle; $\frac{1}{2} e_2 \frac{1 \otimes \gamma_\mu}{r}$ and $\frac{1}{2} e_1 \frac{\gamma_\mu \otimes 1}{r}$, respectively.

The self potentials are nonlinear integral expressions. The arguments of Φ consist of seven variables because $\Phi(x_1, x_2)$ is different from zero only if $(x_1 - x_2)$ is lightlike; only then there is a communication between the particles. This means that we have one time-variable and three space variables for each particle. We see this more cleary if we introduce center of mass and relative variables according to

$$\Pi = \pi_1 + \pi_2 , \quad \pi = \pi_1 - \pi_2$$

$$x = x_1 - x_2 , \quad X = x_1 + x_2$$

(42)

Then equation (39), without the selffield terms for simplicity, becomes

$$\left\{ \Gamma^\mu \Pi_\mu + k^\mu \pi_\mu - \frac{e_1 e_2}{r_\perp} \gamma^\mu \otimes \gamma_\mu - m_1 I \otimes \gamma \cdot n - m_2 \gamma \cdot n \otimes I \right\} \Phi = 0$$

(43)

where we have introduced

$$\Gamma^\mu = \frac{1}{2} (\gamma^\mu \otimes \gamma \cdot n + \gamma \cdot n \otimes \gamma^\mu)$$

and

$$k^\mu = \frac{1}{2} (\gamma^\mu \otimes \gamma \cdot n - \gamma \cdot n \otimes \gamma^\mu)$$

(44)

We see now that $k \cdot n = 0$, i.e., the component of k^μ parallel to n^μ vanishes which means that the component of the relative momentum π_μ parallel to n_μ drops out of the equation automatically. For $n = (1000)$, in particular, we have

$$\left\{ \Gamma^0 \Pi_0 - \vec{\gamma} \cdot \vec{\Pi} - \vec{k} \cdot \vec{\pi} - \frac{e_1 e_2}{r} \gamma^\mu \otimes \gamma_\mu - m_1 I \otimes \gamma_0 - m_2 \gamma_0 \otimes I \right\} \Phi = 0$$

(45)

Thus we have only one time variable conjugate to the center of mass energy Π_0 and three degrees of freedom for the center of mass momentum $\vec{\Pi}$ and three degrees of freedom for the relative momentum $\vec{\pi}^0$; π_0 does not enter, as it should be so on physical grounds. In contrast the Bethe-Salpeter equation has two time coordinates. Since Π_0 is the "Hamiltonian" of the system we obtain, by multiplying (45) by Γ_0^{-1} the Hamiltonian form of the two-body equation

$$\Pi_0 \Phi = \left\{ \vec{\alpha} \cdot \vec{\Pi} + (\vec{\alpha}_1 - \vec{\alpha}_2) \cdot \vec{\pi} + \frac{e_1 e_2}{r} (1 - \vec{\alpha}_1 \cdot \vec{\alpha}_2) + m_1 \beta_1 \cdot I + m_2 I \cdot \beta_2 \right\} \Phi$$

where we have defined

$$\vec{\alpha} \equiv \frac{1}{2}(\vec{\alpha}_1 + \vec{\alpha}_2); \ \vec{\alpha}_i = \gamma_i^0 \vec{\gamma}_i, \beta_i = \gamma_{0i}, \ i = 1, 2 \tag{46}$$

Our two-body equation has the form of a generalized Dirac equation, now a 16-component wave equation. In fact it reduces to the one-body Dirac equation in the limit when one of the particles is heavy.

The above developments are completely relativistic and covariant. The physical results are independent of the vector n although a vector n must appear for manifest covariance. Thus recoil corrections are included to all orders. Further interesting properties of the equation, beside being a one-time relativistic equation, are that relative and center of mass terms in the Hamiltonian are additive, and radial and angular parts of the relative equation are exactly separable. It has also a nonrelativistic limit to the two-body Schrödinger equation. We shall discuss numerical results in Section VIII.

3.6 The Interpretation of Negative Energy States

It is often stated that only in second quantized field theory can one have an adequate description of antiparticles and negative energy solutions where one changes the roles of the creation and annihilation operators for the negative energy solutions. We shall now show that there is also a consistent way of dealing with the negative energy solutions and antiparticles in the Dirac equation as a classical field theory and elaborate how we obtain the annihilation potential in positronium, for example.

There are actually not one but two Diract equations

$$(\gamma \cdot p - m)\psi_I = 0$$
$$(\gamma \cdot p + m)\psi_{II} = 0 \tag{47}$$

obtained from the factorization of the Klein-Gordon operator, for example. By convention we just peak one and work with the complete set of solutions of this equation. Now the negative energy solutions of ψ_I coincide with the positive energy solutions of ψ_{II}. Furthermore in the presence of the electromagnetic field with minimal coupling we have the two equations

$$(\gamma \cdot (p - eA) - m) \psi_I = 0$$
$$(\gamma \cdot (p - eA) + m) \psi_{II} = 0 \tag{48}$$

and we can easily prove that

$$\psi_I(-p, -e) = \psi_{II}(p, e) \tag{49}$$

that is the negative energy momentum solutions of ψ_I coincide with the positive energy solutions of ψ_{II} of opposite charge. Therefore we should consider positive energy solutions of both equations as physical particles. The total number of such physical solutions is the same as the total number of both positive and negative energy solutions of a single Dirac equation.

With this interpretation we obtain quite naturally the annihilation diagrams and annihilation potentials between particles and antiparticles. Consider our interaction action

$$\int dx\,dy\,\bar{\psi}_1(x)\gamma^\mu\psi_1(x)D(x-y)\bar{\psi}_2(y)\gamma_\mu\psi_2(y)$$

Here the classical fields $\psi_i(x)$ contain all positive and negative energy solutions according to our general expansion (22). Separating positive and negative energy solutions as

$$\psi_n(x) = \psi_{E_N>0} + \psi_{E_N<0} \equiv \psi_n^+ + \psi^-$$

and inserting it into the action we get 16 terms. In the limiting case of the lowest order scattering, where we replace the fields by plane wave solutions, we have essentially two distinct types of vertices at each point x or y, namely

$$\bar{\psi}^+(x)\gamma_\mu\psi^+(x)$$

and

$$\bar{\psi}^+(x)\gamma_\mu\psi^-(x)$$

(50)

In the second case we have used our interpretation of the negative energy solutions as the antiparticles with reversed energy-momentum p_μ. The complete interaction action to this order consists of all combinations of these two vertices located at x and y for particles 1 and 2 multiplied with the Green's function $D(x-y)$. Of these 16 terms some cannot be realized because of the overall energy-momentum conserving δ-function, $\delta(p_1 + p_2 - p_3 - p_4)$, and we are left with two disctinct types of terms

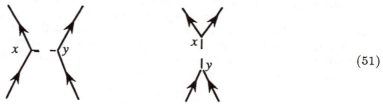

(51)

plus the same terms with particles and antiparticles interchanged. This result agrees with the standard QED. But we shall go a bit further and apply it to bound state problems in Section VII after a discussion of the case of identical particles.

Identical Particles

For two identical particles we use the postulate of the first quantized quantum theory that the field is symmetric or antisymmetric under the interchange of all dynamical variables of identical particles. In our formulation we go back to the original action principle and assume that the current j_μ is antisymmetric in the two fields

$$j_\mu = \frac{1}{2}e\left(\bar{\psi}_1\gamma_\mu\psi_2 - \bar{\psi}_2\gamma_\mu\psi_1\right), \quad e_1 = e_2 = e \tag{52}$$

This implies in the interaction action

$$W_{\text{int}} = \frac{1}{4}e^2 \left[\int dx\,dy\,\bar\psi_1(x)\gamma_\mu\psi_2(x)D(x-y)\bar\psi_1(y)\gamma^\mu\psi_2(y) \right.$$
$$\left. - \int dx\,dy\,\bar\psi_1\gamma_\mu\psi_2 D(x-y)\bar\psi_2\gamma^\mu\psi_1 + (1 \leftrightarrow 2) \right]$$

(53)

and again when the fields are expanded we see that identical particles with exactly the same wave functions i.e., the same quantum numbers or the same state, will not interact and that in the lowest approximation we will get besides the direct interaction also an exchange term as shown in the following diagrams

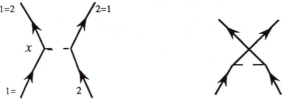

Finally we combine the two effects, identical particles and particle-antiparticle properties, to discuss systems like electron-positron complex and positronium. According to our discussion this system is just a part of the larger electron-electron system taking into account the interpretation of the negative energy levels and the identity of the particles.

3.7 Calculation of the Anomalous Magnetic Moment $(g-2)$

We show now that our basic interaction action also contains besides Lamb shift, spontaneous emission and vacuum polarization also the anomalous magnetic moment in the same single expression. We shall also introduce at this occasion the more general four-dimensional energy-momentum Fourier expansion instead of the energy Fourier expansion (22) which was appropriate for the fixed external field problem.

The interaction action is given by

$$W_{\text{int}} = -\frac{e^2}{2} \int dx\,dy\,j^\mu(x)D(x-y)j_\mu(y)$$

(54)

We expand the fields as four dimensional Fourier integrals

$$\psi(x) = \int dp\,e^{-ipx}\psi(p)$$

(55)

and insert it into the action

$$W_{\text{int}} = -\frac{e^2}{2}\frac{1}{(2\pi)^4} \int dx\,dy\,dk\,dp\,dq\,dr\,ds\,\bar\psi(p)\gamma^\mu\psi(p)\frac{e^{-ik(x-y)}}{k^2+i\varepsilon}\bar\psi(r)\gamma_\mu\psi(s)$$
$$\times e^{i(p-q)x+(r-s)y}$$

which can be written as

$$W_{\text{int}} = -\frac{e^2}{2}(2\pi)^4 \int dp\,dq\,dr\,ds\,j^\mu(p,q)\frac{1}{(r-s)^2+i\varepsilon}j_\mu(r,s)\delta(p-q+r-s) \qquad (56)$$

where $j^\mu(p,q)$ stand for the double Fourier transform

$$j^\mu(p,q) = (2\pi)^8 \int dx\,dy\,\bar\psi(x)\gamma^\mu\psi(y)e^{-ipx}e^{iqy} \qquad (57)$$

The δ-function arises from the x,y and k-integrations. We separate the action into two terms to satisfy the δ-function

$$\begin{aligned}(i) \quad & p=q, \text{ hence } r=s \\ (ii) \quad & p=s, \text{ hence } q=r\end{aligned} \qquad (58)$$

Again as before the first corresponds to vacuum polarization; the second term contains Lamb shift and spontaneous emission as real and imaginary parts of the energy shift ΔE. It also contains the anomalous magnetic moment as the coefficient of the magnetic part of the Lamb shift for any external field. For the calculation of $(g-2)$ it is thus not necessary to solve a problem with an external magnetic field or to solve any external problem for that matter.

The second term with (ii) can be written as

$$W_{\text{int}}^{(ii)} = -\frac{e^2}{2} \int dx\,dy\,dp\,dq\,D(x-y)j^\mu(q,p)j_\mu(q,p)e^{ipx}e^{i(y-x)}q \qquad (59)$$

Here we recognize the c-number electron propagator-function $S(x-y)$

$$\int dq\,\psi(q)\bar\psi(q)e^{i(y-x)q} \equiv S(x-y) \qquad (60)$$

which satisfy the inhomogeneous wave equation

$$[\gamma^\mu(p_\mu - eA_\mu) - m] = i\delta(x-y) \qquad (61)$$

Let us also take the Fourier transform of S

$$S(x-y) = \frac{1}{(2\pi)^4} \int dp\,e^{-ip(x-y)}S(p) \qquad (62)$$

Then the action (59) becomes

$$W_{\text{int}}^{(ii)} = -\frac{e^2}{2}i \int dp\,dP\,\bar\psi(p)\frac{\gamma^\mu S(p)\gamma_\mu}{(p-P)^2+i\varepsilon}\psi(p) \qquad (63)$$

It is related to the energy, more precisely to a mass shift by an overall δ-function and we can write

$$\Delta E = \frac{W_{\text{int}}^{(ii)}}{(2\pi)^4} = \int dp \bar{\psi}(p) \Delta M(p) \psi(p) \tag{64}$$

where we have introduced an effective *mass matrix* by

$$\Delta M(p) = \frac{e^2}{2} \frac{i}{(2\pi)^4} \int ds \frac{\gamma^\mu S(p-s)\gamma_\mu}{s^2 - i\varepsilon} \tag{65}$$

It remains now to evaluate the mass matrix ΔM. First we expand the Green's function or propagator in an external field as follows

$$\frac{1}{\gamma^\mu(p_\mu - eA_\mu) - m} = \frac{\not{p} - e + m}{p^2 - m^2} + 2e \frac{p \cdot A(\not{p}+m)}{(p^2 - m^2)^2} - ie \frac{(\not{p}+m)\gamma^\mu\gamma^\nu F_{\mu\nu}}{(p^2 - m^2)^2} + \cdots$$

where

$$\not{p} \equiv \gamma^\mu p_\mu, \quad \not{A} = \gamma^\mu A_\mu, \quad p \cdot A = p^\mu A_\mu. \tag{66}$$

It turns out that only the third term which is gauge invariant gives a nonvanishing contribution to lowest order in α: the terms containing A_μ give vanishing contributions. The mass operator becomes

$$\Delta M(p) = \frac{e^2}{2} \frac{i}{(2\pi)^4}(-ie) \int ds \frac{\gamma^\mu\left((p-s) - m\right)\gamma^\lambda\gamma^\varphi F_{\lambda\varphi}\gamma_\mu}{s^2\left((p-s)^2 - m^2\right)^2} \tag{67}$$

The integrals can be performed giving[16]

$$\Delta M(p) = \frac{e^2}{2} \frac{i}{(2\pi)^4}(-ie)\frac{2}{m} \int_0^1 dy (1-y)\sigma^{\mu\nu} F_{\mu\nu}$$

and finally

$$\Delta M(p) = -\frac{\alpha}{2\pi} \frac{e}{2m} \sigma_{\mu\nu} F_{\mu\nu} = -\frac{\alpha}{2\pi} \frac{e}{2m}(\vec{\sigma} \cdot \vec{B} + i\vec{\alpha} \cdot \vec{E}) \tag{68}$$

Thus we recognize the anomalous magnetic moment to this order in front of the $\vec{\sigma} \cdot \vec{B}$ term for any field as we have mentioned.

If we insert the mass operator into the energy shift formula (64) we can evaluate it as the expectation value of the operator $\vec{\sigma} \cdot \vec{B} + i\vec{\alpha} \cdot \vec{E}$. For relativistic Coulomb problem for example the magnetic part of the Lamb shift can be analytically evaluated exactly.[17]

This way of calculating the anomalous magnetic moment also shows now how to calculate higher order terms. We must take more terms in the Green's function expansion (66). This may be much simpler than the diagrammatic method of perturbative QED where there are already 891 Feynman diagrams in the order $(\alpha/\pi)^4$.

3.8 Covariant Analysis of Radiative Processes for Two-Body Systems

In this section we discuss how to treat radiative processes, like Lamb shift, etc., for a system like positronium or muonium beyond the naive reduced mass method. As mentioned above the action formalism is more convenient than the equations of motion.

We go back to our covariant 2-body action (37) and separate center of mass and relative coordinates and momenta according to

$$
\begin{array}{lll}
r = x_1 - x_2 & x_1 = R + \dfrac{1}{2}r & P = p_1 + p_2 \\[2mm]
R = \dfrac{1}{2}(x_1 + x_2) & x_2 = R - \dfrac{1}{2}r & p = \dfrac{1}{2}(p_1 - p_2) \\[2mm]
q = z - u & z = Q + \dfrac{1}{2}q & p_1 = \dfrac{1}{2}P + p \\[2mm]
Q = \dfrac{1}{2}(z + u) & u = Q - \dfrac{1}{2}q & p_2 = \dfrac{1}{2}P - p
\end{array}
\tag{69}
$$

All quantities here are four-vectors. Then the action becomes

$$
\begin{aligned}
W = \int dR dq \bar{\Phi}(R,r) &\bigg\{ \left[\gamma \cdot \left(\frac{1}{2}P + p \right) - m_1 \right] \otimes \gamma \cdot n + \gamma \cdot n \otimes \left[\gamma \cdot \left(\frac{1}{2}P - p \right) - m_2 \right] \\
&- e_1 e_2 D(r) - \frac{e_1^2}{2} \int dQ dq \gamma_\mu \otimes \gamma \cdot n D \left(R - Q + \frac{1}{2}(r - q) \right) \bar{\Phi}(Q,q) \gamma^\mu \otimes \gamma \cdot n \Phi(Q,q) \\
&- \frac{e_2^2}{2} \int dQ dq \gamma \cdot n \otimes \gamma_\mu D \left(R - Q - \frac{1}{2}(r - q) \right) \bar{\Phi}(Q,q) \gamma \cdot n \otimes \gamma^\mu \Phi(Q,q) \bigg\} \Phi(R,r)
\end{aligned}
\tag{70}
$$

In the absence of a fixed external field the system is translationally invariant and the generalization of the Fourier expansion (22) is the four dimensional Fourier transform of the composite field $\Phi(R,r)$ which has actually one time variable, $\Phi(R,r_\perp)$, the relative coordinates is a 3-vector $r_\perp$ perpendicular to n.

$$
\Phi(R,r_\perp) = \int \frac{d^4 P}{(2\pi)^4} e^{iPR} \psi(P, r_\perp)
$$

We insert this expansion everywhere in our action and obtain

$$W = \int dR dr_\perp \int \frac{dP_n}{(2\pi)^4} \frac{dP_m}{(2\pi)^4} \bar\psi(P_n, r_\perp) e^{-iP_n R} \left\{ [\Gamma_\mu P^\mu + \mathcal{L}_{rel}(r_\perp, p)] e^{iP_m R} \right.$$

$$- \frac{1}{2} \int \frac{dP_r}{(2\pi)^4} \frac{dP_s}{(2\pi)^4} dQ dq_\perp dk \left[\frac{e_1^2}{2} \frac{e^{-ik[R-Q+\frac{1}{2}(r_\perp - q_\perp)]}}{k^2} \gamma^\mu \otimes \gamma \cdot n\bar\psi(P_r, q_\perp) e^{-iP_r Q} \gamma_\mu \right.$$

$$\otimes \gamma \cdot n e^{iP_s Q} \psi(P_s, q_\perp)$$

$$+ \frac{e_2^2}{2} \frac{e^{-ik[R-Q-\frac{1}{2}(r_\perp - q_\perp)]}}{k^2} \gamma \cdot n \otimes \gamma^\mu \bar\psi(P_r, q_\perp) e^{-iP_r Q} \gamma \cdot n \otimes \gamma_\mu e^{iP_s Q}$$

$$\left. \times \psi(P_s, q_\perp) \right] \left. \vphantom{\frac{e_1^2}{2}} \right\} \psi(P_m, r_\perp)$$

(71)

where $\mathcal{L}_{rel}$ is the Lagrangian of the relative motion and is given by

$$\mathcal{L}_{rel}(r_\perp, p) = (\gamma^\mu p_\mu - m_1) \otimes \gamma \cdot n + \gamma \cdot n \otimes (-\gamma^\mu p_\mu - m_2) - \frac{e_1 e_2}{r_\perp} \gamma^\mu \otimes \gamma_\mu \quad (72)$$

and

$$\Gamma_\mu = \frac{1}{2}(\gamma_\mu \otimes \gamma \cdot n + \gamma \cdot n \otimes \gamma_\mu) \quad (73)$$

The result of performing the R and Q-integrations, letting

$$k_\mu = \gamma_\mu \otimes \gamma \cdot n - \gamma \cdot n \otimes \gamma_\mu \quad (74)$$

is

$$W = \int \frac{dP_n}{(2\pi)^4} dP_m dr_\perp \bar\psi(p_n, r_\perp) \left\{ \left[\Gamma_\mu P^\mu + k^\mu p_\mu - m_1 I \otimes \gamma \cdot n - m_2 \gamma \cdot n \otimes I - \frac{e_1 e_2}{r_\perp} \gamma^\mu \otimes \gamma_\mu \right] \right.$$

$$\times \delta(P_n - P_m) - \frac{1}{2} \int \frac{dP_r}{(2\pi)^4} dP_s \frac{dk}{(2\pi)^4} dr_\perp dq_\perp \delta(P_n - P_m - k) \delta(P_r - P_s - k) \left[e_1^2 e^{-i\frac{1}{2}k(r_\perp - q_\perp)} \right.$$

$$\times \gamma^\mu \otimes \gamma \cdot n\bar\psi(P_r, q_\perp)\gamma_\mu \otimes \gamma \cdot n + e_2^2 e^{i\frac{1}{2}k(r_\perp - q_\perp)} \gamma \cdot n \otimes \gamma^\mu \bar\psi(P_r, q_\perp)\gamma \cdot n \otimes \gamma^\mu \right] \psi(P_s, q_\perp) \left. \vphantom{\frac{1}{2}} \right\}$$

$$\psi(P_m, r_\perp)$$

(75)

Introducing the form factors

$$\overset{(1)}{T^\mu}_{nm}(k) \equiv \int dr_\perp \bar\psi(P_n, r_\perp) e^{\frac{i}{2}k r_\perp} \gamma^\mu \otimes \gamma \cdot n\psi(P_m, r_\perp)$$

and

(76)

$$\overset{(2)}{T^{\mu}}{}_{nm} = \int dr_{\perp}\bar{\psi}(P_n, r_{\perp})e^{-\frac{i}{2}kr_{\perp}}\gamma \cdot n \otimes \gamma^{\mu}\psi(P_m, r_{\perp})$$

for the two particles, we can write the action in the compact form

$$W = \sum_{nm} \int dr_{\perp}\bar{\psi}(P_n, r_{\perp})\left[\Gamma_{\mu}P^{\mu} + \mathcal{L}_{rel}\right]\psi(P_m, r_{\perp})\delta(P_n - P_m)$$

$$-\frac{1}{2}\sum_{nm}\sum_{rs}\int\frac{dk}{(2\pi)^4}\frac{1}{k^2}\left[e_1^2\overset{(1)}{T^k}{}_{nm}\overset{(1)}{T^{rs}}{}_{\mu} + e_2^2\overset{(2)}{T^{\mu}}{}_{nm}\overset{(2)}{T^{rs}}{}_{\mu}\right]\delta(P_n - P_m + k)\delta(P_r - P_s - k)$$

where

(77)

$$\mathcal{L}_{rel}(r_{\perp}) = k^{\mu}p_{\mu} - m_1 I \otimes \gamma \cdot n - m_2\gamma \cdot n \otimes I - \frac{e_1 e_2}{r_{\perp}}\gamma^{\mu} \otimes \gamma_{\mu}$$

Note that $k^{\mu}n_{\mu} = 0$.

Now we can perform the k^0-integration, and without loss of generality set $n = (1000)$, and obtain

$$W = \sum_{nm}\delta(E_n - E_m)\delta(\vec{P}_n - \vec{P}_m)d\vec{r}\bar{\psi}_n(\vec{P}_n, \vec{r})[\Gamma_0 P^0 - \vec{\Gamma}\cdot\vec{P} + \mathcal{L}_{rel}]\psi_m(\vec{P}_m, \vec{r})$$

$$-\frac{1}{2}\sum_{nm}\sum_{rs}\int\frac{d\vec{k}}{(2\pi)^4}\frac{1}{\omega_{nm}^2 - \vec{k}^2}\left[e_1^2\overset{(1)}{T^{\mu}}{}_{nm}(\omega_{nm}, \vec{k})\overset{(1)}{T^{rs}}{}_{\mu}(\omega_{rs}, \vec{k})\right.$$

$$\left. + e_2^2\overset{(2)}{T^{\mu}}{}_{nm}(\omega_{nm}, \vec{k})\overset{(2)}{T^{rs}}{}_{\mu}(\omega_{rs}, \vec{k})\right]\delta(\omega_{nm} + \omega_{rs})\delta(\vec{P}_n - \vec{P}_m + \vec{k})\delta(\vec{P}_r - \vec{P}_s - \vec{k})$$

Now we look at the selfinteraction terms only and expand the denominator

(78)

$$W^{self} = \frac{1}{2}\sum_{nm}\sum_{rs}\int\frac{d\vec{k}}{(2\pi)^4}\left[e_1^2\overset{(1)}{T^{\mu}}{}_{nm}(\omega_{nm}, \vec{k})\overset{(1)}{T^{rs}}{}_{\mu}(-\omega_{nm}, -\vec{k}) + e_2^2\overset{(2)}{T^{\mu}}{}_{nm}(\omega_{nm}, \vec{k})\overset{(2)}{T^{rs}}{}_{\mu}(-\omega_{nm}, -\vec{k})\right]$$

$$\times \delta(\omega_{nm} + \omega_{rs})\left\{(\delta(\omega_{rs} - k) + \delta(\omega_{rs} + k))\frac{2\pi i}{2k} + P\frac{1}{ik}\left(\frac{1}{\omega_{rs} - k} - \frac{1}{\omega_{rs} + k}\right)\right\}$$

$$\delta(\vec{P}_n - \vec{P}_m + \vec{k})\delta(\vec{P}_n - \vec{P}_m + \vec{P}_r - \vec{P}_s)$$

(79)

where P stands for the principal value of the integral and $\displaystyle\sum\!\!\!\!\!\!\int$ means a summation over discrete states and an integration over the continuum states. As in the case of the Coulomb problem and $(g-2)$-calculation, we separate the two terms corresponding to

(a) $n = m$, hence $r = s$

and
$$(b)\ n = s,\ \text{hence}\ m = r$$
and dictated by the δ-functions and obtain finally

$$
W^{\text{self}} = -\frac{1}{2}\sum_{ns}\int\frac{d\vec{k}}{(2\pi)^4}\left\{e_1^2 \overset{(1)}{T^\mu}_{nn}(0,\vec{k})\overset{(1)}{T^{ss}}_\mu(0,-\vec{k}) + e_2^2 \overset{(2)}{T^\mu}_{nn}(0,\vec{k})\overset{(2)}{T^{ss}}_\mu(0,-\vec{k})\right\}
$$

$$
\times\left\{\frac{i\pi}{k}(\delta(k)+\delta(-k)) + \frac{1}{2k}P\left(-\frac{2}{k}\right)\right\}\delta(\vec{P}_n - \vec{P}_m + \vec{k})\delta(\vec{P}_n - \vec{P}_m + \vec{P}_r - \vec{P}_s)
$$

$$
-\frac{1}{2}\sum_{\hbar m}\int\frac{d\vec{k}}{(2\pi)^4}\left\{e_1^2 \overset{(1)}{T^\mu}_{nm}(\omega_{nm},\vec{k})\overset{(1)}{T^{rs}}_\mu(-\omega_{nm},-\vec{k}) + e_2^2 \overset{(2)}{T^\mu}_{nm}(\omega_{nm},\vec{k})\overset{(2)}{T^{\hbar m}}_\mu(-\omega_{nm},-\vec{k})\right\}
$$

$$
\times\left\{\delta(\omega_{nm}-k)+\delta(\omega_{nm}+k)\frac{i\pi}{k} + P\frac{1}{2k}\left(\frac{1}{\omega_{nm}-k} - \frac{1}{\omega_{nm}+k}\right)\right\}
$$

$$(80)$$

We recognize again the following terms:
(i) Term containing $\delta(k) + \delta(-k)$. The contribution of this term to the dk-integral vanishes.
(ii) The term $P1/k$: This term corresponds to vacuum polarization.
(iii) The term with $(i\pi/k)[\delta(\omega_{nm}-k)+\delta(\omega_{nm}+k)]$. This term gives the spontaneous emission or absorption from level n to m or vice versa.
(iv) The term with $P\frac{1}{2k}\left(\frac{1}{\omega_{nm}-k} - \frac{1}{\omega_{nm}+k}\right)$ gives the Lamb shift.

These are our formulas for the radiative processes of the two-fermion system[18]. In the limit they go over to the fixed center Coulomb problem on the one hand, and for free particles to perturbative QED results.

For identical particles and particle-antiparticle system like positronium we have to antisymmetrize our currents as discussed in Sec. VI. Thus the mutual interaction action has two terms. The first is the usual direct interaction term

$$
W^{\text{int}}_{e^-e^+}(1) = -e^2\int dxdy\bar{\psi}_I(x)\overset{(1)}{\gamma^\mu}\psi_I(x)D(x-y)\bar{\psi}_{II}(x)\overset{(2)}{\gamma_\mu}\psi_{II}(y)
$$
$$
= -e^2\int dxdy\bar{\phi}_I(x,y)\overset{(1)}{\gamma^\mu}D(x-y)\overset{(2)}{\gamma_\mu}\phi(x,y) \tag{81}
$$

corresponding to the potential

$$
V = -e^2\overset{(1)}{\gamma^\mu}\frac{1}{r}\overset{(2)}{\gamma_\mu} \tag{82}
$$

The second term is

$$
W^{\text{int}}_{e^-e^+}(2) = e^2\int dxdy\bar{\phi}(x,y)\overset{(1)}{\gamma^\mu}D(x-y)\overset{(2)}{\gamma_\mu}\phi(y,x)
$$
$$
= \frac{e^2}{2\pi}\int d\vec{r}\bar{\phi}_E(p_0,-\vec{r})\overset{(1)}{\gamma^\mu}\frac{e^{i\vec{k}\cdot\vec{r}}}{(p_0+p_0')^2 - \vec{k}^2}\overset{(2)}{\gamma_\mu}\phi_E(p_0,-\vec{r})\frac{d\vec{k}}{(2\pi)^3} \tag{83}
$$

where p_0 and p_0' are the initial and final state energies and E is total conserved center of mass energy of the whole system. In the positronium the relative momentum is approximately zero so that we can set

$$p_\mu \cong (m, \vec{o})$$

and the action becomes

$$W_{e^-e^+}^{\text{int}}(\text{annihilation}) \cong \frac{2\alpha}{m^2} \int d\vec{r}\, \bar{\phi}(\vec{r}) \overset{(1)}{\gamma^\mu} \delta(\vec{r}) \overset{(2)}{\gamma_\mu} \phi(-\vec{r}) \tag{84}$$

Now we show that this term gives correctly the annihilation contribution to the hyperfine splitting in the $n = 1$ state of positronium, for example. The effective potential above (84), when inserted into our wave equations gives an energy shift only for the levels $j = l \cong 0$ and for $j - 1 = l \cong 0$.

$$\delta E(j = \ell = 0) \cong \frac{m\alpha^4}{2n^3}$$

and

$$\delta E(j - 1 = \ell = 0) \cong \frac{m\alpha^4}{4n^3}$$

The difference is the annihilation contribution in the hyperfinesplitting

$$\delta E_{Hfs}(\text{annihilation}) \cong \frac{3m\alpha^4}{12n^3} \tag{85}$$

To this order it agrees with perturbative QED. It is however obtained here in first quantized QED with selffields.

3.9 Further Results

There are still discrepancies between theory and experiment in almost all tests of QED. The Table III summarizes all measured levels in positronium, muonium and Hydrogen, positronium lifetimes, the anomalous magnetic moments a_e and a_μ, and some theoretical values in parenthesis. In reviewing some of these discrepancies, W-Lichten[18] writes "It seems likely that the problem lies in the difficulty of QED calculations which have not been carried out to a high order enough, perhaps a totally new type of calculation is needed". The self field approach to QED provides a new type of calculation. It's important to have a complementary or alternate method to perturbative QED, for a theory is tested not only against experiment but also against other theories in order to clarify the basic assumptions and concepts, specially in view of recent results that perturbative QED might be inconsistent or a trivial theory. Self-field QED modifies our notion of the quantized radiation field and the interpretation of quantum theory. The emphasis is shifted from the field to the source of radiation, an electronic charge distribution which objectively and deterministically evolves as a

TABLE III. BOUND STATE TESTS OF QED

POSITRONIUM

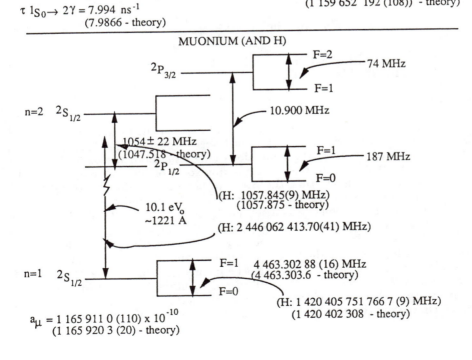

$\tau\ 3_{S_1} \to 3\gamma = 7.0514\ \mu s^{-1}$
     (7.0383 - theory)

$\tau\ 1_{S_0} \to 2\gamma = 7.994\ ns^{-1}$
     (7.9866 - theory)

$a_e = 1\ 159\ 652\ 188.4\ (4.3) \times 10^{-12}$
    $(1\ 159\ 652\ 192\ (108))$ - theory)

MUONIUM (AND H)

$2P_{3/2}$ F=2   74 MHz, F=1

$n=2$ $2S_{1/2}$ 10.900 MHz

1054 ± 22 MHz
(1047.518 - theory)
$2P_{1/2}$ F=1   187 MHz, F=0

(H: 1057.845(9) MHz)
(1057.875 - theory)

10.1 eV$_o$
~1221 Å

(H: 2 446 062 413.70(41) MHz)

$n=1$ $2S_{1/2}$ F=1   4 463.302 88 (16) MHz
(4 463.303.6 - theory)   F=0

(H: 1 420 405 751 766 7 (9) MHz)
(1 420 402 308 - theory)

$a_\mu = 1\ 165\ 911\ 0\ (110) \times 10^{-10}$
  (1 165 920 3 (20) - theory)

classical field and produces a selffield which acts back on the charge itself. Quantized properties of the light reflect the discrete frequencies of the oscillating charge distribution.

The two-body relativistic equation discussed in Sections V and VII gives us a possibility to make improved calculations for positronium and muonium, in particular. In positronium, the experiments seem to be more accurate than the theory and the perturbative calculations remain incomplete[19]. Considerable analytical work has been done on the study of the two-body equation (39)ff: separation of radial and angular parts and further reduction of the radial equations[20]. It turns out that the two-body equation, when the electromagnetic potentials are kept to order α^4, is exactly soluble with an energy spectrum

$$E^2 = \frac{M^2 + \Delta m^2}{2} + \frac{M^2 - \Delta m^2}{2}\left[1 + \frac{\alpha^2}{(n_r + \ell)^2}\right]^{-1/2} \tag{86}$$

with $M = m_1 + m_2$, $\Delta m = m_1 - m_2$, generalizing the Dirac spectrum. We have treated the remaining potentials of order α^5 and higher as perturbations. But having tested the equation in this way, one can now make direct nonperturbative numerical calculations. The treatment of the negative energy states and the covariance of the equation has also been discussed[22] according to the methods outlined in Section VI. We give here some of the results[20].

1) For parapositronium, eq. (86), is exact including terms of the order α^4 since normal and anomalous magnetic moment terms do not contribute to this order. It gives

$$E^{\text{para ps}} = 2m - \frac{m\alpha^2}{4n^2} - \frac{m\alpha^4}{2n^3(2j+1)} + \frac{11}{64}\frac{m\alpha^4}{n^4} + 0(\alpha^6) \tag{87}$$

2) Introducing the anomalous magnetic moments a_1, a_2, which are in the selfenergy term, as a Pauli-coupling we obtain the ground-state hyperfine splitting

$$\Delta E^{\text{Hfs}} = \frac{8}{3}\frac{\zeta}{(1+\zeta)^2}m\alpha^4\left[(1+a_1)(1+a_2) - \frac{3}{4}\zeta a_2^2 + \frac{3}{4}a_1\frac{a_2}{(1+\zeta)^2}\right]$$

with $\zeta = m_1/m_2$. Numerically this gives 1420.348 MHz for H, and 4.463.060 MHz for muonium, compared to the experimental values 1420.405752 and 4.463302, respectively.

3) Positronium hyperfinesplitting including the annihilation term, eq. (85), gives

$$\Delta E^{\text{Hfs}} = \frac{7}{12}m\alpha^4 + \frac{5}{12}m\alpha^4\left(\frac{\alpha}{2\pi}\right).$$

This "Lambshift" term $-\frac{\alpha}{2\pi}(\frac{16}{9} + \ell n2)m\alpha^4$ has to be added perturbatively, but we hope to calculate these terms and more eventually numerically.

4) Positronium ($n = 2, n = 1$) splitting, including annihilation and anomalous magnetic moment contributions

$$\Delta E_{21} = \frac{3}{8}Ry - 0.468098\alpha^2 Ry - \frac{\alpha^2 Ry}{2\pi}\frac{35}{96}.$$

5) Positronioum fine structure

$$\Delta E(2^3 S_1 - 2^3 P_2) = -\frac{1}{12}\alpha^2 Ry + \frac{7}{48}\alpha^2 Ry - \frac{7}{480}\alpha^2 Ry + 0(\alpha^5)$$

$$\underset{(\text{recoil})}{} \quad \underset{(\text{annihilation})}{} \quad \binom{\text{normal magnetic}}{\text{moment}}$$

$$= \frac{23}{480}\alpha^2 Ry + 0(\alpha^5)$$

6) H or muonium ($n = 2$, $n = 1$) splitting

$$\Delta E_{21} = \frac{3}{8}\mu\alpha^2 + \frac{8}{128}\mu\alpha^4 + \frac{15}{128}\mu\alpha^4 \frac{\zeta}{1+\zeta}$$

$$- \frac{7}{16}\mu\alpha^4 \left(1 + \frac{2(a_1 + \zeta^2 a_2)}{(1+\zeta)^2} - 2a_1 a_2 \frac{\zeta}{1+\zeta}\right)$$

$$- \frac{7}{12}\frac{\mu^2}{M}\alpha^4(1 + a_1 + a_2 + a_1 a_2)$$

where $M = m_1 + m_2$, $\zeta = m_1/m_2$, $\mu = \frac{m_1 m_2}{M}$.

For other details and applications of self-field QED we refer to the literature listed in the Appendix.

References to Chapter 3.

1. See e.g. W. Thirring, *Principles of Quantumelectrodynamics* (Academic Press, N.Y. 1958).
   J.D. Bjorken and S.D. Drell, *Quantum Fields* (McGraw Hill, N.Y. 1965).
   R.P. Feynman, *Quantum electrodynamics* (Benjamin, N.Y. 1961).

2. A.O. Barut, in *Quantumelectrodynamics*, ed. P. Urban, Suppl. Acta Physica Austriaca, Vol. 2 (Springer, Berlin 1965); and *The Theory of the Scattering Matrix*, (Macmillan, N.Y. 1962), Ch. 13.

3. G. Scharf, *Finite Quantumelectrodynamics*, Springer 1989, and These Proceedings.

4. See e.g. C. Itzykson and J.-B. Zuber, *Quantum Field Theory* (McGraw Hill, N.Y. 1964).

5. A.O. Barut and I.H. Duru, *Phys. Reports*, **172**, 1 (1989).

6. J. Schwinger, *Particles, Sources and Fields*, Vol. II (Addison Wesley, Reading, MA 1973).

7. H.A. Lorentz, *The Theory of Electrons*, (Dover, N.Y. 1952).

8. A.O. Barut, *Phys. Rev.*, **D10**, 3335 (1974).

9. A.O. Barut and N. Ünal, *Phys. Rev.*, **A40**, 5404 (1989).

10. A. Lees, *Philos. Mag.*, **2S**, 385 (1939).

11. D.A.M. Dirac, *Proc. Roy. Soc.*, (London), **A268**, 57 (1962).

12. A.O. Barut and Y. Salamin, *Phys. Rev.*, **A37**, 2284 (1988); **A39**, (1990).

13. A.O. Barut and Y. Salamin, *Zeits. f. Physik D*, (in press).

14. A.O. Barut and N. Ünal, A Regularized Analytic Evaluation of Vacuum polarization in Coulomb field, *Phys. Rev. D*, **41**, 3822 (1990).

15. A.O. Barut, J. Kraus, Y. Salamin and N. Ünal (to be published).

16. For more details on $(g-2)$ calculations see A.O. Barut and J.P. Dowling, *Zeits. f. Naturf.*, **44a**, 1051 (1989) and A.O. Barut, J.P. Dowling and J.F. van Huele, *Phys. Rev.*, **A38**, 4405 (1988).

17. For more details see A.O. Barut and N. Ünal, Complete QED of the electron-positron system and positronium (to be published).

18. W. Lichten, in *The Hydrogen Atom*, edit. by G.F. Bassani *et al* (Springer-1989), p. 39.

19. A. Rich *et al*, in *"Frontiers of Quantumelectrodynamics and Quantumoptics"*, edit by A.O. Barut (Plenum Press, 1991).

20. A.O. Barut and N. Ünal, *J. Math. Phys.*, **27**, 3055 (1986), *Physica*, **142A**, 457 an 488 (1987).

21. A.O. Barut, *Physica Scripta*, **36**, 493 (1987), and in *"Constraint Theory and Rela tivistic Dynamics"*, L. Lusanna *et al*, edit., (World Scientific, 1987), p. 122.

Appendix: References on Selffield QED

1. Nonperturbative QED: The Lamb shift, A.O. Barut and J. Kraus, *Found. of Physics*, **13**, 189 (1983).

2. QED based on selfenergy, A.O.Barut and J.F. van Huele, *Phys. Rev. A*, **32**, 3887 (1985)

3. QED based on selfenergy vs. quantization of fields: Illustration by a simple model. A.O. Barut, *Phys. Rev. A*, **34**, 3502 (1986)

4. An exactly soluble relativistic quantum two fermion problem. A.O. Barut and N. Ünal, *J. Math. Phys.*, **27**, 3055 (1986)

5. A new approach to bound state QED. I. Theory, *Physica Scripta*, **142A**, 457 (1987), II. Spectra of positronium, muonium and Hydrogen. A.O. Barut and N.Ünal, *Physica*, **142A**, 488 (1987)

6. An approach to finite non-perturbative QED. A.O. Barut in "Proc. 2nd Intern. Symposium on Foundations of Quantum Mech.", *Phys. Soc. of Japan*, 1986, p. 323.

7. On the treatment of Møller and Breit potentials and the covariant 2-body equation for positronium and muonium, A.O. Barut, *Physica Scripta*, **36**, 493 (1987)

8. On the covariance of two-fermion equation, A.O. Barut, in "Constraint theory and Relativistic dynamics", (L. Lusanna *et al*, edit.) World Scientific, 1987; p. 122

9. Formulation of nonperturbative QED as a nonlinear first quantized classical field theory, A.O. Barut in "Differential Geometric Methods in Theoretical Physics", (H. Doebner *et al*, edit's), World Scientific 1987; p. 51

10. QED based on selfenergy: Spontaneous emission in cavities, A.O. Barut and J.P. Dowling, *Phys. Rev. A*, **36**, 649 (1987)

11. QED based on self energy: The Lamb shift and long-range Casimir-Polder forces near boundaries, A.O. Barut and J.P. Dowling, *Phys. Rev. A*, **36**, 2550 (1987)

12. QED based on selfenergy, A.O. Barut, *Physica Scripta*, **T21**, 18 (1988)

13. Relativistic spontaneous emission, A.O. Barut and Y. Salamin, *Phys. Rev. A*, **37**, 2284 (1988)

14. QED based on selfenergy: A nonrelativistic calculation of $(g-2)$, A.O. Barut, J.P. Dowling and J.F. van Huele, *Phys. Rev. A*, **38**, 4405 (1988)

15. QED based on selfenergy: Cavity dependent contributions of $(g-2)$, A.O. Barut and J.P. Dowling, *Phys. Rev. A*, **38**, 2796 (1989)

16. QED based on self fields, A Relativistic calculation of $(g-2)$, A.O. Barut and J.P. Dowling, *Zeits. f. Naturf*, **44A**, 1051 (1989)

17. Path integral formulation of QED from classical particle trajectories, A.O. Barut and I.H. Duru, *Phys. Reports*, **172**, 1–32 (1989)

18. Problème relativiste á deux corps en électyrodynamique quantique, *Heelv. Phys. Acta*, **62**, 436 (1989)

19. Selffield QED: The two-level atom, A.O. Barut and J.P. Dowling, *Phys. Rev. A*, **41**, 2284 (1990)

20. QED based on selffields: On the origin of thermal radiation detected by an accelerated observer, A.O. Barut and J.P. Dowling, *Phys. Rev. A*, **41**, 2227 (1990)

21. Relativistic Spontaneous emission in Heisenberg representation. A.O. Barut and Y. Salamin, *Z. f. Physik D*, (in press)

22. Relativistic $2S_{1/2} \rightarrow 1S_{1/2} + \gamma$ decay rates of H-like atoms for all Z, A.O. Barut and Y. Salamin, *Phys. Rev. A*, (in press)

23. QED-The unfinished business, A.O. Barut in "Proc. III. Conf. Math. Physics" (World Scientific, 1990) (F. Hussain, ed.) p. 493

24. A regularized analytic evaluation of vacuum polarization in Coulomb field, A.O. Barut and Ünal, *Phys. Rev. D*, **41**, 3822 (1990)

25. The Foundations of Self-field Quantumelectrodynamics, A.O. Barut in "New Frontiers of Quantumelectrodynamics and Quantumoptics", (edit. by A.O. Barut), Plenu Press, NY 1990

26. QED Based on Selffields: Cavity effects, J.P. Dowling, *ibid*,

27. Fundamental Symmetries of Quantumelectrodynamics in *Symmetry in Science III*, Plenum Press 1989 (ed. B. Gruber), p. 3–13

28. QED in Non-Simply Connected Regions, A. O. Barut and I. H. Duru, *Quantum Optics*

29. Contribution of the individual discrete levels to the Lamb shift in hydrogenic atoms, A. O. Barut, B. Blaive and R. Boudet *J. Phys. B*.

30. Interpretation of selffield QED, A. O. Barut and J. P. Dowling *Phys. Rev. A*, (in press)

31. Is Second Quantization Necessary? in *Quantum Theory and the Structure of Space and Time*, Vol. 6 (edit. by L. Castell and C. F. vm Weizsäcker), Hanser Verlag, Munchen 1986; p. 83–90.

THE EXPLICIT NONLINEARITY OF QUANTUM ELECTRODYNAMICS

W.T. Grandy, Jr.
Department of Physics and Astronomy
University of Wyoming
Laramie, Wyoming 82071 USA

ABSTRACT. The underlying presumptions and principal features of major semiclassical theories of radiation are examined critically. There are significant physical and mathematical difficulties leading to the conclusion that, in their present state of development, these theories are incapable of competing with QED either qualitatively or quantitatively. Some possible means of salvaging the semiclassical approach are suggested, the motivation being to seek a more physically satisfying picture than that provided by quantum field theory.

For over fifty years conventional wisdom has held that the *only* correct description of the electromagnetic interaction of matter is with the quantized radiation field. Yet, for almost as long numerous writers have considered this view not entirely acceptable, at best, and the attempt to explain the data both quantitatively and qualitatively without field quantization remains an active topic of interest. This article constitutes a review and analysis of some contemporary efforts to construct a viable theory of leptons interacting with the *classical* electromagnetic field (generically referred to as 'semiclassical' theories).

It is not unreasonable to ask the question immediately, Why study this approach? Briefly, to help in answering this, we recall that the fundamental expression of electrodynamics stems from the linearity of Maxwell's equations, which suggests that the total electromagnetic field is the sum of any external field A_μ^{ext} acting on a source, plus the field of the source itself, A_μ^{self}. Thus, in the covariant gauge $\partial^\mu A_\mu^{\text{self}} = 0$, the system of electron and fields, say, is described by the coupled Dirac and Maxwell equations:

$$\left(\gamma^\mu \pi_\mu - mc\right)\psi(x) = \frac{e}{c}\gamma^\mu A_\mu^{\text{self}}(x)\psi(x)\,, \tag{1a}$$

$$\Box A_\mu^{\text{self}}(x) = 4\pi e\overline{\psi}(x)\gamma_\mu\psi(x)\,, \tag{1b}$$

with $\pi_\mu \equiv (i\hbar\partial_\mu - \frac{e}{c}A_\mu^{\text{ext}})$. These highly nonlinear equations have proved extraordinarily resistant to closed-form solution, and only quantum field theory (QFT) has succeeded in solving them to the high order of accuracy required by experiment. This is achieved by replacing the classical fields with operator fields, introducing the associated commutation relations, and adopting the notions of a physical vacuum and all-pervasive radiation field. The result, in practice, is a perturbative theory commonly referred to as QED—with or without the Latin connotation—which in its agreement with observation is the most successful physical theory known.

149

D. Hestenes and A. Weingartshofer (eds.), The Electron, 149–164.

But this quantitative success bears a qualitative price in the form of what some would consider bizarre physical features. Primary among these is the interpretation of the vacuum as a vibrant fluctuating reality which is a literal source and sink of real particles and fields. Vacuum fluctuations lead necessarily to the notion of virtual states, and all the radiative effects previously associated with the self-interaction of an electron with its own field are now attributed to the fluctuating vacuum field. As Jaynes (1973, 1978, 1990) has noted frequently, however, one can be forgiven some skepticism about the reality of vacuum fluctuations upon examining actual numerical values: the energy density associated with the Lamb shift would produce a Poynting vector about three times the total power output of the sun, and a gravitational field disrupting the entire solar system! Although operator fields provide a mathematical means for describing creation and annihilation, there is not yet any understanding of a *physical* mechanism for these processes. In addition, the mathematical description itself breaks down in the form of various infinities and, for example, acausal behavior of the Feynman propagator, which is also highly singular on the lightcone. These difficulties have been 'resolved' in part by the mathematically-questionable, but enormously clever techniques of the renormalization program (*e.g.*, Itzykson and Zuber, 1980).

Although several authors have suggested that radiative effects should be described by a mixture of vacuum fluctuations and radiation reaction associated with the classical self-field (*e.g.*, Milonni, 1984, and references therein; Davies and Burkitt, 1980), we shall here focus entirely on the question of field quantization itself. Moreover, there is very little argument for the need for field quantization in scattering processes *per se*, so that the issues reside principally with radiative processes such as the Lamb shift, spontaneous emission, and anomalous magnetic moments, to which we restrict the following discussion. The generic semiclassical approach to quantum electrodynamics thus begins with a formal integration of Eq.(1b) to obtain coupled nonlinear differential and integral equations:

$$(\gamma^\mu \pi_\mu - mc)\psi(x) = \frac{e}{c}A_\mu(x)\gamma^\mu\psi(x),\tag{2a}$$

$$\begin{aligned}A_\mu(x) &= \frac{4\pi e}{c}\int D(x-y)j_\mu[y;A(y)]\,d^4y\\ &= 4\pi e\int D(x-y)\overline{\psi}(y)\gamma_\mu\psi(y)\,d^4y,\end{aligned}\tag{2b}$$

up to a solution of the homogeneous equation, which is here taken as zero. (From now on A_μ will refer to the self-field.) It is just prior to this step that QFT and semiclassical theories part company (*e.g.*, Källén, 1958).

Among the general features of these equations yet to be decided are the boundary conditions on the Green function D. If A_μ is to be associated with a classical field, then one would certainly expect it to be retarded (unless one were particularly attracted to the Wheeler-Feynman absorber theory). Below, however, we shall see that arguments are sometimes made for Feynman-Stückelberg boundary conditions, despite the difficulties already noted regarding the so-called causal propagator D_F. We remark here only that D_F implies that A_μ is complex, and thus $\partial_\mu j^\mu \neq 0$. This is not a difficulty for QFT, because the self-field does not appear—the vacuum field itself is not observable, only its effects.

A second feature of Eqs.(2) of some importance is that they are *never* uncoupled. That is, A_μ can be zero only if $j_\mu[y;0] \equiv 0$ (sometimes called Hammerstein's theorem in the theory of nonlinear integral equations), but that is never the case here. It is therefore difficult to know just what kind of approximation it is to set the left-hand side of Eq.(2a) to zero and obtain stationary-state solutions for the Coulomb problem, say. On the one hand it is obvious that there can be no such thing as exactly stationary states, while on the other the approximation is evidently rather accurate in practice. In the absence of external fields it is a theorem that no *plane-wave* solutions for ψ exist (*e.g.*, Das and Kay, 1989), but there *are* unique global solutions to the initial-value problem described by Eqs.(2) for $t \geq 0$ when $\pi_\mu \to p_\mu$ (Flato, *et al*, 1987).

The general Maxwell-Dirac problem is exceptionally complicated and remains poorly understood. The quantum field-theoretic viewpoint leads to solution of the equations of motion through a tortuous string of arguments often defying common sense, and the perturbative result constitutes an effective linearization. If the nonlinearity is a fundamental feature of the system, then the excellent agreement with experiment suggests that the nonlinearity must be subsumed in the vacuum. The meaning of 'understanding' varies greatly among personalities and contexts, but on one definition QFT can be considered superb in that regard.

By way of contrast, the semiclassical approach described by Eqs.(2) alone constitutes a first-quantized classical field theory for both ψ and A_μ. It is not necessary at this point to interpret ψ in other than the conventional way, as a probability amplitude, but the major semiclassical developments of recent years tend to follow Schrödinger's interpretation (Schrödinger, 1926,1927). The wavefunction is interpreted literally as the matter field, in which $e|\psi|^2$ constitutes the actual charge distribution, and it is deemed sensible to focus on a single particle and the single event. A thoughtful and detailed discussion of objections to this interpretation, and their possible resolutions, has been provided by Dorling (1987). Because the issue actually goes to the heart of the so-called Copenhagen interpretation of quantum mechanics itself, space limitations prohibit much further discourse here. Nevertheless, the interpretation is central to the approaches discussed below, and we shall have a few more relevant comments later.

1. Semiclassical Theory I

It is important to emphasize that Eq.(2b) does not exhibit a solution for A_μ, but rather is a nonlinear integral equation for the self-field. As is customary in treating coupled equations, one can combine Eqs.(2) into a single integro-differential equation for ψ:

$$\left(\gamma^\mu \pi_\mu - mc\right)\psi(x) = \frac{4\pi e^2}{c} \int D(x-y)\overline{\psi}(y)\gamma_\mu\psi(y)\,d^4y \cdot \gamma^\mu\psi(x) . \tag{3a}$$

Although A_μ no longer appears here explicitly, it is by no means eliminated, and the validity of this procedure in the nonlinear case generally depends on what is done next with Eq.(3a). At best the right-hand side is a *time-dependent* perturbation. Note that one could derive this expression by including an additional term in the Dirac action:

$$W_{\text{int}} = -\frac{4\pi e^2}{2c} \int d^4x \int d^4y\, \overline{\psi}(x)\gamma^\mu\psi(x)D(x-y)\overline{\psi}(y)\gamma_\mu\psi(y)$$

$$= \int d^4x \int d^4y\, \overline{\psi}(x)\left[-\frac{4\pi e^2}{2c}\gamma^\mu\psi(x)\overline{\psi}(y)\gamma_\mu D(x-y)\right]\psi(y) , \tag{3b}$$

where the second line is written as such to facilitate later comment.

Equation (3a) has been studied to modest extent by a number of authors, including Finkelstein (1949), Lloyd (1950), Kaempffer (1955), and Finkelstein, *et al* (1956). To examine it in somewhat more detail we first rewrite it in a noncovariant Schrödinger form:

$$i\hbar\partial_t\psi(x) = (\mathsf{H}_0 + \mathsf{H}_{\text{self}})\psi(x)\,, \tag{4a}$$

with

$$\mathsf{H}_0 \equiv c\boldsymbol{\alpha}\cdot\boldsymbol{\pi} + eA_0^{\text{ext}} + \gamma^0 mc^2\,, \tag{4b}$$

$$\mathsf{H}_{\text{self}} \equiv 4\pi e^2 \int D(x-y)\overline{\psi}(y)\gamma_\mu\psi(y)\,d^4y \cdot \gamma^0\gamma^\mu\,. \tag{4c}$$

Because H_{self} is a time-dependent addition to the Dirac Hamiltonian, and never vanishes, it appears that there are no precisely-stationary solutions, except for the ground state.

The nonlinearity of the equations of motion (4) prohibits superposition of solutions, but $\psi(x)$ can still be expanded in terms of a complete set. Surely the Hilbert space remains realized by L_2 and is spanned by the set $\{\varphi_\ell\}$ satisfying $\mathsf{H}_0\varphi_\ell(x) = \epsilon_\ell\varphi_\ell(x)$. It will be convenient to define 'frequencies' $\omega_\ell \equiv \epsilon_\ell/\hbar$, $\omega_{ij} \equiv \omega_i - \omega_j$, in terms of the stationary-state energies ϵ_ℓ. Hence,

$$\psi(x) = \sum_\ell a_\ell(t)e^{-i\omega_\ell t}\,\varphi_\ell(x)\,, \qquad \sum_\ell |a_\ell|^2 = 1\,, \tag{5}$$

and the continuous portion of the spectrum is included implicitly. Substitution into Eqs.(4) yields the equations of motion for the coefficients:

$$i\hbar\dot{a}_j(t) = \sum_\ell a_\ell(t)e^{i\omega_{j\ell}t} H_{j\ell}(t)\,, \tag{6a}$$

where

$$H_{j\ell}(t) = 4\pi e^2 c \sum_{m,n} \int_{-\infty}^{\infty} K_{j\ell,mn}(t-t')a_m^*(t')a_n(t')e^{i\omega_{mn}t'}\,dt'\,. \tag{6b}$$

The entire problem has thus been reduced to an integro-differential equation in a *single* variable. Although the kernel $K_{j\ell,mn}$ is completely known, its precise form will depend sensitively on the boundary conditions chosen for the Green function $D(x-y)$.

Equations (6) had been studied years ago by Lanyi (1973), who derived them from considerations of energy-momentum conservation with kernel defined by the radiation Green function D_r (half the difference between retarded and advanced solutions). Through a series of expansions and approximations, including the dipole approximation, he obtained a reduced set of nonlinear equations for a two-level atom that are roughly equivalent to those of the neoclassical theory to be discussed presently. Owing to his use of the homogeneous Green function, his results were restricted to the problem of pure spontaneous emission. Equations (6), however, are exact within the standard paradigm of first-quantized quantum mechanics, for there are as yet no approximations, nor has there been a choice made for D.

But boundary conditions must be specified eventually, so let us first consider the retarded Green function $D_R = D_P + \frac{1}{2}D_r$, in terms of principal-value and radiation Green functions:

$$D_P(z) = \tfrac{1}{2}\varepsilon(z_0) \int \frac{d^3k}{(2\pi)^3} e^{i\mathbf{k}\cdot\mathbf{z}} \frac{\sin(kz_0)}{k}$$

$$= \tfrac{1}{2}\varepsilon(z_0)D_r(z), \tag{7}$$

where

$$\varepsilon(z_0) \equiv \begin{cases} +1, & z_0 > 0 \\ -1, & z_0 < 0 \end{cases}. \tag{8}$$

If Feynman boundary conditions are chosen instead we obtain the causal Green function $D_F = D_P + \frac{1}{2}D_1$, where D_1 is obtained from D_r by the replacement $\sin \rightarrow \cos$. The utility of writing the Green functions in this way will become apparent presently, but here we simply note that the splitting has the same form as the classical decomposition into velocity and acceleration fields.

Let us now return to Eqs.(6) and introduce the convenient notation

$$b_j(t) \equiv a_j(t)e^{-i\omega_j t}, \tag{9}$$

so that

$$i\hbar \dot{b}_j(t) = \epsilon_j b_j(t) + 4\pi e^2 c \sum_\ell b_\ell(t) \sum_{m,n} \int_{-\infty}^{\infty} K_{j\ell,mn}(t - t')b_m^*(t')b_n(t')\, dt'. \tag{10}$$

When the self-field term is neglected one regains the expected stationary-state solution,

$$b_j(t) = b_j(0)e^{-i\omega_j t}. \tag{11}$$

The next approximation is obtained by substitution of this solution into the integral in Eq.(10), presumably corresponding to the one-photon approximation of QFT. In making such an approximation, of course, we have linearized the equations, and it is not clear at this point what violence has therefore been done to the physical description. The $O(\alpha)$ form of Eq.(10) is now

$$i\hbar \dot{b}_j(t) \simeq \epsilon_j b_j(t) + 4\pi e^2 c \sum_{\ell,m,n} b_\ell(t)b_m^*(0)b_n(0) \int_{-\infty}^{\infty} K_{j\ell,mn}(t - t')e^{i\omega_{mn}t'}\, dt'. \tag{12}$$

Define 'transition currents'

$$J^\mu(\mathbf{k})_{j\ell} \equiv \int e^{i\mathbf{k}\cdot\mathbf{x}} \overline{\varphi}_j(x)\gamma^\mu \varphi_\ell(x)\, d^3x$$

$$= \int e^{i\mathbf{k}\cdot\mathbf{x}} J^\mu(x)_{j\ell}\, d^3x, \tag{13}$$

and recall the Green functions discussed in connection with Eq.(7). A short calculation
yields for the integral in Eq.(12)

$$\int_{-\infty}^{\infty} K_{j\ell,mn}(t-t')e^{i\omega_{mn}t'}\,dt' = \int \frac{d^3k}{(2\pi)^3}\frac{J^\mu(\boldsymbol{k})_{j\ell}J_\mu(-\boldsymbol{k})_{mn}}{k}e^{i\omega_{mn}t}A_{mn}(k,\omega)\,,\qquad (14a)$$

where

$$A_{mn}(k,\omega) \equiv \mathrm{P}\frac{ck}{c^2k^2-\omega_{mn}^2} + i\frac{\pi}{2}\left[\delta(ck+\omega_{mn})\mp\delta(ck-\omega_{mn})\right]\,.\qquad (14b)$$

The $\mp$ signs correspond to retarded and Feynman boundary conditions, respectively; indeed,
these are the *only* differences arising from the choice of Green function. Moreover, both
the δ-functions and the sign differences come entirely from the homogeneous solutions to
Maxwell's equations, and the principal value is related to D_P only.

One now retains only the term $\ell = j$ in Eq.(12), in a sort of Wigner-Weisskopf approximation, and extracts the j-terms from the remaining sums by discarding 'counter-rotating'
terms. The resulting approximate equation of motion is readily integrated to obtain

$$b_j(t) = b_j(0)e^{-\frac{i}{\hbar}E_j t}\,,\qquad\qquad \sum_j |b_j|^2 = 1\,.\qquad (15)$$

With $E_j \equiv \epsilon_j + \Delta E_j$, we have

$$\begin{aligned}
\Delta E_j = {}&\frac{e^2}{2\pi^2}\sum_n |b_n(0)|^2\,\mathrm{P}\int\frac{d^3k}{k^2}J^\mu(\boldsymbol{k})_{jj}J_\mu(-\boldsymbol{k})_{nn} \\
&+ \frac{e^2}{2\pi^2}\mathrm{Re}\left\{\sum_n{}' b_j^*(0)b_n(0)\int\frac{d^3k}{k^2}J^\mu(\boldsymbol{k})_{jj}J_\mu(-\boldsymbol{k})_{jn}\right. \\
&\left.\qquad\times\mathrm{P}\left[\frac{1}{ck-\omega_{jn}}+\frac{1}{ck+\omega_{jn}}\right]\frac{e^{i\omega_{jn}t}-1}{i\omega_{jn}t}\right\} \\
&+ i\pi\frac{e^2}{2\pi^2}\left\{\begin{matrix}\mathrm{Re}\\\mathrm{Im}\end{matrix}\right\}\left\{\sum_n{}' b_j^*(0)b_n(0)\int\frac{d^3k}{k^2}J^\mu(\boldsymbol{k})_{jj}J_\mu(-\boldsymbol{k})_{jn}\right. \\
&\left.\qquad\times\left[\delta(ck-\omega_{jn})\mp\delta(ck+\omega_{jn})\right]\frac{e^{i\omega_{jn}t}-1}{i\omega_{jn}t}\right\}\,,\qquad (16)
\end{aligned}$$

where the prime on the sums indicates that $n \neq j$, and the coefficients $b_\ell(0)$ are to be
determined from initial conditions. Thus, except for very short times, the energy shift is
time dependent, or *chirped*. The first two (real) terms on the right-hand side of Eq.(16) have
the interpretation of vacuum-polarization and self-energy contributions, respectively, and
thus might be thought to constitute the Lamb shift. The imaginary part of E_j corresponds
to spontaneous emission and, unfortunately, spontaneous absorption, which is not observed.
(These identifications are actually not that simple to make, and we shall discuss them
further below.) Owing to the primes on the sums in Eq.(16), the initial state cannot be

exactly an eigenstate of H_0, in agreement with the earlier observation that there are no truly stationary states.

Equations (6) contain as a very special case the neoclassical theory developed by Jaynes and his students (Crisp and Jaynes, 1969; Stroud and Jaynes, 1970), though the approximations leading to Eqs.(15) and (16) are *not* part of that theory. Rather, Jaynes restricted his studies to a two-level atom and dipole interaction, and for this model was able to maintain the basic nonlinearity in the subsequent equations of motion. For pure spontaneous emission the radiated energy is proportional to $\tanh \beta(t-t_0)$, where β is twice the Einstein A-coefficient, and the line shape has a sech^2-structure. This, of course, is a signature of soliton behavior and is the kind of pulse one might hope for in an unquantized, non-photon theory, and for long times one regains the familiar exponential decay of QED. The latter was only a presumption of the original theory of Weisskopf and Wigner (1930) at any rate, and has remained such. One would naturally expect a more complicated structure in a more detailed theory.

As mentioned earlier, similar results were obtained by Lanyi (1973), although that work was restricted to pure spontaneous emission owing to the absence of the D_P-portion of the Green function. No such restriction applies to the neoclassical theory, and a 'dynamical' Lamb shift containing the chirp emerges. Unfortunately, there is good evidence that the chirp does not exist (Citron, *et al*, 1977), thereby casting a serious cloud over the neoclassical theory and the result of Eq.(16). On the one hand it might be expected that the full nonlinear theory applied beyond two-level atoms might possess solutions in closer correspondence to the observed physical behavior. Support for this expectation can be found in the semiclassical calculations of Mahanty (1974), who was able to reproduce most of Power's results (Power, 1966), including the Bethe logarithm. On the other hand, it is not difficult to prove rigorously from the full nonlinear equations (6) that the first term on the right-hand side of Eq.(16) is the *exact* static contribution to the energy shift—simply presume that E_j in Eq.(15) is independent of the time and substitute into Eqs.(6). The outstanding question is whether or not the time dependence predicted in this approach is experimentally determinable.

2. Semiclassical Theory II

Earlier we noted that Eq.(3a) can be derived from an action principle, and this has been the point of departure for Barut and collaborators in developing an alternate semiclassical approach (*e.g.*, Barut and Salamin, 1988; see, also, Barut's article in these Proceedings). With $\hbar = c = 1$, the total action follows from Eqs.(3):

$$W = W_0 + W_1,$$

$$W_0 = \int \overline{\Psi}(x)\left[\gamma^\mu \pi_\mu - m\right]\Psi(x)\,d^4x,$$

$$W_1 = -2\pi e^2 \int\int \overline{\Psi}(x)\gamma^\mu \Psi(x) \int \frac{d^4k}{(2\pi)^4}\frac{e^{-ik\cdot(x-y)}}{k^2}\overline{\Psi}(y)\gamma_\mu \Psi(y)\,d^4x\,d^4y.$$

$$(17)$$

Rather than carry out variations on W, Barut focuses directly on radiative corrections to bound-state energies. Although the self-field remains implicit, the time evolution of processes of interest is eliminated completely by working entirely within the action. Thus, one is already committed to small corrections to stationary-state quantities and any possible time dependence of the energy shifts is suppressed.

The next and most far-reaching presumption in the development is that the state vector Ψ can be expanded in a temporal Fourier series:

$$\Psi(x) = \sum_n \psi_n(x) e^{-iE_n x_0} , \tag{18}$$

where the coefficients ψ_n remain to be determined but are taken to form a complete set, and the notation now includes explicitly all portions of the Dirac spectrum. It is difficult to understand this *ansatz* under any interpretation of quantum theory, for clearly Ψ is *not* normalizable. Barut asserts that Ψ is simply an unnormalizable solution to the equations of motion, and that ψ_n represents an actual quantum-mechanical wavefunction. This seems rather disingenuous, however, for if Ψ has no physical meaning neither does W. In light of this view, a secondary feature of Eq.(18) which is puzzling is the subsequent interpretation of E_n as a system energy level, for which no arguments are presented.

It is useful to suspend these objections momentarily and investigate further the consequences of the *ansatz* (18). As a next step substitute the expansion (18) into the action and carry out the k_0, y_0, and x_0 integrations (in that order). The Feynman propagator D_F is chosen on grounds that this is a necessary choice to treat correctly both particle and antiparticles in the theory. Why this criterion is applied to the 'photon' propagator in the context of an unquantized electromagnetic field is not clear—below it will be seen that the choice is effectively irrelevant. Substitution from Eq.(18) results in four generalized sums over states in W_1, and the time integrations reduce these to two by introducing a factor $\delta(E_n - E_m + E_r - E_s)$, a *$\delta$-function*. Subsequently the quantities $\{E_n\}$ are found to be complex (see below), so that at this point the meaning of the δ-function is a bit obscure. In addition, this is actually treated as a Kronecker-δ in collapsing the sums, and arguments are made that the only choices implied are $n = m$, $r = s$, and $n = s$, $r = m$. It is not clear, for example, why $E_n = \frac{1}{2}(E_m + E_s)$, $E_r = \frac{1}{2}(E_m + E_s)$ are not equally valid choices. Bialynicki-Birula (1986) has also objected to this kind of 'voodoo' mathematics.

The final developmental step is to establish an iterative scheme, in which the first approximation consists of replacements $\psi_n \rightarrow \psi_n^{(0)}$, $E_n \rightarrow E_n^{(0)} + \Delta E_n$, in terms of the unperturbed solutions valid when self-fields are omitted. These are Coulomb wavefunctions and energies in the present case of interest. After some algebra one finds the $O(\alpha)$ radiative energy shift to be

$$\Delta E_n = \Delta E_n^{VP} + \Delta E_n^{LS} + \Delta E_n^{SE} , \tag{19}$$

where the first two terms on the right-hand side are real and comprise the *vacuum polarization*,

$$\Delta E_n^{VP} = \frac{e^2}{2} \sum_s \int d^3x \int d^3y \, J^\mu(x)_{nn} J_\mu(y)_{ss} \, \mathrm{P} \int \frac{d^3k}{(2\pi)^3} \frac{e^{i\boldsymbol{k}\cdot(\boldsymbol{x}-\boldsymbol{y})}}{k^2} , \tag{20}$$

and *self-energy* ,

$$\Delta E_n^{LS} = 2\pi e^2 \sum_s \int d^3x \int d^3y \, J^\mu(x)_{ns} J_\mu(y)_{sn} \int \frac{d^3k}{(2\pi)^3} \frac{e^{i\boldsymbol{k}\cdot(\boldsymbol{x}-\boldsymbol{y})}}{2k}$$

$$\times \mathrm{P}\left[\frac{1}{E_s - E_n - k} - \frac{1}{E_s - E_n + k} \right] , \tag{21}$$

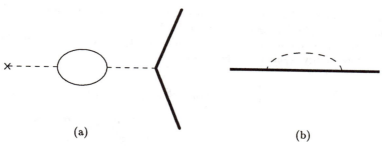

(a) (b)

Fig. 1. Feynman diagrams contributing in leading order to the Lamb shift: (a) vacuum polarization, and (b) electron self-energy.

contributions to the Lamb shift, the third (imaginary) term describes *spontaneous emission*,

$$\Delta E_n^{SE} = 2\pi e^2 \sum_s \int d^3x \int d^3y J^\mu(x)_{ns} J_\mu(y)_{sn} \int \frac{d^3k}{(2\pi)^3} \frac{e^{ik\cdot(x-y)}}{2k}$$

$$\times \left\{ i\pi \left[\delta(E_s - E_n + k) \pm \delta(E_s - E_n - k) \right] \right\}, \quad (22)$$

and $J^\mu(x)_{ns}$ is defined in Eq.(13). The $\pm$ sign in this last result exhibits the *only* difference in choosing Feynman or retarded boundary conditions, respectively. As found earlier, the real energy shift is independent of this choice. Note carefully that the sum-integral in each expression is to range over all discrete and continuum states for both positive and negative energies, although in the present model the negative-energy states are just the continuum states of the positron.

One now attempts to interpret these expressions as done in the preceding section, which is facilitated by reference to the perturbative results of conventional QED. The leading-order contributions to the Lamb shift correspond to the Feynman diagrams of Figure 1, referring to vacuum polarization and electron self-energy. In our notation the original evaluation by French and Weisskopf (1949) yields $\Delta E_n = \Delta E_n^{(a)} + \Delta E_n^{(b)}$, where

$$\Delta E_n^{(a)} = \frac{\alpha}{4\pi^2} \int \frac{d^3k}{k^2} \sum_s (1 - \delta_s) J^\mu(k)_{nn} J_\mu(-k)_{ss}, \quad (23a)$$

$$\Delta E_n^{(b)} = \frac{\alpha}{4\pi^2} \int \frac{d^3k}{k} \sum_s \frac{J^\mu(k)_{ns} J_\mu(-k)_{sn}}{E_n - E_s - k\delta_s}, \quad (23b)$$

and $J^\mu(k)_{ns}$ is defined in Eq.(13). The quantity $\delta_s = \pm 1$, depending on whether s denotes a positive- or negative-energy state. Although these bear a superficial resemblance to Eqs.(20) and (21), their content and origins are rather different.

The first difference is that Eq.(23a) specifically excludes positive-energy states. In QFT it arises conceptually as a modification of the photon propagator, rather than directly as part of the energy shift, but perhaps neither view should be viewed as written in stone. Secondly, the principal values in Eq.(21) are both summed over both positive- and negative-energy states, and do *not* correspond to the two values taken by δ_s in Eq.(23b). These remarks imply extra terms in Eqs.(20) and (21) over and above those of Eqs.(23), and the latter are known to yield close agreement with experiment. It would be remarkable if these extra terms made a negligible contribution to $\mathrm{Re}(\Delta E_n)$.

In the nonrelativistic limit Eq.(21) has been shown by Barut and Van Huele (1985) to reduce to the low-energy result of Bethe (1947):

$$\Delta E_n^{LS} \xrightarrow{\ NRL\ } -\frac{2}{3}\frac{1}{m^2}\frac{\alpha}{\pi}\sum_m (E_n - E_m)\int \frac{dk}{E_n - E_m - k},$$
$$\times\ _nT_m(\boldsymbol{k})\cdot{}_mT_n(-\boldsymbol{k}), \qquad (24a)$$

with

$$_nT_m(\boldsymbol{k}) \equiv \int \psi_n^*(\boldsymbol{x})(-i\nabla)\psi_m(\boldsymbol{x})e^{i\boldsymbol{k}\cdot\boldsymbol{x}}\, d^3x, \qquad (24b)$$

In this limit there are no differences, because Eqs.(21) and (23b) reduce to the same thing. But eventually this result must be matched to the corresponding high-energy expression at the cutoff implied in Eq.(24a), and we have seen that this will not result in continued agreement. Indeed, it is rather surprising that one obtains Eq.(24a) correctly.

Equation (22) has been studied at length by Barut and Salamin (1988), in which they note that the δ-functions are mutually exclusive—the first implies $E_n > E_s$, and the second vice-versa. The first has the obvious interpretation of spontaneous emission from an excited state ψ_n to a set $\{E_s\}$ of lower levels, and clearly explains why the ground state is stable. But the second δ-function predicts spontaneous absorption, which apparently is not observed. This extra 'absorption' term seems to be a difficulty with semiclassical theories in general (e.g., Davies and Burkitt, 1980), one not shared by QFT. Barut and Salamin make the dubious claim that one can simply select the 'correct' term, and so retain only the first. In that case the $\pm$ sign is irrelevant, and the spontaneous decay is also independent of boundary conditions on the Green function D. Barut argues elsewhere in these Proceedings that omission of the absorption term follows from energy conservation. The term is there in the mathematics, however, and energy conservation has little to do with its presence or absence. By way of analogy, there is nothing in energy conservation forbidding an automobile from spontaneously cooling itself and jumping to the top of the nearest building—it is rendered highly improbable by an additional selection rule, called the second law of thermodynamics. Nor is the difficulty avoided by moving the basic level of discussion from the action to the equations of motion themselves, for both δ-functions still appear (e.g., Barut, 1988).

With the notation $\exp(-iE_n t) = \exp(iE_n^{(0)}t)\exp(-\frac{1}{2}\Gamma_n t)$, the decay rate becomes

$$\Gamma_n = -2\mathrm{Im}(E_n) = -e^2 \sum_{s<n}\int d^3x \int d^3y\, J^\mu(\boldsymbol{x})_{ns}J_\mu(\boldsymbol{y})_{sn}$$
$$\times \int \frac{d^3k}{(2\pi)^3}\frac{e^{i\boldsymbol{k}\cdot(\boldsymbol{x}-\boldsymbol{y})}}{k}\frac{\pi}{2}\delta(E_s - E_n + k). \qquad (25)$$

The requisite integrals have been carried out and for various scenarios agree well with some other theoretical calculations and some experimental results—the latter, apparently, are sparse. If the dipole approximation is made and one identifies an electron velocity as

$v = c\alpha$, then introduction of the Heisenberg equations of motion and a sum over 'photon' polarization vectors yields a more familiar approximate form:

$$\Gamma_n \simeq \tfrac{2}{3}\alpha \sum_{s<n} \omega_{ns}^3 |r_{ns}|^2 \,, \tag{26}$$

where $\omega_{ns} \equiv E_n - E_s$. These matrix elements still contain relativistic wavefunctions, so that the summation over the electron spin states will yield a factor of 2, thereby providing the correct factor of 4/3 and Einstein's A-coefficient.

An additionally puzzling feature of these calculations is that the negative-continuum states appear to be ignored in Eq.(25), though they satisfy the indicated inequality. They *are* included in the conventional expressions of QFT, and the latter yield reasonably accurate results (*e.g.*, Goldman and Drake, 1981). It is therefore a bit surprising that Eq.(25) might produce some numerical agreement with other work.

Finally, the expression (20) has been evaluated to leading order in $(Z\alpha)$ for s-states by Barut and Ünal (1990), in which they 'almost' reproduce the known results of QED (see, *e.g.*, Grandy, 1991). It should be noted also that this formalism has been applied to other problems, including the Casimir-Polder effects (Barut and Dowling, 1987), and 'derivation' of the electron anomalous magnetic moment (Barut and Dowling, 1989). We shall examine this latter calculation in some detail in the Appendix.

3. Summary and Conclusions

Unquestionably, semiclassical calculations avoiding field quantization are capable of indicating the presence of all radiative effects of interest. Classical-mechanical versions of the Lamb shift have been illustrated more than once by Jaynes (1978, 1990), and we have seen above the appearance of spontaneous emission in the neoclassical theory. What we now call vacuum polarization was first derived semiclassically by Uehling (1935), Schwinger, *et al*, (1978) derived the Casimir-Polder effects without field quantization, and an anomalous magnetic moment for the electron of $\alpha/2\pi$ can be found by appropriate choice of relativistic cutoff parameter either classically (Grandy and Aghazadeh, 1982) or semiclassically (Barut and Dowling, 1988). But, as Jaynes (1978) has observed, "The mere fact of getting the right numerical magnitude of $\Delta\omega_{\text{Lamb}}$ cannot be claimed as a valid 'derivation' of the Lamb shift if we do not get also the correct qualitative behavior." While applauding this standard, in extending it to others one might wish also to broaden it by including an insistence on mathematically unimpeachable procedures.

The neoclassical theory has not been applied extensively, and has focused primarily on two-level atoms (though see Mahanty, 1974). It does, however, possess the considerable merit of retaining the essential nonlinearity of the problem, and one consequence of this is a line shape which is possibly a more realistic picture of the emitted radiation. Schrödinger's interpretation of the wavefunction may or may not prove central to semiclassical theories, but there are more glaring obstacles to further progress that seem intrinsic to both the neoclassical theory and its parent model of Section A. A first such obstacle is the appearance of 'absorption' terms in both real and imaginary parts of the energy shift—their emergence in Barut's approach as well suggests they are intrinsic. A second is the chirp in the Lamb shift, for whose absence in the data there is good evidence. No ready resolution of these questions appears to be in sight.

Aside from the mathematical objections raised above, Barut's approach linearizes the theory as a first approximation, thereby losing the essential feature which may be thought to enable an avoidance of second quantization. Of course, this is also a shortcoming of the semiclassical calculation in Section A (but not of the two-level neoclassical theory). In addition, there is still an infinite renormalization term which one would have thought might at least be finite when Coulomb wavefunctions are used for the unperturbed quantities.

There does remain the fact that both of these major semiclassical developments make *some* contact with other theoretical calculations, as well as with some of the data. As Khalatnikov (1989) reminds us, though, "... no coincidence of theory with experiment can justify logical gaps in the theorist's work." It is the supposed logical difficulties with QFT, after all, that have motivated the work under consideration here, and there is no point to falling into the same trap. There is qualitative encouragement, nevertheless, that the basic semiclassical approach to the radiation field is sound. Indeed, devotees of QFT themselves have shown a recent inclination to shift in subtle ways the specifics of the issue. For example, Mandel (1986), and others (Kimble and Mandel, 1975; Wódkiewicz, 1980) have argued that the radiative effects do *not* provide good tests of QFT as opposed to semiclassical theories. Rather, it is argued that the quantum nature of the electromagnetic field is really manifest only in photon correlation experiments. Although a number of these appear capable of interpretation in terms of the classical field and its first-quantized sources, no serious calculations along these lines are known to the author. Moreover, in such experiments the issues of quantized fields and the interpretation of quantum mechanics itself, in connection with Bell's theorem, are deeply intertwined. No doubt the advocates of a semiclassical theory of radiation would settle for merely the radiative effects at this time!

Jaynes has repeatedly emphasized that his motivation in constructing a neoclassical theory was not so much the development of a specific replacement for QFT as it was to uncover experiments which might suggest directions in which that development might proceed (Jaynes, 1973, 1978). Investigation of the Maxwell-Dirac theory by Barut and others, however, has been more ambitious and has been aimed at providing just that direction. Although both approaches seem to have run into serious difficulty, it can be argued that the development simply has not proceeded far enough—principally because of mathematical intractability. The truly essential feature of the Maxwell-Dirac equations is their non-linearity, which has yet to be fully exploited. Owing to the complexity of the equations in 3+1 dimensions, soliton-like solutions have not been found, but the possibility surely exists. Such solutions would have enormous bearing on an understanding of the general elementary-particle problem itself.

Unmentioned—at least explicitly—in semiclassical treatments to date is the *desideratum* of finding a *physical* mechanism for pair creation and annihilation. Quantum field theory has eliminated the problem by fiat, but this procedure at least has the merit of accommodating the phenomena. At the heart of any truly viable alternative to QFT must be a natural description of the entire e^+e^- spectrum, and in turn this suggests a need to study the entire electromagnetic *system* as a single entity. The evidence for myopia in looking at only individual pieces of the total $e^+e^-\gamma$-system has been before us for a long time, but no one has yet seen how to put it together. Quite possibly it will be necessary to include a $\nu\bar{\nu}$-component as well. Earlier neutrino theories of light met insurmountable obstacles, but perhaps only because the complete system was not being considered.

In the author's view some variant of these generalizations is absolutely required if the quest for a non-second-quantized theory of matter and radiation is to be sustained. At a minimum, a first step might be to obtain solutions to the Maxwell-Dirac equations in the absence of external fields and employ these as the zero-order input for investigation of radiative corrections, rather than the intrinsically linear Coulomb solutions. Although such solutions exist (Flato, *et al*, 1987), finding them explicitly has not proved an easy task. Nevertheless, we have noted earlier that, though there are no plane-wave solutions, unique global solutions to the initial-value problem do exist. From the work of Mathieu and Morris (1984) one can infer that there are no localized *stationary* solutions, even for the Coulomb problem. In agreement with earlier discussion of the energy shifts, then, we conjecture that there do exist localized, nonstationary solutions. Once again, the major issue will no doubt be the precise nature of the time dependence.

Appendix.

In a recent article Barut and Dowling (1989) claim to provide a derivation of the electron anomalous magnetic moment. They consider the interaction term in the action, differing by a factor -4π from our Eq.(3b), along with what appears to be a Fourier representation

$$\Psi(x) = \int e^{-ip\cdot x}\, \Psi(p)\, d^4 p\,. \qquad (A\text{--}1)$$

This, however, is taken to be understood in the sense of Eq.(18) above, so that the p_0-integral may contain a sum over the discrete part of the spectrum, including spin. Substitution of Eq.(A–1) for each of the four functions occurring in the action, followed by evaluation of three of the integrals, yields

$$W_{\text{int}} = \frac{e^2}{c}(2\pi)^4 \int d^4 p \int d^4 q \int d^4 r \int d^4 s\, \overline{\Psi}(p)\gamma^\mu \Psi(q)$$

$$\times \frac{1}{(r-s)^2 + i\epsilon}\overline{\Psi}(r)\gamma_\mu \Psi(s)\delta(p-q+r-s)\,. \qquad (A\text{--}2)$$

The next step defies understanding, for the δ-function is employed to evaluate *two* of the integrals in Eq.(A–2). That is, the values $s = p$, $r = q$ are selected as satisfying the stated condition—which they do—and the continuum of other choices is ignored. Though the procedure is clearly unacceptable mathematically, the claim is now made that the physically interesting piece of the action with respect to $g - 2 \neq 0$ is that corresponding to these special values:

$$W'_{\text{int}} = \frac{e^2}{2} \int d^4 p \int d^4 x \int d^4 y\, D(x-y)\overline{\Psi}(p)e^{ip\cdot x}$$

$$\times \gamma^\mu \left\{ \int d^4 q \Psi(q)\overline{\Psi}(q)e^{-i(x-y)\cdot q} \right\} \gamma_\mu \Psi(p)e^{-ip\cdot y}\,. \qquad (A\text{--}3)$$

Although we must consider a factor of ∞ to have been omitted on the right-hand side, let us follow the further development by Barut and Dowling anyway.

Denote the function within the braces in Eq.(A–3) as $S(x-y)$, whose Fourier transform is $\Psi(p)\overline{\Psi}(p)$, where $\Psi(x)$ is to satisfy the *complete* equations of motion obtained from

Eqs.(17) above through variation with respect to $\overline{\Psi}(x)$. The authors now claim that this function S is the propagator, or Green function solution for the *complete* nonlinear equations, subject to the following boundary condition:

$$S(x-y) = \theta(y_0-x_0)\int_{(+)} \Psi(q)\overline{\Psi}(q)e^{-i(x-y)\cdot q}\,d^4q$$

$$-\theta(x_0-y_0)\int_{(-)} \Psi(q)\overline{\Psi}(q)e^{-i(x-y)\cdot q}\,d^4q\,. \quad \text{(A-4)}$$

The notation $(\pm)$ on the integrals indicates integration over positive- and negative-energy states, so that the right-hand side is now independent of energies. In first approximation Ψ is replaced by the solution to the equations of motion for zero self-field, in which case $S(x-y)$ is asserted to become the familiar function satisfying $(\gamma^\mu\pi_\mu - m)S(x-y) = i\delta(x-y)$. One's skepticism about the meaning attributed to S begins with the implication from its definition that the general propagator is always a function of $(x-y)$ alone. This is not generally true. Moreover, the expression (A-4) is not at all equivalent to similar decompositions found in, say, Bjorken and Drell (1964). Nevertheless, we follow the authors along this path and adopt $S(p) = (\gamma^\mu\pi_\mu - m)^{-1}$ as the leading-order approximation to the Fourier transform of $S(x-y)$. [The reader may find it challenging to try to demonstrate that $\Psi(p)\overline{\Psi}(p) = (\gamma^\mu p_\mu - m)^{-1}$ for free particles, let alone anything else.] They then rewrite Eq.(A-3) as

$$W'_{\text{int}} = -\frac{ie^2}{2}\int d^4p\int d^4P\,\overline{\Psi}(p)\frac{\gamma^\mu S(p)\gamma_\mu}{(p-P)^2+i\epsilon}\Psi(p)$$

$$= \int d^4p\,\overline{\Psi}(p)\,\delta M(p)\,\Psi(p)\,, \quad \text{(A-5)}$$

which, after a change of integration variables, defines a mass shift

$$\delta M(p) \equiv \frac{e^2}{2}\frac{i}{(2\pi)^4}\int \frac{\gamma^\mu S(p-s)\gamma^\mu}{s^2+i\epsilon}\,d^4s\,. \quad \text{(A-6)}$$

One notes that the interaction part of the action is proportional to the energy shift: $W_{\text{int}} = (2\pi)^4\delta E$.

If the above procedure were correct this would be a remarkable result, for Eq.(A-6) exhibits precisely the $O(\alpha)$ approximation to Schwinger's mass operator (*e.g.*, Berestetskii, *et al*, 1982; p.477). But if one reviews the 'derivation' beginning with Eq.(3b) above, then the implication is that the mass operator $M(x,y)$ can be identified as the term in brackets in the second line of Eq.(3b). This cannot be true even though Eq.(A-6) provides a first approximation in momentum space. For one thing, the 'identification' in Eq.(3b) contains no classical version of the vertex operator, which is absolutely essential in higher orders. For another, when one actually works through the 'derivation' of Eq.(A-5) it is found that a factor $\delta(0)$ has been omitted on the right-hand side. Thus, one can present anything one pleases if it is multiplied by infinity! The observation by the authors that their expression is equivalent to that found similarly by Babiker (1975) is simply irrelevant, for the latter was derived in a mathematically consistent way employing field operators for the electron. Appealing to Babiker's result is akin to merely cancelling the sixes in the fraction 64/16 and claiming the end to justify the means!

It would be very pleasing to see an unimpeachable derivation of the electron anomaly that is independent of notions regarding the vacuum and field quantization. The work under discussion, however, does not accomplish this.

REFERENCES

Babiker, M.: 1975, 'Source-Field Approach to Radiative Corrections and Semiclassical Radiation Theory', *Phys. Rev. A* **12**, 1911.

Barut, A.O.: 1988, 'Quantum-Electrodynamics Based on Self-Energy', *Physica Scripta* **T21**, 18.

Barut, A.O., and J.P. Dowling: 1987, 'Quantum Electrodynamics Based on Self-Energy, without Second Quantization: The Lamb Shift and Long-Range Casimir-Polder van der Waals Forces Near Boundaries', *Phys. Rev. A* **36**, 2550.

Barut, A.O., and J.P. Dowling: 1989, 'QED Based on Self-Fields: A Relativistic Calculation of $g-2$', *Z. f. Naturf.* **44a**, 105.

Barut, A.O., and Y.I. Salamin: 1988, 'Relativistic Theory of Spontaneous Emission', *Phys. Rev. A* **37**, 2284.

Barut, A.O., and N. Ünal: 1990, 'Regularized Analytic Evaluation of Vacuum Polarization in a Coulomb Field', *Phys. Rev. D* **41**, 3822.

Barut, A.O., and J.F. Van Huele: 1985, 'Quantum Electrodynamics Based on Self-Energy: Lamb Shift and Spontaneous Emission without Field Quantization', *Phys. Rev. A* **32**, 3187.

Berestetskii, V.B., E.M. Lifshitz, and L.P. Pitaevskii: 1982, *Quantum Electrodynamics*, Pergamon Press, Oxford.

Bethe, H.A.: 1947, 'The Electromagnetic Shift of Energy Levels', *Phys. Rev.* **72**, 339.

Bialynicki-Birula, I.: 1986, 'Comment on "Quantum Electrodynamics Based on Self-Energy: Lamb Shift and Spontaneous Emission without Field Quantization', *Phys. Rev. A* **34**, 3500.

Bjorken, J.D., and S.D. Drell: 1964, *Relativistic Quantum Mechanics*, McGraw-Hill, New York.

Citron, M.L., H.R. Gray, C.W. Gabel, and C.R. Stroud: 1977, 'Experimental Study of Power Broadening in a Two-Level Atom', *Phys. Rev. A* **16**, 1507.

Crisp, M.D., and E.T. Jaynes: 1969, 'Radiative Effects in Semiclassical Theory', *Phys. Rev.* **179**, 1253.

Das, A., and D. Kay: 1989, 'A Class of Exact Plane Wave Solutions of the Maxwell-Dirac Equations', *J. Math. Phys.* **30**, 2280.Davies, B., and A.N. Burkitt: 1980, 'On the Relationship between Quantum Random and Semiclassical Electrodynamics', *Aust. J. Phys.* **33**, 671.

Dorling, J.: 1987, 'Schrödinger's Original Interpretation of the Schrödinger Equation: A Rescue Attempt', in C.W. Kilmister (ed.), *Schrödinger*, Cambridge Univ. Press, Cambridge, p.16.

Finkelstein, R.J.: 1949, 'On the Quantization of a Unified Field Theory', *Phys. Rev.* **75**, 1079.

Finkelstein, R., C. Fronsdal, and P. Kaus: 1956, 'Nonlinear Spinor Field', *Phys. Rev.* **103**, 1571.

Flato, M., J. Simon, and E. Taflin: 1987, 'On Global Solutions of the Maxwell-Dirac Equations', *Commun. Math. Phys.* **112**, 21.

French, J.B., and V.F. Weisskopf: 1949, 'The Electromagnetic Shift of Energy Levels', *Phys. Rev.* **75**, 1240.

Goldman, S.P., and G.W.F. Drake: 1981, 'Relativistic Two-Photon Decay Rates of $2S_{1/2}$ Hydrogenic Ions', *Phys. Rev. A* **24**, 183.

Grandy, W.T., Jr.: 1991, *Relativistic Quantum Mechanics of Leptons and Fields*, Kluwer, Dordrecht.

Grandy, W.T., Jr., and A. Aghazadeh: 1982, 'Radiative Corrections for Extended Charged Particles in Classical Electrodynamics', *Ann. Phys.* **142**, 284.

Itzykson, C., and J.-B. Zuber: 1980, *Quantum Field Theory*, McGraw-Hill, New York.

Jaynes, E.T.: 1973, 'Survey of the Present Status of Neoclassical Radiation Theory', in L. Mandel and E. Wolf (eds.), *Coherence in Quantum Optics*, Plenum, New York.

Jaynes, E.T.: 1978, 'Electrodynamics Today', in L. Mandel and E. Wolf (eds.), *Coherence in Quantum Optics IV*, Plenum, New York.

Jaynes, E.T.: 1990, 'Probability in Quantum Theory', in W.H. Zurek (ed.), *Complexity, Entropy and the Physics of Information*, Addison-Wesley, Reading, MA.

Kaempffer, F.A.: 1955, 'Elementary Particles as Self-Maintained Excitations', *Phys. Rev.* **99**, 1614.

Källén, A.O.G.: 1958, 'Quantenelektrodynamik', in S. Flügge (ed.), *Handbuch der Physik, Band V, Teil 1*, Springer-Verlag, Berlin.

Khalatnikov, I.M.: 1989, 'Reminiscences of Landau', *Physics Today*, May, p.34.

Kimble, H.J., and L. Mandel: 1975, 'Problem of Resonance Fluorescence and the Inadequacy of Spontaneous Emission as a Test of Quantum Electrodynamics', *Phys. Rev. Lett.* **34**, 1485.

Lanyi, G.: 1973, 'Classical Electromagnetic Radiation of the Dirac Electron', *Phys. Rev. D* **8**, 3413.

Lloyd, S.P.: 1950, 'Elimination of the Self-Electromagnetic Field', *Phys. Rev.* **77**, 757(A).

Mahanty, J.: 1974, 'A Semi-Classical Theory of the Dispersion Energy of Atoms and Molecules', *Nuovo Cimento* **22B**, 110.

Mandel, L.: 1986, 'Non-Classical States of the Electromagnetic Field', *Phys. Scripta* **T12**, 34.

Mathieu, P., and T.F. Morris: 1984, 'Existence Conditions for Spinor Solitons', *Phys. Rev. D* **30**, 1835.

Milonni, P.W.: 1984, 'Why Spontaneous Emission?', *Am. J. Phys.* **52**, 340.

Power, E.A.: 1966, 'Zero-Point Energy and the Lamb Shift', *Am. J. Phys.* **34**, 516.

Schwinger, J., L.L. DeRaad, Jr., and K.A. Milton: 1978, 'Casimir Effect in Dielectrics', *Ann. Phys.* **115**, 1.

Schrödinger, E.: 1926, 'Quantisierung als Eigenwertproblem', *Ann. d. Phys.* **79**, 361.

Schrödinger, E.: 1927, 'Energieaustauch nach der Wellenmechanik', *Ann. d. Phys.* **83**, 956.

Stroud, C.R., Jr., and E.T. Jaynes: 1970, 'Long-Term Solutions in Semiclassical Radiation Theory', *Phys. Rev. A* **1**, 106.

Uehling, E.A.: 1935, 'Polarization Effects in the Positron Theory', *Phys. Rev.* **48**, 55.

Weisskopf, V., and E.P. Wigner: 1930, 'Linienbreite auf Grund der Diracschen Lichttheorie', *Z. Phys.* **63**, 54.

Wódkiewicz, K.: 1980, 'Resonance Fluorescence and Spontaneous Emission as Tests of QED', in A.O. Barut (ed.), *Foundations of Radiation Theory and Quantum Electrodynamics*, Plenum, New York.

ON THE MATHEMATICAL PROCEDURES OF SELFFIELD QUANTUMELECTRODYNAMICS

A. O. Barut
Department of Physics
University of Colorado
Boulder, CO 80309, USA

In this Note I clarify some of the mathematical developments and procedures of Self-field QED which are different from those in the usual quantum field theory and has caused some misunderstandings to those deeply attached or used to the techniques of the latter. I think the critical remarks[1] are more on the way and the style we extract observable quantities directly from the action rather than on the substance or the philosophy of our approach which has its own different but complete interpretation.

By eliminating the vector potential A_μ from the coupled Maxwell-Dirac equations we found that the ψ-field satisfies a nonlinear integro-differential equation[2]. Here ψ is a complex scalar field, not a probability amplitude. Then we look for possible frequencies of such a field by expanding it into a Fourier series

$$\psi(x) = \sum_n \psi_n(\boldsymbol{x})e^{-iE_n t} \tag{1}$$

For some reasons there has been objections[1] to such an expansion, although this is the proper way of treating any wave field, linear or nonlinear, which we do all the time for time-dependent Schrödinger equation, for Navier-Stokes equation and so on. Apparently they think of another expansion in perturbation theory where a probabilistic, normalized wave function is expanded as

$$\psi(x) = \sum_n C_n(t)\psi_n(\boldsymbol{x})e^{-iE_n t} \tag{2}$$

in which both ψ and ψ_n are normalized according to quantum mechanics, $\{\psi_n\}$ being further an orthonormal set. We have no condition that our field (unfortunately labelled by the same symbol ψ) in (1) should be normalized. The ψ_n and E_n in (1) have not the same meaning as those in (2), again by custom the same symbols have been used, regretable, as we realize. It is not clear at all that the series (1) diverges, as claimed, because the unknown coefficients ψ_n satisfy coupled nonlinear equations, they are not orthonormal, and these coupled equations provide us with the physical interpretation

165

D. Hestenes and A. Weingartshofer (eds.), The Electron, 165–169.
© *1991 Kluwer Academic Publishers. Printed in the Netherlands.*

of ψ_n and E_n, the only quantities we need. At any rate as Heaviside put it :"This series is divergent, therefore we shall be able to do something with it.". In contrast the ψ_n and E_n in (2) are known for a given problem. This is the big difference. Every approach has its own interpretation and one should not confuse a language used in one approach with the other.

Self-field QED has the logical structure of classical electrodynamics and the current j_μ is the most important quantity . In both cases what we actually use , as developed in an early version of the theory[3], is an expansion of the current j into its possible frequencies. We shall see that all observables are expressed in terms of the Fourier coefficients of the current

$$j_\mu(\boldsymbol{x},t) = \sum_{nm} j_\mu^{nm}(\boldsymbol{x})e^{-\omega_{nm}t} \tag{3}$$

Here ω_{nm} are the observed frequencies of the atomic system. Schrödinger already observed at the very beginning that one should be able to formulate atomic processes in terms of the Fourier spectrum of matter and of radiation in space and time[4]. Inserting (3) and the Fourier expansion of the Green's function D into the interaction action

$$W_{\text{int}} \sim \int j_\mu(x)D^{\mu\nu}(x-y)j_\nu(x)dxdy$$

and performing dx^0, dy^0, dk^0-integrations we obtain

$$W_{\text{int}} \sim \int dk \sum_{nmrs} \delta(\omega_{nm}+\omega_{rs})j_\mu^{nm}(\boldsymbol{k})D^{\mu\nu}(\omega_{nm},\boldsymbol{k})j_\nu^{rs}(-\boldsymbol{k}) \tag{4}$$

At this point we can also answer the critical queries about the δ-function in (4) which in later work appeared as $\delta(E_n - E_m + E_r - E_s)$. Now we could have developed the whole theory in terms of the equations of motion without such a δ-function. But we found that it is more direct and more interesting to calculate all the observables in a unified form from the interaction action itself. One obtains so immediately the decay rates, for example, instead of first finding the amplitudes and then squaring it to get the rates. But the action being an integral over all space and time is infinite. However this is a well known situation in S-matrix theory: one factorizes a factor $\delta(0)$ from W_{int} to arrive at observables.

The next query concerns the ways the δ-function in (4) can be satisfied. In the set of all frequencies $\{\omega_{nm}\}$ there are only two ways we can satisfy the δ-function: (a) $\omega_{nm} = 0$ and $\omega_{rs} = 0$, (b) $\omega_{nm} = -\omega_{rs}$. We write then these two types of terms separately and identify them with the contributions of vacuum polarization and Lamb shift plus spontaneous emission, respectively. There are no other terms. In the form $\delta(E_n - E_m + E_r - E_s)$ there seems to be at first more ways of satisfying the δ-function, but these take one outside the set of frequencies $\{\omega_{nm}\}$. One further query was what happens to the argument of the δ-function when due to spontaneous emission the energies get a complex shift (to which we shall come back immediately). The answer is that these imaginary parts always cancel. At any rate the argument of δ-function is always zero for observables as stated above. The summations go over all indices, discrete or continuous, with their degeneracies, that satisfy the δ-function constraint. We shall now see how this is done in the iterative solutions of the nonlinear equations.

Because the total action W vanishes when the equations of motion are inserted, we obtain a condition to evaluate the energy shifts due to selffields iteratively. They are given by

$$E^{\text{int}} \sim \int dk \sum_{ns} j_\mu^{nn}(k) j_\nu^{ss}(-k) D^{\mu\nu}(0, k) + \int dk \sum_{nm} j_\mu^{nm}(k) j_\nu^{mn}(-k) D^{\mu\nu}(\omega_{nm}, k) \quad (5)$$

We may use the Fourier components of the Green's function in Coulomb gauge with nonvanishing components

$$D_{00} = \frac{1}{k^2}, \quad D_{0k} = D_{k0} = 0$$

$$D_{j\ell}(\omega.k) = \frac{1}{\omega^2 - k^2} \left(\delta_{j\ell} - \frac{k_j k_\ell}{k^2} \right) \quad (6)$$

to obtain

$$E_{\text{int}} \sim \int dk \Bigg[\sum_{ns} j_0^{nn}(k) j_0^{ss}(-k) + \sum_{\substack{nm \\ n \neq m}} j_0^{nm}(k) j_0^{mn}(-k)$$

$$+ \sum_{nm} j_j^{nm}(k) j_\ell^{mn}(-k) \frac{k^2}{\omega_{nm}^2 - k^2} \left(\delta_{j\ell} - \frac{k_j k_\ell}{k^2} \right) \Bigg] \quad (7)$$

In the last term we use the identity

$$\frac{k^2}{\omega^2 - k^2} = \frac{1}{2} \left(\frac{\omega}{\omega - k} + \frac{\omega}{\omega + k} - 2 \right) \quad (8)$$

and, because we have a sum over all n and m and the integrand is invariant under the exchange $n \leftrightarrow m$, the two terms in (8) lead to the same contribution. Hence (7) becomes

$$E_{\text{int}} \sim \int dk \Bigg[\sum_{ns} j_0^{nn}(k) j_0^{ss}(-k) + \sum_{\substack{nm \\ n \neq m}} j_0^{nm}(k) j_0^{mn}(-k)$$

$$+ \sum_{nm} j_j^{nm}(k) j_\ell^{mn}(-k) \left[\frac{\omega_{nm}}{\omega_{nm-k}} - 1 \right] \left(\delta_{j\ell} - \frac{k_j k_\ell}{k^2} \right) \Bigg] \quad (9)$$

This is the calorimetric interaction energy of all levels. We consider a fixed level n and evaluate the energy shift of this level due to all others. In the last term of (9) because of the pole of the integrand we use the standard formula

$$\frac{1}{\omega - k} = P \frac{1}{\omega - k} - i\pi\delta(\omega - k) \quad (10)$$

There is only one imaginary part to the energy shift. In some papers we wrote at this stage two δ- functions : $\delta(\omega - k) + \delta(\omega + k)$, but we actually worked with one of them. The first one gives the spontaneous decay from n to m, if m is a lower state, the second from m to n, if n is lower. We combine these terms as explained above before we single out a particular state n. This answers the remarks about "spontaneous absorption"[1] interpretation of the second term. At any rate the causal Green's function has only one imaginary part, and spontaneous absorption cannot occur by energy conservation. Our formula for spontaneous emission,although it looks quite different and has been obtained by an entirely different reasoning, has been shown recently to be exactly equivalent, in the lowest iteration, to the QED formula[5]. Similarly, our formulas for vacuum polarization and Lamb shift, although they look at first to be different, are equivalent in lowest order of iteration to the formulas of QED, the first to the expression used by Wichmann and Kroll[6], the second to the more recent and more complete work of Mohr[7]. Thus apparent looks should not be attributed to a difference in substance.

Finally two other remarks on the nature of Selffield-QED may help to clarify the approach. The first is that it is not a "semiclassical" theory. "Semiclassical" usually means that we keep both the matter and electromagnetic fields side by side, but quantizing only the matter field and not the electromagnetic field, Here we have no separate degrees of freedom for the electromagnetic field A_μ; it has been eliminated But we have a nonlinear field theory for the matter field alone. Therefore we do not think that the selffield-QED is an approximation to QED. We view it as an exact theory of the Maxwell-Dirac system, and as far as we can see sofar in the lowest order of iteration, it gives identical results to QED, but with a possibility that it may give nonperturbative results and can be extrapolated to short distances. The question of quantization of A_μ does not arise. As we have often stressed, the quantized properties of the field reflect the quantized properties of the source and there are no new degrees of freedom in the field besides those of the source.

The second remark concerns the interpretation of negative energy states. Although as mentioned above, the Selffield-QED has the logical structure of classical electrodynamics, the source current is now the Dirac current which has a more complicated frequency spectrum. Only the positive frequency solutions of the Dirac equation refer to the electron, the negative frequency solutions have to be interpreted consistently even in first quantized theory, as the states of the antiparticle, the positron. However we often use in the calculations, with great advantage,the completeness relation which involves both the positive and negative energy states. But then the correct physical interpretation must be implemented. For example, in the calculation of vacuum polarization contribution of the electron, we extend the contour of integration to negative energy cut as well in order to use the Green's function of the Dirac equation , but for the contribution of negative energy states we change $e \rightarrow -e$, and then take one half of the result[8]. For the same reason we use the causal Green's function $D(x - y)$ which picks only one pole for $t' > t$, another for $t' < t$, thus indirectly controls the positive end negative frequency states of the current source.

In conclusion, concepts, ideas and calculational techniques used in quantum theory and QED should not be translated unchanged into another approach, the Self-field QED. But with the correct interpretation the latter provides an efficient, unified and, we think, a fully consistent framework for the whole field of radiative processes with new directions of extrapolation.

References

1. T.Grandy, Jr.,These Proceedings. See also the earlier paper by I. Bialynicky-Birula, *Phys. Rev. A* **34**, 3500(1986) and reply to it, A.O.Barut, *Phys. Rev.A* **34**, 3502 (1986)

2. A.O.Barut, These Proceedings, and references therein.

3. A.O.Barut, in *Quantum Theory and Structure of Space and Time*, Vol.6 (edited by L.Castell et al), C.Hanser Verlag, Munchen 1986, p.83

4. E.Schrödinger, *Die Naturwissenschaften*, **17**, 326 (1929)

5. A.O.Barut and Y.Salamin, *Phys. Rev. A* (in press)

6. E.H.Wichmann and N.Kroll, *Phys. Rev.* **101**, 83 (1956)

7. P.Mohr, *Ann. of Physics*, **88**, 26 (1974)

8. A.O.Barut and N.Unal, *Phys. Rev. D* **41**, 3822 (1990)

NON-LINEAR GAUGE INVARIANT FIELD THEORIES OF THE ELECTRON AND OTHER ELEMENTARY PARTICLES

F.I. COOPERSTOCK
Department of Physics and Astronomy
University of Victoria
P.O. Box 3055
Victoria, B.C. Canada   V8W 3P6

ABSTRACT. We review the Einstein-Rosen program of building elementary particles in solitonic structures in singularity-free non-linear gauge invariant field theories. The role of gravity via general relativity is discussed. It is found that a zone of negative energy density surrounds the particle core which is indicated to be much larger than 10^{-33} cm. A model that encompasses the electron, muon and tau is found with particle sizes ~10^{-16} cm, within experimental limits. Spin and magnetic moment have the potential to be incorporated with fields of axial symmetry. The quarks can also be modelled, but thus far, two additional coupling constants have been required. The new approach of modelling the electron as a quantum soliton in Dirac-Maxwell theory is described. Preliminary results indicate an emerging wave function with characteristic spread of the order 10^{-16} cm.

1. Introduction

Field theory has been one of the crowning achievements of modern physics. It has described electromagnetism in Maxwell's equations, gravitation in Einstein's equations of general relativity and in more recent times, it has merged with quantum theory. Its compelling elegance and logic have led many researchers to believe that field theory holds the key to the fundamental description of physical phenomena.

In the eyes of some of the greatest achievers in the history of physics such as Einstein, there was a conviction that a proper field theory should be free of singularities. Moreover, the concept of "particle" should not be separate from the field. From the well-known paper of Einstein and Rosen [1]:

A complete field theory knows only fields and not the concepts of particle and motion. For these must not exist independently of the field but are to be treated as part of it. On the basis of the description of a particle without singularity one has the possibility of a logically more satisfactory treatment of the combined problem: The problem of the field and that of motion coincide.

D. Hestenes and A. Weingartshofer (eds.), The Electron, 171–181.

Singularities signal the breakdown of physics and their avoidance has served as a useful constraint in the construction of physical theory. Another useful constraint has been that of gauge invariance. Indeed, in recent years, the concept of gauge invariance has come to play an increasingly important role in the theories of fundamental interactions [2].

In terms of structure, nonlinearity has been a vital element in our efforts to model the complexity of nature with mathematics. With regard to field theory, it is through the medium of nonlinearity that structures emerge which we identify as fundamental particles such as the electron. The problem is to identify the ideal field theory (or theories) which describes the physical world. Its success would be measured not only by its completeness and accuracy, but also by its capacity to predict new phenomena which are experimentally verifiable. These are the fundamentals of good science. While esthetic appeal and elegance might be deemed desirable by most researchers including ourselves, it is not our task to prejudge the workings of nature, which might follow a course that is not in accord with our predilections or expectations. Of even less concern is conformity to existing popular trends: progress in science is best served by an openness to new ideas.

It is my goal in this paper to describe the field theory of elementary particles which Rosen and I [3] developed, as well as more recent work, both completed and in progress, which my collaborators and I have pursued. This includes a more detailed treatment of the role of general relativity [4], an attempt to encompass the quarks [5], spin and magnetic moment [6], and work in progress to incorporate the Dirac theory in the formation of quantum solitons.

2. Historical Background

In Maxwell's theory of the electromagnetic field, we deal with the Lagrangian

$$L = \frac{-1}{8\pi} F^{\mu\nu} F_{\mu\nu} - \frac{1}{c} J^{\mu} A_{\mu} \tag{2.1}$$

where the Maxwell tensor, $F_{\mu\nu}$, is related to the four-vector potential A_{μ} as

$$F_{\mu\nu} = A_{\nu,\mu} - A_{\mu,\nu} \tag{2.2}$$

and J^{μ} is the current four-vector. The variational principle, or Lagrange's equations, yields the source set of Maxwell's equations

$$F^{\mu,\nu},\nu = - \frac{4\pi}{c} J^{\mu} \quad . \tag{2.3}$$

In pure electromagnetic theory, elementary charged particles appear as point singularities. As we discussed earlier, this is unsatisfactory as it signals the breakdown of physical theory.

Poincaré had created finite non-singular "elementary" particles, but because of the Coulomb repulsion of interacting elements of the finite structure, stresses had to be adjoined to maintain the integrity of the particle structure. These "Poincaré stresses" inject a phenomenological element to the theory whereas a fundamental, unified structure is preferable, if attainable. Einstein showed that with electromagnetic and gravitational fields alone, such a unified theory could not be realized.

Gauge invariance provides the avenue by which new fields are added to build particles with fields. It is well-known that a gauge transformation,

$$A_{\mu} \rightarrow A'_{\mu} = A_{\mu} + \lambda_{,\mu} \tag{2.4}$$

where λ is a scalar function, retains the value of $F_{\mu\nu}$ by virtue of the defining equation (2.2). While Mie [7] developed a non-singular finite elementary particle by modifying Maxwell theory, he did so with A_μ appearing explicitly in the Lagrangian. Thus, Mie's theory was not gauge invariant.

Born and Infeld [8] created a new non-linear electrodynamic theory which, unlike Mie's theory, had A_μ appear in the Lagrangian only in the form of (2.2). Thus, their theory was gauge invariant and it did produce finite particle states. However, Rosen [9] showed later that this theory still contained singularities.

Rosen noted that Mie's theory with A_μ explicitly appearing in the Lagrangian did avert the singularity problem and he succeeded in retaining A_μ while maintaining gauge invariance by introducing a new complex scalar field ψ. This was achieved by having ψ undergo a phase rotation

$$\psi \rightarrow \psi' = \psi e^{i\epsilon\lambda}$$

(2.5)

whenever A_μ undergoes a gauge transformation (2.4). In this manner, A_μ is introduced explicitly and gauge invariance is retained with derivatives of ψ entering into the Lagrangian only in the form of a "gauge derivative",

$$D_\mu\psi \equiv \partial_\mu\psi - i\epsilon A_\mu\psi .$$

(2.6)

Rosen [9] constructed his Lagrangian with the simplest scalar combinations, $\psi\bar{\psi}$ and $D_\mu\psi \overline{D^\mu\psi}$, added to the free electromagnetic term,

$$L = - F_{\mu\nu} F^{\mu\nu}/8\pi - D_\mu\psi\overline{D^\mu\psi} + \sigma^2\psi\bar{\psi}$$

(2.7)

with one non-trivial constant σ introduced as shown. The particular choice of signs was the only one which yielded particle solutions.

Variations with respect to A_μ and $\bar{\psi}$ yield the field equations

$$F^{\mu\nu},\nu = -\frac{4\pi}{c}J^\mu , \quad J^\mu = \frac{i\epsilon}{2}(\bar{\psi}D^\mu\psi - \psi\overline{D^\mu\psi})$$

$$\partial_\nu (D^\nu\psi) - i\epsilon A_\nu D^\nu\psi + \sigma^2\psi = 0 .$$

(2.8)

Note that the current source for the Maxwell equations derives from ψ coupled to A_μ itself via the gauge derivative. The field equation for ψ is a modified Klein-Gordon equation.

Rosen found a continuum of non-singular, stationary, spherically symmetric particle solutions of the form

$$\psi = \theta(r) \exp(-i\epsilon\mu t), \quad \mu = constant,$$

(2.9)

but the particle energies were all negative.

Years later, Finkelstein, Lelevier and Ruderman [10] and Rosen, Rosenstock [11] examined a purely scalar field Lagrangian with an additional quartic term, of the form

$$L = (\partial^\mu\psi)(\partial_\mu\bar{\psi}) - \sigma^2\psi\bar{\psi} + \frac{1}{2}g\psi^2\bar{\psi}^2, \quad g = constant .$$

(2.10)

Because of the quartic term, the sign structure in (2.10) could be chosen differently from that in

(2.7) and still yield particle solutions. However, because of the new choice of signs, the particle energies are positive. Moreover, only discrete particle solutions of the form

$$\theta(r) \ (r \to \infty) \ = Ae^{-\alpha r}/_r \ , \qquad \alpha^2 = \sigma^2 - \omega^2 > 0 \qquad (2.11)$$

are possible. These discrete states are readily plotted in the θ, r plane with the lowest energy "ground state" having no nodes, the first excited state having one node, the second excited state having two nodes, etc.

3. Lepton Modelling

Clearly, the charged versions of such states are realized by replacing ∂_μ by D_μ and adding the free electromagnetic part, $-F_{\mu\nu}F^{\mu\nu}/8\pi$ to the Lagrangian of (2.10). This does indeed yield charged quantized energy particle states with positive mass. Numerical integration reveals that the particle sizes which these states represent are of the order of 10^{-13} cm, the classical electron radius. However, in recent years, experiments have shown that the upper limit to the size of an electron is of the order of 10^{-16} cm, and hence this approach is inadequate.

Rosen and I [3] also considered a Lagrangian which couples the scalar ψ of his original paper [9] and the scalar (now called ψ_1) of the quantized energy states [10,11] to electromagnetism:

$$L = -F_{\mu\nu}F^{\mu\nu}/_{8\pi} - D^\mu\psi\overline{D_\mu\psi} + \sigma^2\psi\overline{\psi} +$$

$$(\partial^\mu\psi_1)(\partial_\mu\overline{\psi}_1) - \sigma^2\psi_1\overline{\psi}_1 + \tfrac{1}{2}g\psi_1^2\overline{\psi}_1^2 - f\overline{\psi}\psi\overline{\psi}_1\psi_1 \ . \qquad (3.1)$$

We found that with the proper choice of parameters, it is possible with this Lagrangian to find particle solutions whose sizes are even well below the experimental upper limit. At that point, we endeavoured to realize our greater ambition, namely to model not only the electron but also the other clearly ponderable lepton masses, the muon and the tau. However, in adjusting the parameters to model the first excited state of ψ_1 as the muon, the particle sizes were found to be of order 10^{-15} cm, which is beyond the experimental upper limit. Also, regardless of the coupling parameters, it was not possible to adjust the second excited state to have the mass of the tau.

A successful model [3] was constructed by coupling the original [9] Rosen scalar, ψ, with two scalars ψ_1, ψ_2 of the quantized positive energy states [10,11] to electromagnetism:

$$L = -F_{\mu\nu}F^{\mu\nu}/_{8\pi} - (D^\mu\psi)(\overline{D_\mu\psi}) + \sigma^2\psi\overline{\psi} +$$

$$\sum (\partial^\mu\psi_i)(\partial_\mu\overline{\psi}_i) - \sigma^2\psi_i\overline{\psi}_i + \tfrac{1}{2}g_i\psi_i^2\overline{\psi}_i^2 - f_i\psi\overline{\psi}\psi_i\overline{\psi}_i \ . \qquad (3.2)$$

We found that we could choose parameters to fit the charge and the masses of e, μ and τ to within 0.06% of their experimentally determined values and the particle sizes which emerge are of the order of 10^{-16} cm.

An attractive element which is revealed in these models is that of confinement: as we consider successively higher excitation states, the energies increase because the energy gains from the uncoupled terms in the energy density consistently outweigh the energy losses from the coupled negative binding energy terms. However, because the latter increase in absolute value with successive excitations, there is never a dissolution of the particle. This is attractive for two reasons: firstly, if these fields could become decoupled, a particle with negative energy, stemming from the original Rosen scalar ψ, would exist independently, contrary to our experience. Secondly, this confinement mechanism could conceivably be the protoype of that which is responsible for confinement of quarks in hadrons.

With regard to the question of stability, T.D. Lee and his collaborators [16] (see also [6]) considered a class of scalar field soliton solutions which included those generated by the Lagrangian of (2.10). They found that the ground-state solution is classicially stable but quantum-mechanically metastable. The excited states were found to be unstable. Since this Lagrangian forms the base of our particle models generated by scalar fields coupled to electromagnetism, it would be expected that the Lee *et al* stability results would hold for these as well. Moreover, the stability of the ground state and instability of the higher states are what we seek in the modelling of elementary particles.

4. The Role of Gravity

We recall that Einstein showed that electromagnetism and gravitation were not sufficient to model elementary particles. With the weakness of gravitation relative to the other interactions in nature, there is a natural tendency to dismiss gravitation out of hand insofar as its capacity to influence elementary particle structure. However, regardless of how small a mass may be, given sufficient compactification, gravity can assume an important or even dominant role. This is readily seen from the Reissner-Nordström metric

$$ds^2 \; = \; (1 - \tfrac{2m}{r} + \tfrac{e^2}{r^2})dt^2 \; - \; (1 - \tfrac{2m}{r} + \tfrac{e^2}{r^2})^{-1} dr^2 - r^2 d\Omega^2,$$

$$(4.1)$$

$$d\Omega^2 \; = \; d\theta^2 \; + \; sin^2\theta d\phi^2$$

(where we now use geometrical units in which G=c=1 and all quantities are measured in centimeters). For m ~ r or e ~ r, gravity assumes vital proportions as deviations from Minkowski space become important.

The coefficient of 2/r, namely m - e^2/2r, in the g_{00} component of the metric in (4.1) is the energy which is localized within a sphere of coordinate radius r. The entire mass, m, is seen as r $\rightarrow \infty$ and this is understandable because of the electromagnetic field energy e^2/2r which is stored in the field from a radius r to infinity. An observer at radius r does not perceive this energy which lies at radii beyond him.

The most interesting situation to consider is when r is sufficiently small to render

$$m(r) \; \equiv \; m - e^2 \,/\, 2r \qquad\qquad (4.2)$$

negative. In this case, a neutral test particle at such an r value would be gravitationally repelled rather than attracted. This phenomenon [12], often referred to as "Reissner-Nordström repulsion", has hitherto been regarded as a curiosity of general relativity, with no manifestation in the physical world. However, when we consider the values of m and e for that most ubiquitous of particles, the electron, namely 6.77 x 10^{-56} cm and 1.38 x 10^{-34} cm respectively, we find that m(r) becomes negative as we pass below r ~ 10^{-13} cm, the classical electron radius. If the electron were indeed of this magnitude, this phenomenon would not be realized because the metric (4.1) would not be valid within the particle itself. However, experiments tell us that the electron is no larger than 10^{-16} cm, and hence there is a zone from at least 10^{-16} cm to 10^{-13} cm where Reissner-Nordström repulsion would actually exist.

What are the important consequences of these results? Firstly, they establish that anti-gravity, however limited, would appear to be a part of physics rather than merely science fiction. Secondly, with the existence of a zone of negative energy density, a key condition of the well-known Hawking-Penrose singularity theorems [13] is removed, as is the inevitability of the onset of singularities in nature.

There are various points that are to be discussed in this subject. To begin, it is important to emphasize that the phenomenon is very limited, confined to the zone up to the classical electron radius. We are not saying that there are negative masses. Indeed, as one examines m(r) for $r > 10^{-13}$ cm, one finds that 99% of the positive observed mass of the electron is already contained at $r \sim 10^{-11}$ cm. Although the positive m is what we perceive in experiments to the present day, it would be most interesting if an experiment could be designed to detect the negative energy inner core. Unfortunately, the Coulomb interaction overwhelms the gravitational interaction and hence charged probes would not be expected to be viable in this regard.

There is also the question of the applicability of classical general relativity in this domain. Since we are considering such small distance scales, in fact within a Compton wavelength, inevitably the question of quantum effects arises. However, according to references contained in [13], the manifold structure of space-time remains intact to scales of at least 10^{-15} cm. Also, the quantization of gravity is believed to be required with certainty only at the considerably more extreme Planck scale of 10^{-33} cm, and the phenomena under consideration are safely beyond this, in the zone where the manifold structure is secure. Thus, quantum considerations, insofar as gravitation is concerned, would not appear to be relevant.

Another issue is that of the appropriateness of the Reissner-Nordström metric (4.1) for the gravitational description of elementary particles such as the electron. The electron has both spin and magnetic moment and hence the Kerr-Newman [14] metric, which describes a spinning charge, would appear to be a more appropriate choice. This is particularly underlined by the fact that the Kerr-Newman metric reveals a gyromagnetic ratio which agrees with that of the electron.

Unfortunately, energy localization is more difficult to rationalize for these rotational states as opposed to the purely static states with spherical symmetry such as (4.1). However, recently, Virbhadra [15] has succeeded in demonstrating that the integration of the energy up to coordinate radius r within the Kerr-Newman field gives the same result to third order in the spin parameter, using two different pseudotensors, that of Tolman and that of Landau-Lifshitz. For the distances of concern in the present modelling of elementary particles, the Virbhadra expression is not adequate and S. Richardson and I are currently attempting to extract the exact integrated energy within r. With this expression in hand, we will be able to compare the onset of "Kerr-Newman repulsion" to that of Reissner-Nordström repulsion.

Gravitation theory can tell us more about elementary particles. We recall that with sufficient compactification, any body, however small its mass, can reveal significant or even dominant effects of gravity. We have considered a variety of non-singular field theoretic models of elementary particles in which the compactification is important. The results were:

a) whenever gravity played a significant role, the charge-to-mass ratio e/m was of order unity or less, $e/m \lesssim 1$.

b) whenever gravity played a dominant role, e/m approached 1.

However, for the known elementary particles, e/m is very far from unity. For example, $e/m \sim 10^{21}$ in the case of the electron. Thus, if the model results have general validity, and if these results

are not significantly altered by the inclusion of spin, we may conclude that gravity is not a significant factor in the structure of elementary particles, at least for those particles with which we are
familiar.

These results in turn suggest further information about the sizes of elementary particles. For consider the metric of (4.1).   With the known e and m of the electron, for example, the metric components show major deviations from unity when r is of the order 10^{-33} cm. Hence, gravitation becomes a dominant force at this, the Planck radius. However, if gravity is dominant only for particles with e/m ~ 1, as suggested by all of our models, then it would necessarily follow that the
known elementary particles are much larger than 10^{-33}cm. Thus, experiments provide an upper
limit of 10^{-16} cm and theory suggests a lower limit much greater than 10^{-33} cm. The challenge for
the future is one of narrowing these widely separated limits.

5. Spin and Magnetic Moment

To this point, the essential focus has been on models without spin. There is support for such an
approach at the level of classical modelling from the point of view that intrinsic spin may be seen as
a strictly quantum-mechanical attribute and that Dirac theory, which describes the interaction of the
electron with other particles and fields, successfully incorporates both spin and magnetic moment.
We will consider the electron with Dirac theory in the following section. However, before doing
so, it is of interest to consider an alternative approach, namely one of extending the classical field
theory in an attempt to encompass spin and magnetic moment [6] at the level of classical field theory.

We consider, for the sake of illustration, the simplest Lagrangian which yields charged positive
energy particle states:

$$L = - F_{\mu\nu}F^{\mu\nu}/8\pi + (D_\mu\psi)(\overline{D^\mu\psi}) - \sigma^2\psi\bar\psi + \tfrac{1}{2}g\psi^2\bar\psi^2 \tag{5.1}$$

This yields field equations

$$(D^\mu\psi)_{,\mu} - i\epsilon A_\mu D^\mu\psi + \sigma^2\psi - g\psi^2\bar\psi = 0 \tag{5.2}$$

$$F^{\mu\nu}{}_{,\nu} = - 4\pi J^\mu , \quad J^\mu = \tfrac{1}{2}i\epsilon(\bar\psi D^\mu\psi - \psi\overline{D^\mu\psi}) . $$

If we now consider solutions which are only axially symmetric, i.e. functions of r and θ, then it can
be shown that this requires at least an additional azimuthal component of the four-vector potential,
A_μ, i.e.,

$$A_\mu = (A_0,0,0,A_\phi) \equiv (\Phi,0,0,A) . \tag{5.3}$$

The most general form of the scalar field ψ is

$$\psi = \xi(r,\theta) \exp(i\omega t + is\phi), \quad s = 2\pi n, \quad n = integer . \tag{5.4}$$

The s has been introduced as shown to assure single-valuedness of ψ.

When the energy-momentum tensor components are constructed from the fields in the usual manner, we find that the x and y components of angular momentum vanish identically, as expected and

the z component is

$$L^z = \int (xT^{yo} - yT^{xo})dV = -\int T_{o\phi}dV =$$

$$= -\int dV[\frac{1}{4\pi}(\Phi_{,r} A_{,r} + \Phi_{,\theta} A_{,\theta}/r^2 +$$

$$+ \xi^2 (\omega - \epsilon\Phi)(s - \epsilon A)] \ .$$

(5.5)

The magnetic moment is constructed from the current which, again, yields identically vanishing x and y components and

$$M^z = \frac{1}{2}\int (xJ^y - yJ^x)dV = \frac{1}{2}\int J_\phi dV = \epsilon/2 \int (\epsilon A - s)\xi^2 dV \ .$$

(5.6)

Thus, provided solutions exist which render non-vanishing integrals in (5.5) and (5.6), there is scope for modelling both spin and magnetic moment in these classical field-theoretic structures. Unfortunately, it is far more difficult to find axially symmetric solutions of (5.2) because they are partial differential equations of considerable complexity. It would be interesting to see if solutions could be found and if so, whether the new degree of freedom could lead to a modelling of elementary particles with a simpler Lagrangian than that of (3.2).

Although the present emphasis is upon the leptons, in particular the electron, we conclude this section by noting that we have had a partial success in modelling the quarks [5]. We were not able to model the clearly massive particles of the three families: the leptons, e, μ, τ and the quarks, up, charm, top and down, strange, bottom with a single set of coupling constants. Two additional constants were required to fit all of the masses and charges (of which the latter two families are fractionally charged) of the three families of particles. However, as before in the case of the lepton modelling, the theory and the solutions predict particle sizes which are now found to lie in the experimentally acceptable range 10^{-18} to 10^{-17} cm.

6. Quantum Solitons

We now consider an alternative approach to the modelling of the electron, and possibly the other leptons as well. While the earlier discussion indicated that non-spherically symmetric solutions have the potential to build spin and magnetic moment into the particle, there is a more direct route. We recall that Dirac theory [17, 18] builds spin and magnetic moment via the spinorial structure of the theory and the theory is very successful in describing the fine structure of the energy spectrum of hydrogen. All that remains are the minute corrections from quantum electrodynamics to explain the Lamb shift. In hydrogen, the Dirac equation is solved by coupling the electron to the Coulomb field of the proton. In this treatment, the particles are treated as points. We have now embarked on the following generalization: we treat the electron as a Dirac soliton built by the coupling of its own electromagnetic field with its wave function, the Dirac spinor ψ. This is an extension of the Einstein-Rosen program to the quantum domain. In contrast to the problem of the hydrogen atom where the electromagnetic field is imposed via the Coulomb potential, we now have the Coulomb potential enter as the boundary condition. The coupled Dirac-Maxwell equations now determine the structure of both fields, subject to conditions of regularity.

The Dirac Lagrangian becomes locally gauge invariant by the inclusion of the four-vector potential with minimal coupling as

$$L = i\hbar c\bar{\psi}\gamma^\mu \partial_\mu \psi - mc^2\bar{\psi}\psi - \tfrac{1}{16\pi} F_{\mu\nu} F^{\mu\nu} - e\bar{\psi}\gamma^\mu \psi A_\mu \tag{6.1}$$

where γ^μ are expressed in terms of the perhaps more familiar α and β matrices as

$$\gamma^\circ = \beta , \quad \underset{\sim}{\gamma} = \beta\underset{\sim}{\alpha} , \tag{6.2}$$

These satisfy

$$\alpha_x^2 = \alpha_y^2 = \alpha_z^2 = \beta^2 = 1 \tag{6.3}$$

and $\alpha_x,\alpha_y,\alpha_z,\beta$ all anticommute in pairs. For a central field, the Dirac equation can be separated exactly in spherical coordinates [17,18]. We define

$$p_r \equiv (\underset{\sim}{r}\cdot\underset{\sim}{p} - i\hbar)/r , \quad \alpha_r \equiv \underset{\sim}{\alpha}\cdot\underset{\sim}{r}/r , \quad k \equiv \tfrac{\beta}{\hbar}(\underset{\sim}{\sigma'}\cdot\underset{\sim}{L} + \hbar) \tag{6.4}$$

where

$$\underset{\sim}{L} = \underset{\sim}{r} \times \underset{\sim}{p} , \quad \sigma' = \begin{pmatrix} \underset{\sim}{\sigma} & 0 \\ 0 & \underset{\sim}{\sigma} \end{pmatrix} \tag{6.5}$$

and $\underset{\sim}{\sigma}$ represents the three Pauli matrices. From (6.3-5), it follows that

$$\underset{\sim}{\alpha}\cdot\underset{\sim}{p} = \alpha_r p_r + i\hbar\alpha_r \beta k r^{-1} , \tag{6.6}$$

and the Dirac Hamiltonian H is

$$H = c\alpha_r p_r + i\hbar c\alpha_r \beta k r^{-1} + \beta mc^2 + V \tag{6.7}$$

whence k commutes with H and is a constant. k^2 has eigenvalues $(j + \tfrac{1}{2})^2$, $j = 1/2, 3/2, 5/2, \ldots$ and hence

$$k = \pm 1, \pm 2, \ldots \tag{6.8}$$

We choose a representation in which H and k are diagonal and represented by E and k (numbers). The angular and spin parts of ψ are fixed by ψ being an eigenfunction of k.

We now consider the electron as described by ψ generated by the coupling to its own electromagnetic field. Writing ψ with spinor components F_1/r, F_2/r, G_1/r, G_2/r with the F's and G's all functions of r, the field equations are expressed as

$$G_2' + \tfrac{k}{r}G_2 - \left[\tfrac{mc^2}{\hbar c}(1-\lambda) + \tfrac{V}{\hbar c}\right] F_1 = 0$$

$$G_1' + \tfrac{k}{r}G_1 + \left[\tfrac{mc^2}{\hbar c}(1-\lambda) + \tfrac{V}{\hbar c}\right] F_2 = 0$$

$$F_2' - \frac{k}{r} F_2 + \left[\frac{mc^2}{\hbar c} (1 + \lambda) - \frac{V}{\hbar c} \right] G_1 = 0 \tag{6.9}$$

$$F_1' - \frac{k}{r} F_1 + \left[\frac{mc^2}{\hbar c} (1 + \lambda) - \frac{V}{\hbar c} \right] G_2 = 0$$

$$\nabla^2 V = - \frac{4\pi e^2}{r^2} (F_1^2 + F_2^2 + G_1^2 + G_2^2)$$

$$\lambda \equiv E / mc^2, \qquad V = eA_0$$

with limit conditions.

$$L_{r \to o} \{F_1, F_2, G_1, G_2\} = \{0\}, \quad L_{r \to o} \{V'\} = 0,$$

$$L_{r \to \infty} \{V\} \sim \frac{C_1}{r} + C_2, \qquad C_1, C_2 \; constants \; .$$

Preliminary results have indicated an emerging Dirac soliton with characteristic spread of the wave function of the order 10^{-16} cm. However, the numerical integration of (6.9) is delicate and much work remains to be done. The non-linearity of this system of equations renders it much more diffi-cult than the corresponding linear problem of the hydrogen atom.

References

1. Einstein, A. and Rosen, N. (1935) Phys. Rev. **48**, 73.
2. Quigg, C. (1983) "Gauge Theories of the Strong, Weak and Electromagnetic Interactions" (Benjamin, Reading, Mass).
3. Cooperstock, F.I. and Rosen, N. (1989) Int. J. of Theor. Phys. **28**, 423.
4. Bonnor, W.B. and Cooperstock, F.I. (1989) Phys. Lett. A. **139**, 442.
5. Sharman, P.H. and Cooperstock, F.I. (1990) Can. J. Phys. **68**, 531.
6. Cooperstock, F.I. (1989) Foundations of Physics Lett. **2**, 553.
7. Mie, G. (1912) Ann. der Physik **37**, 511; (1913), **40**, 1.
8. Born, M. and Infeld, L. (1934) Proc. Roy. Soc. (London) **A144**, 425; (1934) **A147**, 522; (1935) **A150**, 141. See also Hoffman, B. and Infeld, L. (1937) Phys. Rev. **51**, 765.
9. Rosen, N. (1939) Phys. Rev., **55**, 94.
10. Finkelstein, R., LeLevier, R. and Ruderman, M. (1951) Phys. Rev. **83**, 326.
11. Rosen, N. and Rosenstock, H.B. (1952) Phys. Rev. **85**, 257.
12. Papapetrou, A. (1974) Lectures on General Relativity, D. Reidel, Dordecht, Holland.
13. Hawkins, S.W. and Ellis, G.F.R. (1973) The Large-Scale Structure of Space-Time, Cambridge University Press, Cambridge.
14. Newman, E.T. *et al* (1965) J. Math. Phys. **6**, 918.
15. Virbhadra, K.S. (1990) Phys. Rev. D. **41**, 1086.
16. Freidberg, R., Lee, T.D. and Sirlin A. (1976) Phys. Rev. **13**, 2739.
17. Dirac, P.A.M. (1947) The Principles of Quantum Mechanics, 3rd ed., Oxford, New York.
18. Schiff, L.I. (1955) Quantum Mechanics, 2nd ed., McGraw-Hill, New York.

HOW TO IDENTIFY AN ELECTRON IN AN EXTERNAL FIELD

A. Z. CAPRI
Theotretical Physics Institute
Department of Physics
University of Alberta
Edmonton, AB.
T6G 2J1
Canada

ABSTRACT. We show that a construction previously obtained by us leads to a Lorentz invariant vacuum even in the presence of an external field. This allows an unambiguous definition of an instantaneous one-particle state.

1. INTRODUCTION

The title may suggest that the problem I want to discuss is simply an academic problem. After all, we all know what an electron is. It is a particle with the following properties:

$$\text{mass} \quad m = 9.109\ 389\ 7(54) \times 10^{-31} \text{ kg}$$

$$\text{charge} \quad e = 1.602\ 177\ 33(49) \times 10^{-19} \text{ C}$$

$$\text{spin} \quad s = 1/2\ \hbar.$$

So why is there such a vast literature on this subject? (A good list of references to recent work is to be foumd in the book by Fulling as well as in the somewhat older book by Birrel and Davies reference[1].) What is the problem about identifying an electron in a given external field? The answer is that there are many problems. For example, if the external field varies in time then there is a non-zero probability for pair creation and one can no longer be sure that the state one is looking at is a single-particle state. Furthermore, two observers not at rest relative to each other will see different external fields. If one of these observers sees an electron, does it follow that the other observers also sees the same thing, namely one electron? Or at an even lower level, if one of them sees a vacuum (zero-particle) state, what does the other observer see?

This problem is not simply theoretical, although the experimentalist can frequently avoid these difficulties by having the external field vary so slowly in time that for practical purposes it is constant. Nevertheless if two experimentalists zip past each other at some considerable speed they encounter the same difficulty. Still, even if the problem is not of immediacy to experimentalists it certainly involves many important questions of principle, especially if the external field is a gravitational field. Here a solution of this pr5oblem is viewed as a first step toward quantizing gravity. What i am trying to point out is that the vacuum for an external field problem is a complicated entity. In fact, there is at present no concensus among theorists on how to define it.

The reason for all these difficulties is that an external field destroys the fundamental kinematic symmetries all relativistic theories are supposed to have, namely Lorentz and

D. Hestenes and A. Weingartshofer (eds.), The Electron, 183–189.

translation invariance of the field equations. As already stated, the external field looks different in every Lorentz frame.

In this paper I shall to discuss only smooth (differentiable) external fields that vanish rapidly for very large (positive or negative) distances or times. This guarantees that asymptotically we have free fields and avoids the necessity for technical discussions that, although important, are not germane to the physical problem of interest.

2. WHAT IS THE PROBLEM WITH THE VACUUM?

In a quantum field theory with time-translation invariance it is always possible to separate the field into positive and negative frequency parts in a straightforward causal manner; the future does not influence the past. If the theory also has Lorentz invariance, this decomposition is also Lorentz invariant; positive frequency is positive frequency in every Lorentz frame. In these cases one writes:

$$\phi(x) = \phi^{(+)}(x) + \phi^{(-)}(x) \tag{2.1}$$

Here $\phi^{(\pm)}(x)$ are defined as follows:

$$\phi^{(\pm)}(x) = \frac{1}{2\pi} \int_{-\infty}^{\infty} \theta(\pm\omega) \, d\omega \int_{-\infty}^{\infty} dx^0 \; e^{i\omega x^0} \phi(x^0) \tag{2.2}$$

The vacuum state $|0\rangle$ is then defined as the state annihilated by the positive frequency part of the field, that is:

$$\phi^{(+)} |0\rangle = 0 \tag{2.3}$$

To illustrate what can go wrong when we have a time-dependent external field, we consider a Dirac field interacting with a time-dependent electromagnetic field. The field (Heisenberg) equation now reads:

$$[-i\gamma^\mu(\partial_\mu - ieA_\mu) + m] \, \phi(x) = 0 . \tag{2.4}$$

This equation may be rewritten as an integral equation which incorporates the boundary condition that at very early times the field is a free field which we call $\phi_{in}(x)$.

$$\phi(x) = \phi_{in}(x) + e\int S_R(x-y) \, \gamma^\mu A_\mu(y) \, \phi(y) \, d^4y \tag{2.5}$$

Here S_R is the retarded Green's function for the free Dirac equation. That is,

$$[-i\gamma^\mu\partial_\mu + m] \, S_R(x - y) = \delta(x-y) \tag{2.6}$$

and

$$S_R(x - y) = 0 \quad \text{if} \quad x^0 < y^0 . \tag{2.7}$$

The obvious thing to do now seems to try and separate $\phi(x)$ into positive and negative frequency parts as in equation (2.2). This yields:

$$\phi^{(+)}(x) = \phi^{(+)}{}_{in}(x) + e\int S^{(+)}{}_R(x-y)\, \gamma^\mu A_\mu(y)\, \phi(y)\, d^4y \tag{2.8}$$

One of the major difficulties is now displayed in this equation. The positive frequency part $S^{(+)}{}_R$ of the retarded Green's function S_R no longer satisfies equation (2.7). Thus, even if the the external vector potential $A_\mu(x)$ vanishes for $x^0 < T$,where T is some finite time, we find that $\phi^{(+)}(x)$ depends on $A_\mu(x)$ for all times, including times $x^0 < T$. Thus, the present depends not only on the past but also on the future.

The problem just described was recognized quite some time ago. In fact, back in 1972 G. Labonte' and I found a way to evade this difficulty [2]. What we did not realize then was just how good this method is. I shall first review our construction and then go on to show that it leads to a definition of an instantaneous vacuum that is seen as a vacuum state by all observers connected by a Lorentz transformation to the frame in which the vacuum was originally defined. Thus, although the theory no longer has space-time translation invariance nor Lorentz invariance, we can nevertheless obtain a Lorentz invariant, instantaneous vacuum state. Furthermore, the instantaneous one-particle states will also be seen as one-particle states by all Lorentz-equivalent observers. Thus, there is no ambiguity as to what constitutes an instantaneous state of one electron. The construction used reduces to what would be an obvious procedure for a very slowly varying external field. A more primitive version of these results was presented by us in reference [3].

3. THE VACUUM

We now list certain minimal properties that we believe any reasonable vacuum state should possess.

The instantaneous vacuum state at time τ say, $|0_\tau\rangle$ should satisfy:

a) $|0_\tau\rangle$ is a state of zero particles, where particles are defined as the quanta of the instantaneous Heisenberg field.

b) for $\tau \to \pm\infty$, the vacuum $|0_\tau\rangle \to |0_{out,\ in}\rangle$.

c) The vacuum $|0_\tau\rangle$ may not depend on times $x^0 > \tau$. It may only depend on times $x^0 \le \tau$. This is simply a requirement of causality.

We now show how to accomplish this for a Dirac field interacting with an external electromagnetic field. The Heisenberg equation for this field is:

$$[-i\gamma^\mu(\partial_\mu - ieA_\mu(x^0,\mathbf{x})) + m\,]\,\phi(x) = 0 \tag{3.1}$$

To begin, we define an auxiliary field $\phi_\tau(x)$ by the following equation and initial condition:

$$[-i\gamma^\mu(\partial_\mu - ieA_\mu(\tau,\mathbf{x})) + m\,]\,\phi_\tau(x) = 0 . \tag{3.2}$$

with initial condition:

$$\phi_\tau(x)\,|_{x^0=\tau} = \phi(x)\,|_{x^0=\tau} \tag{3.3}$$

This condition guarantees that at time $t = \tau$ the auxiliary field $\phi_\tau(x)$ coincides with the Heisenberg field $\phi(x)$. The field equation for $\phi_\tau(x)$ has time translation invariance and can be decomposed into positive and negative frequency parts in a causal manner according to equation (2.2).

$$\phi_\tau(x) = \phi^{(+)}_\tau(x) + \phi^{(-)}_\tau(x) \tag{3.4}$$

The reason for introducing the auxiliary field $\phi_\tau(x)$ is to introduce time-translation invariance which is essential if causality (condition c) is to be achieved. A direct decomposition of the Heisenberg field (whose field equation is not time translation invariant) leads, as we have already seen, to acausal behaviour.

It is now possible to define the vacuum $|0_\tau\rangle$ as usual as the state annihilated by the positive frequency part of the auxiliary field:

$$\phi_\tau^{(+)} (x) |0_\tau\rangle = 0 \tag{3.5}$$

This is the vacuum state defined by Labonte' and me. It is the state of interest to us. Incidentally, the construction makes it obvious that at time $t = \tau$ the Heisenberg field and the auxiliary field coincide. They also both satisfy the same equation at this instant and, as we have shown, have the same instantaneous hamiltonian. So at $t = \tau$ the auxiliary field can replace the Heisenberg field. What I want to show next is that the vacuum state defined above is Lorentz invariant in spite of all the non-covariant things in the theory. To discuss this Lorentz invariance requires a slight change of notation. But first I have to explain what exactly is meant by Lorentz invariance of a theory with an external field.

By Lorentz invariance of the Heisenberg equation (3.1) we mean that this equation remains form invariant under a proper orthochronous Lorentz transformation if the external field $A_\mu(x)$ is replaced by:

$$A'_\mu(x') = \Lambda_\mu{}^\nu A_\nu(x) \tag{3.6}$$

where

$$x'_\mu = \Lambda_\mu{}^\nu x_\nu \tag{3.7}$$

This is easily verified to be the case if $\phi(x)$ is replaced by $\phi'(x')$ where

$$\phi'(x') = S(\Lambda)\phi(x) . \tag{3.8}$$

Here $S(\Lambda)$ is the $(1/2,0) \oplus (0,1/2)$ (non-unitary) representation of the Lorentz transformation Λ and we furthermore have,

$$S(\Lambda)\gamma^\mu S^{-1}(\Lambda) = \Lambda^\mu{}_\nu\gamma^\nu . \tag{3.9}$$

The notation we use here is the same as in the book by Bjorken and Drell [4].

To examine the transformation properties of the auxiliary field $\phi_\tau(x)$ we now introduce some more notation. Let n be a time-like unit vector

$$n^2 = 1 \qquad (3.10)$$

We next define a space-like surface σ by:

$$\sigma: \quad n.x = \tau \qquad (3.11)$$

If the points x are restricted to lie on the spacelike surface σ we write x/σ. Thus, if $f(x)$ is some field and its argument (field point) is restricted, to lie on σ, we write $f(x/\sigma)$. This simply means that $f(x^0,\mathbf{x})$ is replaced by $f((\tau+\mathbf{n.x})/n^0,\mathbf{x})$

Under the Lorentz transformation (3.7), the surface σ is transformed into $\Lambda\sigma$ where

$$\Lambda\sigma: \quad n'.x' = \tau. \qquad (3.12)$$

The auxiliary field is now labelled with the surface σ rather than τ. Thus, the field equations for the auxiliary field now read:

$$[-i\gamma^\mu(\partial_\mu - ieA_\mu(x/\sigma)) + m]\,\phi_\sigma(x) = 0. \qquad (3.13)$$

with initial data

$$\phi_\sigma(x/\sigma) = \phi(x/\sigma). \qquad (3.14)$$

We are finally ready to prove the Lorentz invariance of the vacuum corresponding to this instantaneous field.

4. PROOF OF LORENTZ INVARIANCE

Consider the c-number corresponding to the quantized theory defined above. The extension back to a q-number (second quantized)theory is easy and will be indicated at the end of this section.

The Heisenberg field ϕ has associated with it a self-adjoint hamiltonian $h(A)$. The corresponding hilbert space is $L^2(R^3)$ with the usual Dirac scalar product. Associated with $h(A)$ is a Møller operator $U(A)$ which is unitary since the electromagnetic field A dies out rapidly in time. This allows us to write the Heisenberg field ϕ as

$$\phi(x) = U(A)\,\phi_{in}(x) \qquad (4.1)$$

Here we have already introduced the incoming asymptotic (free) field ϕ_{in}. We also define the transformed electromagnetic field $^\Lambda A$ by

$$^\Lambda A^\mu(x) = \Lambda^\mu{}_\nu A^\nu(\Lambda^{-1}x) \qquad (4.2)$$

and a corresponding hamiltonian $^\Lambda h$ and Møller operator $U(^\Lambda A)$. These correspond to the respective quantities in the Lorentz transformed frame. Furthermore we have the transformation properties of the free asymptotic fields:

$$(^\Lambda\phi_{in})(x) = S(\Lambda)\ \phi_{in}\ (\Lambda^{-1}x) = (V(\Lambda)\phi_{in})(x)\ . \tag{4.3}$$

Now because the transformed Møller operator is again unitary we have:

$$^\Lambda\phi(x) = U(^\Lambda A)\ (^\Lambda\phi_{in})(x) = (U(^\Lambda A)\ V(\Lambda)\phi_{in})(x) \tag{4.4}$$

Thus, we have unitary evolution from the same asymptotic field ϕ_{in} to the Lorentz transformed Heisenberg field $^\Lambda\phi$.

We next define the self-adjoint hamiltonian $h_\sigma(A)$ obtained from $h(A)$ by restricting the electromagnetic field A to lie on the surface σ. The auxiliary field ϕ_σ evolves with the dynamics specified by this hamiltonian by starting from the surface σ with the initial data specified by equation (3.14).

$$\phi_\sigma(x/\sigma) = \phi(x/\sigma)\ . \tag{4.5}$$

Corresponding to this evolution we have again a unitary operator $W_\sigma(A)$ such that the field ϕ_σ evolves from the surface σ according to this operator:

$$\phi_\sigma(x) = W_\sigma(A)\ \phi(x/\sigma) = (W_\sigma(A)\ U_\sigma(A)\ \phi_{in})\ (x) \tag{4.6}$$

Here we have also introduced the Møller operator $U_\sigma(A)$ that evolves the Heisenberg field $\phi(x)$ from the asymptotic incoming field ϕ_{in} to the surface σ. It is important to notice that we again have unitary evolution from ϕ_{in} to ϕ_σ.

Finally, we define the hamiltonian $^\Lambda h_\sigma$ for the auxiliary field corresponding to the Lorentz transformed electromagnetic field. Associated with this we have a unitary operator $^\Lambda W_{\Lambda\sigma}\ (^\Lambda A)$ such that :

$$^\Lambda\phi_\sigma(x) = {}^\Lambda W_{\Lambda\sigma}\ (^\Lambda A)\phi(x/\Lambda\sigma)$$

$$= ({}^\Lambda W_{\Lambda\sigma}\ (^\Lambda A)\ U_{\Lambda\sigma}(^\Lambda A)\ {}^\Lambda\phi_{in})(x)$$

$$= ({}^\Lambda W_{\Lambda\sigma}\ (^\Lambda A)\ U_{\Lambda\sigma}(^\Lambda A)\ V(\Lambda)\ \phi_{in})(x). \tag{4.7}$$

From equation (4.6) we find:

$$\phi_{in}\ (x) = (U^\dagger{}_\sigma(A)\ W^\dagger{}_\sigma(\Lambda)\ \phi_\sigma)\ (x) \tag{4.8}$$

Thus,

$$^\Lambda\phi_\sigma(x) = ({}^\Lambda W_{\Lambda\sigma}\ (^\Lambda A)\ U_{\Lambda\sigma}(^\Lambda A)\ V(\Lambda)\ U^\dagger{}_\sigma(A)\ W^\dagger{}_\sigma(\Lambda)\ \phi_\sigma)\ (x) \tag{4.9}$$

This shows that under a Lorentz transformation the fields ϕ_σ and $^\Lambda\phi_\sigma$ are connected by a unitary transformation. It only remains to translate all of this into a q-number theory. But this is easy.

Consider the quantized field φ smeared with a test function f. We write this as $\varphi(f)$. A unitary transformation of the test function f say Uf induces a unitary transformation V on the field φ according to:

$$V\varphi(f)V^\dagger = \varphi(Uf) .$$

(4.10)

Thus, we have unitary transformations between the auxiliary field as defined in one Lorentz frame and a second Lorentz frame. This means that a vacuum in one Lorentz frame is also a vacuum in any other Lorentz frame and a one-particle state in one frame is also a one-particle state in all other Lorentz frames. This is as much as one can expect, although it is amazing that one can get even this much.

A natural question that arises is, "What has happened to all the interesting phenomena that the relativists have found, such as the background of thermal particles in an accelerating frame?" Preliminary calculations with a model consisting of a real scalar field coupled to a c-number source indicate that these effects are still there, but arise in a totally different manner. In that model it is the fact that the source for the auxiliary field is static that leads to interesting features for the auxiliary field ϕ_σ It is shifted from the Heisenberg field by a time-dependent c-number as well as a time-independent c-number. This second part acts like a zero energy mode and can be taken with the annihilation (positive frequency) part of the field or with the creation (negative frequency) part. Another way to handle this time-independent term is to introduce a representation of the auxiliary field ϕ_σ in which it has a non-zero vacuum expectation value both in the asymptotic vacuum $|0_{in}>$ and the instantaneous vacuum $|0_\sigma>$. In this case the auxiliary field ϕ_σ would no longer be unitarily equivalent to the asymptotic field ϕ_{in} . These various possibilities remain to be investigated.

5. REFERENCES

[1] Birrel N. D. and Davies P. C. W. (1982) *Quantum Fields in Curved Space*, Cambridge University Press, Cambridge/ Fulling S. A. (1989) *Aspects of Quantum Field Theory in Curved Space-Time*, Cambridge University Press, Cambridge.

[2] Labonte' G. and Capri A. Z. (1972) 'Vacuum for external-field problems', Il Nuovo Cimento, 10B, 583 - 591.

[3] Capri A. Z. , Kobayashi M. and Takahashi Y. (1990) 'Lorentz invariance of the vacuum for external field problems', Class. Quantum Grav., 7, 933 - 938.

[4] Bjorken J. D. and Drell S. D. (1964) *Relativistic Quantum Mechanics*, McGraw-Hill, New York.

THE ELECTRON AND THE DRESSED MOLECULE

A.D. BANDRAUK
Département de chimie
Faculté des sciences
Université de Sherbrooke
Sherbrooke, Québec, J1K 2R1, Canada

ABSTRACT. Interactions of molecules with intense laser fields produce highly nonlinear effects mediated by the bonding electrons. Recent experiments in nonlinear photoelectron spectroscopy of diatomics such as H_2 have shown the presence of new electronic states and thus new boundstates of the molecular ions, which are manifestations of the dressing of the diatomic ion by the large number of photons present at high intensities as predicted by theory. In this chapter we will expose the theory of dressed molecules and present numerical calculations by coupled equations of these new phenomena which occur whenever intense laser fields interact with molecules.

1. INTRODUCTION

Electrons are the "gluons" in molecules which prevent nuclei from exploding apart. Thus electrons provide the necessary bonding to create stable multinuclear species called molecules. Great progress in our understanding of the electronic structure of molecules has come from the introduction of the molecular orbital concept by Mulliken in the 1950's and 60's. Thus as in atoms, electrons in molecules occupy orbitals which envelope the whole nuclear space, creating stable molecular species if the molecular orbitals are bonding and unstable species if these are antibonding [1].

The bonding characteristics of molecular orbitals can be inferred from photoelectron spectroscopy [2]. Recent improvements in this method has even led to determination of the electron momentum distribution in these orbitals [3]. A concomitant structure which appears often in the photo electron spectrum is the vibronic structure of the remaining molecular ion after photoionization. This structure which is created by the coupling of the ionized electron to the core of the ion reveals the vibrational structure of the molecular ion and the

D. Hestenes and A. Weingartshofer (eds.), The Electron, 191–217.

degree of coupling between both electron and ion [4]. We conclude therefore that the electron serves as an essential probe in understanding molecular structure.

The advent of intense lasers has revealed some singular aspects of the nonlinear behaviour of atoms in intense laser fields [5-7]. Recently, similar nonlinear phenomena (e.g., above threshold ionization, ATI), have been observed in molecules [8-11]. In particular, experiments on the nonlinear photoionization of H_2 have revealed that the vibronic structure of the molecular ion is considerably altered with respect to the free ion [9-10]. It is the goal of this chapter to examine a theoretical model, the <u>dressed</u> <u>molecule</u>, which can help us understand nonlinear molecule-laser interactions, which interactions we reiterate are induced by multiphoton transitions (real and virtual) of the electrons in the molecule.

One can classify the regime of coupling between the laser and the molecular system according to the nature of the process they induce. The first regime is that corresponding to low-intensity lasers which couple weakly with the system. As a result, the excitation processes are well described by leading order perturbation theory, such as Fermi's Golden rule. For molecules, this leads to a Franck-Condon picture of electronic (radiative) transitions [12]. At intermediate to high intensities, one encounters a domain in which multiphoton processes begin to take effect. This is signalled by nonlinear behaviour of the transition probabilities as a function of intensity. In particular two or more states may be strongly coupled together as a result of being near resonant. An example of this is the Rabi oscillations of a two level atom [5-6] or an n-level molecule [13]. Another example which this chapter discusses in detail, is the nonlinear interaction between rovibrational manifolds of different electronic molecular states induced by intense laser fields. Judging from atomic experience, [14], one can establish the upper limit of the intensity I of this regime at 10^{12} W/cm^2 (terawatt/cm^2), since for $I \geq 10^{13}$ W/cm^2, ionization rates exceed dissociation rates for many molecules. Finally one has the very high intensity limit available with current superintense lasers ($I > 10^{13}$ W/cm^2), where Rabi frequencies ($\omega_R = d\mathcal{E}/\hbar$, d = transition moment, $\mathcal{E}$ = electric field) are comparable to the laser frequency, and highly nonresonant transitions compete with resonant processes. Thus in the case of the nonlinear photoelectron spectroscopy of H_2 mentioned above [8-11], the photoionized electron continues to absorb photons creating ATI peaks with a vibronic structure which has no relation to the vibrational structure of free H_2^+. We will show in the present chapter, that the H_2^+ core is dressed by the intense field and that the structure of the ATI peaks reflects the nonlinear interaction of the ion core with the laser while at the same time remaining coupled to the dressed photoionized electron.

In particular we will show that intense lasers can create dressed <u>adiabatic</u> states which are degenerate with the excited field-molecule <u>diabatic</u> states, as a result of a laser induced avoided crossing between the ground bonding state $(^2\Sigma_g^+)$ of H_2^+ and the dissociative anti-bonding state $(1\Sigma_u^+)$ of that ion. From a semiclassical analysis of the problem [15-16], one can predict a <u>stabilization</u> of new dressed molecular states. This stabilization stems from the molecule resonating between the two bound states, adiabatic (unperturbed) and adiabatic (perturbed) of the molecule. Such stabilization of electronic states at high intensities is currently being discussed extensively in the atomic case [17-18]. In the molecular case, the nuclear degrees of freedom offer the possibility of creating stable new electronic states by the laser induced coherent superposition of bonding and antibonding states of the free molecule. In the following, we will show the realization of this effect within a more realistic close-coupling calculation involving many electronic-field states, as befits such a highly nonperturbative problem. We also point out that at high intensities, where Rabi frequencies exceed rotational spacings, laser-induced orientational effects or alignment are expected to predominate in the angular distribution of photodissociation fragments [19].

2. ELECTROMAGNETIC FIELD-PHOTON STATES.

In describing the field-molecule states of a radiation-molecule system, one encounters the dichotomy of a classical description in terms of the classical electric field $\mathcal{E}(x,t)$ and the n photon particle quantum states $|n\rangle$. Maxwell's equations allow one to express free electric field plane waves of frequency ω and wave vector k, (one dimension) as,

$$\mathcal{E}(x,t) = \mathcal{E}_o \sin (kx - \omega t) \quad , \tag{1}$$

with total energy in a region of dimension L,

$$E = \int_o^L \mathcal{E}^2 \, dx = \frac{1}{2} (\omega^2 Q^2 + P^2) \quad , \tag{2}$$

where P and Q are canonical variables satisfying the relation [20-22],

$$P = \dot{Q} = (L)^{1/2} \mathcal{E}_o \sin \omega t \quad . \tag{3}$$

Equation (2) illustrates the harmonic oscillator analogy for the field. Through a further definition of amplitudes a(t) and the complex conjugate a(t)*,

$$a(t) = (2\hbar\omega)^{-1/2} (\omega Q + iP) = ae^{-i\omega t} \quad , \tag{4}$$

one can formulate the real field as

$$\mathcal{E} = \mathcal{E}^{(-)} + \mathcal{E}^{(+)} \quad , \tag{5}$$

$$\mathcal{E}^{(+)} = (\mathcal{E}^{(-)})^* = - i \left(\frac{\hbar\omega}{2L} \right)^{1/2} a(t) \, e^{ikx} \quad . \tag{6}$$

This renders the classical Hamiltonian (2) of the electromagnetic field in the form

$$E = \hbar\omega a^* a \tag{7}$$

Quantization of the classical Hamiltonian transforms the positive $(\mathcal{E}^{(+)})$ and negative $(\mathcal{E}^{(-)})$ parts of the classical field defined in terms of the amplitudes a and a^* into quantum field operators $\hat{a}$ and $\hat{a}^*$. Thus for a cavity of length L we obtain the <u>electric field operators</u>

$$\hat{\mathcal{E}}^{(+)} = - i \left(\frac{\hbar\omega}{2} \right)^{1/2} \hat{a}(t) \, e^{ikx} \quad , \tag{8}$$

$$\hat{\mathcal{E}}^{(-)} = i \left(\frac{\hbar\omega}{2} \right)^{1/2} \hat{a}^+(t) \, e^{-ikx} \quad , \tag{9}$$

The operator $\hat{\mathcal{E}}^{(-)}$ <u>creates</u> photons of frequency ω and wave number k while the operator $\hat{\mathcal{E}}^{(+)}$ <u>annihilates</u> photons. Taken together as an ordered product of operators they yield the quantum field Hamiltonian operator

$$\hat{H} = 2 \, \hat{\mathcal{E}}^{(-)} \, \hat{\mathcal{E}}^{(+)} = \hbar\omega \hat{a}^+ \hat{a} \quad , \tag{10}$$

where the expressions

$$\hat{a}^+|n\rangle = (n+1)^{1/2} \, |n+1\rangle \quad , \quad \hat{a}|n\rangle = n^{1/2} \, |n-1\rangle \quad , \tag{11}$$

define the effect of these operators in the quantum photon number states. The expectation value of the Hamiltonian operator (10) in the eigenstate $|n\rangle$ of photon number $|n\rangle$ is

$$\langle\hat{H}\rangle = \langle n|\hbar\omega \hat{a}^+ \hat{a}|n\rangle = n\hbar\omega \quad . \tag{12}$$

Thus the average energy of the field equals the energy corresponding to an exact number of photons for that state.

Since either part $\mathcal{E}^{(-)}$ or $\mathcal{E}^{(+)}$ of the classical field $\mathcal{E}(x,t)$ is a solution of Maxwell's equations, one could expect that the operators (8-9) have eigenvalues corresponding to some aspect of the classical field, especially in the limit of large photon numbers. Thus rather than quantize the field energy, which leads to number states and eigenvalues of the energy operator H, one can also quantize the field itself and search for eigenvalues and states of the field operators $\hat{\mathcal{E}}^{(\pm)}$.

Thus the quantization of the field which can be expressed by,

$$\hat{\mathcal{E}}^+|\alpha> = \mathcal{E}^{(+)}|\alpha> \quad , \quad <\alpha|\hat{\mathcal{E}}^{(-)} = <\alpha|\mathcal{E}^{(-)} \quad , \qquad (13)$$

produces field eigenvalues $\mathcal{E}^{(\pm)}$ provided we can find the eigenstates of $\hat{a}$, and $\hat{a}^+$, i.e.,

$$\hat{a}|\alpha> = \alpha|\alpha> \quad , \quad <\alpha|\hat{a}^+ = <\alpha|\alpha^* \quad . \qquad (14)$$

For the harmonic oscillator, these eigenstates are well known and are called the <u>coherent</u> states of oscillator defined by, [21], [23],

$$|\alpha> = \sum_n \frac{|\alpha|^n \; e^{-(|\alpha|^2)/2}}{(n!)^{1/2}} |n> \quad . \qquad (15)$$

Combining equations (8), (9), (13) and (14) gives the real classical amplitude as

$$\mathcal{E} = \mathcal{E}^{(+)} + \mathcal{E}^{(-)} = -i \left(\frac{\hbar\omega}{2}\right)^{1/2} (\alpha e^{ikx} - \alpha^* e^{-ikx})$$

$$= (2\hbar\omega)^{1/2} |\alpha| \sin(kx - \omega t) \quad . \qquad (16)$$

Comparing this amplitude with that obtained from Maxwell's equation (1), shows that the classical field $\mathcal{E}$ has an amplitude related to the eigenvalue of the coherent quantum oscillator state $|\alpha>$, equation (15). This is the coherent wave of controlled phase capable of being produced by an ideal laser, and for this reason is called the coherent state of light.

We can now evaluate the energy of this state obtaining from (10) and (13),

$$E = <\alpha|\hat{\mathcal{E}}^{(-)} \; \mathcal{E}^{(+)}|\alpha> = \hbar\omega|\alpha|^2 = \hbar\omega<N> \quad . \qquad (17)$$

Thus from equation (15) we have the result that a coherent classical wave is expressible as a superposition of an infinite number of photon states $|n>$. The wave has an average number of photons $<N>$ defined by (17) and the probability p_n of finding the nth photon state is given by the Poisson distribution,

$$P_n = |<n|\alpha>^2 = \frac{|\alpha|^{2n}}{n!} e^{-|\alpha|^2} \quad . \qquad (18)$$

Thus in the ensuing discussion we shall use the number state represen-

tation of the electric field which interacts with molecules. For high
intensities, i.e. large photon numbers, the Poisson distribution (18)
peaks at the average photon number <N> defined by (17) so that this
will be taken as the most probable photon state for which absorptions
and emissions of photons will occur. This photon number representation of the electric field will enable us to define field molecular
states for the proper description of the dressed states of the molecu-
le-radiation system.

3. THEORETICAL METHOD - COUPLED EQUATIONS

We shall elaborate in the present section on the coupled equations in the field-molecular representation which leads to a proper
and accurate description of dressed molecular states at high intensities [24-25].

For the present, let us consider the general case of photodissociation of a simple diatomic. The Hamiltonian for the system may be
partitioned into four components, namely,

$$\hat{H} = \hat{H}_m + \hat{H}_{na} + \hat{H}_f + \hat{H}_{mf} \quad , \qquad (19)$$

in which the molecular interactions are denoted by $\hat{H}_m$, the Hamiltonian
of the Born-Oppenheimer approximation, and $\hat{H}_{na}$, the nonadiabatic perturbation. The quantized radiation fields were defined in section 2
and are represented by the term

$$\hat{H}_f = \sum_k \omega_k \hat{a}_k^+ \hat{a}_k \quad , \qquad (20)$$

in which the summation is over the frequencies ω_k and wave vector k of
the modes. The creation and annihilation operators $(\hat{a}_k^+, \hat{a}_k)$ have been
defined in equation (11). Lastly the radiative interaction between
the molecules and the fields is denoted by the term H_{mf} and takes the
form in the quantized field representation and dipole approximation
[5-6], [26],

$$\hat{H}_{mf} = \vec{d} \cdot \vec{\mathcal{E}} = \sum_k \left(\frac{2\pi\omega}{V} k \right)^{1/2} \vec{e}^{(k)} \cdot \vec{d}^{(k)} (\hat{a}_k + \hat{a}_k^+) \quad , \qquad (21)$$

in which $\vec{e}$ denotes the polarization vector, V is the volume of the
cavity, and $d^{(k)}$ designates the dipole moment of the molecule for the
k^{th} transition. The effect of nonadiabaticity can be treated simultaneously and can play an important role as in the multiphoton infrared dissociation of ionic molecules [27].

A measure of the various interstate couplings involved will help
in understanding the dynamics. Radiative couplings can usually be

expressed as a Rabi frequency

$$\omega_R \ (cm^{-1}) = \vec{d} \cdot \vec{\mathcal{E}}/\hbar = 1.17 \times 10^{-3} \ d \ (a.u.) \ I \ [W/cm^2]^{1/2} \quad , \quad (22)$$

$$I = \frac{8\pi}{c} \ \mathcal{E}_o^2 \quad , \quad\quad\quad\quad\quad (23)$$

where a.u. denotes atomic units, c is the velocity of light, the intensity I is reported in watts/cm^2 and $\mathcal{E}_o$ is the maximum field amplitude. For a dipole transition moment d ~ 1 a.u., and an intensity I = 10^{11} W/cm^2, one obtains a radiative interaction of ~ 400 cm^{-1}. This is to be compared with the nonradiative (nonadiabatic) interaction between the covalent and ionic state of LiF, $\langle\psi(LiF)|H_{el}|\psi(Li^+F^-)\rangle \simeq$ 600 cm^{-1}, as an example [26] whereas the vibrational frequency of LiF is $\omega(LiF) = 300$ cm^{-1}. It is clear that at high intensities (I > 10^{10} W/cm^2), radiative interactions are nonperturbative and will compete with the nonradiative interactions, hence influencing considerably the photodissociation ratios of branching into various product excited atomic states.

We will endeavour to show in this section that the model of the dressed molecule and the Born-Oppenheimer approximation [1], [12], lead to the determination of the dressed or field-molecule eigenstates as solutions to coupled differential equations that describe the nuclear motion in the presence of the laser field. Thus bound-discrete, bound-continuum, radiative and nonradiative (nonadiabatic) can all be treated simultaneously for any coupling strength, thus allowing us to go beyond the usual perturbative treatments. Since we shall be dealing with bound states as initial conditions, the presence of dissociative (continuum) nuclear states presents a problem, which is curcumvented through the use of a scattering formalism that encompasses all possibilities. Thus, by introducing the technique of artificial channels for entrance [28] and generalized to include exit channels also [25], one can simultaneously treat bound and continuum states. It is thus possible by the present method to calculate rigorously transition amplitudes for any radiative or nonradiative interaction strength in the presence of bound and continuum states, thus covering both perturbative (Fermi-Golden rule) and nonperturbative regimes.

We rewrite the total Hamiltonian (19) by separating the radiative and nonradiative perturbation, H_{mf} and H_{na},

$$\hat{H} = \hat{H}_o + \hat{V} \quad ; \quad \hat{H}_o = \hat{H}_m + \hat{H}_f \quad ; \quad \hat{V} = \hat{H}_{mf} + \hat{H}_{na} \quad . \quad (24)$$

Thus $\hat{H}_o$ is the zeroth-order field-molecule Hamiltonian, and $\hat{V}$ is the total, radiative and nonradiative interaction. We now try to express the field molecule eigenstates of the total Hamiltonian in terms of the eigenstates of $\hat{H}_o$, which are therefore direct products of the unperturbed (Born-Oppenheimer) molecular eigenstates of $\hat{H}_m$, and the unperturbed field eigenstates of $\hat{H}_f$, equation (12). We can therefore define the field-electronic states

$$|e,n> = |e> |n> , \qquad (25)$$

where e is a collective quantum number (symmetry, spin, etc.) for molecular Born-Oppenheimer electronic states, and n is the photon number defined in equations (9-10). We now look for solutions of the total Schrödinger equation: $H|\psi_E> = E|\psi_E>$ with the total wave function expanded in terms of the basic field-electronis states defined in (25),

$$|\psi_E> = \frac{1}{R} \sum_{e,n} F_{en}(R) |e,n> . \qquad (26)$$

$F_{en}(R)$'s are appropriate nuclear radial functions propagating on the potential of the photon-electronic state $|e,n>$. By substituting into the total Hamiltonian defined in equation (24), and premultiplying by a particular state $|e,n>$, one obtains the set of one-dimensional second-order differential equations for $F_{en}(R)$:

$$\left\{ \frac{d^2}{dR^2} + \frac{2m}{\hbar^2} \left[E - V_e(R) - n\hbar\omega \right] \right\} F_{en}(R)$$

$$= \frac{2m}{\hbar^2} \sum_{e'n'} V_{en,e'n'}(R) F_{e'n'}(R) , \qquad (27)$$

where m is the reduced mass of the molecule, $V_e(R)$ is the field free electronic potential of electronic state $|e>$ obtained from ab-initio quantum chemical calculations or from spectroscopic measurements [1], [12]. We treat here rotationless molecules, although in principle both rotational quantum numbers (J,M) can be included rigorously [19], [28].

Equation (27) for the field-molecule problem can be more succinctly expressed in matrix form as,

$$F''(R) + W(R) F(R) = 0 , \qquad (28)$$

where the diagonal energy matrix elements are

$$W_{en,en}(R) = \frac{2m}{\hbar^2} [E - V_e(R) - n\hbar\omega] . \qquad (29)$$

The nondiagonal elements that describe the couplings, i.e.,

$$W_{en,e'n'}(R) = \frac{2m}{\hbar^2} \left[V^m_{en,e'n}(R) + V^r_{en,e',n\pm1} \right] \quad , \tag{30}$$

are of two types: nonradiative ($V^m = H_{na}$) and radiative ($V^r = H_{mf}$).
Since each electronic potential $V_e(R)$ appears in the diagonal matrix
elements (29), we are able to sum numerically over all bound vibrational and continuum (unbound) states of the same potential. Thus only
the electronic and photon states need be specified explicitly in any
numerical calculation. Finally the radiative coupling V^r are nondiagonal in the photon quantum numbers reflecting annihilation (absorption, $\Delta n = -1$) or creation (emission, $\Delta n = +1$) of a photon, equations
(21) and (30). The nonradiative (nonadiabatic) couplings remain diagonal in the photon number n since they do not involve the field.
    All numerical calculations are performed using a Fox-Goodwin
method, which has proved to be very accurate for molecular problems
(errors are of sixth order in the integration step [29]). The asymptotic numerical radial functions are projected onto asymptotic field-
molecule states $|e,n\rangle$ and are expressed as,

$$F_{en}(R) = \sum_{e'n'} F^{e'n'}_{en}(R) \quad ,$$

$$F^{e'n'}_{en}(R) = k^{-1/2}_{en} \{\delta_{ee'} \, \delta_{nn'} \, \exp[-i(k_e R + \delta)] \tag{31}$$

$$- S_{en,e'n'} \, \exp[i(k_e R + \delta)]\} \quad ,$$

$$k^2_{en} = \frac{2m}{\hbar^2} (E - V_e(R_\infty) - n\hbar\omega) \quad .$$

δ is an elastic scattering phase factor, which is zero for neutral
dissociating products but needs to be modified for charged products
[27].
    The coefficients $S_{en,e'n'}$ are defined as the scattering, S-matrix
elements, and the function $F^{e'n'}_{en}(R)$ corresponds to the nuclear radial
functions of the molecule in the final state $|e,n\rangle$ for initial states
$|e',n'\rangle$. In practice one usually projects the real numerical functions onto real asymptotic states, i.e.,

$$F^{e'n'}_{en}(R) = k^{-1/2}_e [\delta_{ee'}\delta_{nn'} \, \sin(k_e R + \delta) + R_{en,e'n'} \, \cos(k_e R + \delta)]. \tag{32}$$

This projection enables one to obtain, from the numerical procedure,
the R matrix, which is related to the S matrix by the expression [30]

$$S = (1 - iR)^{-1} (1 + iR) \quad , \tag{33}$$

and thus one obtains the transition amplitude matrix T,

$$S = 1 - 2 \pi i T \quad . \tag{34}$$

In the molecular problems we shall encounter, invariably the initial state is a bound state, so that one encounters the problem of bound-bound transitions, or one has to calculate the probability of transition from initial bound states to final continuum photodissociation states.   One method, such as encountered in the complex-coordinate method [31], calculates linewidths Γ which are squares of the transition amplitudes $\Gamma = \dfrac{2\pi}{\hbar} |T|^2$. We have shown previously [24-25], [28], that it is possible to obtain transition amplitudes directly from the coupled equations (28), i.e., one can transform all transition amplitude problems, including bound-bound transitions, into a scattering problem by introducing additional __artificial__ channels, continua, as entrance and exit channels.   The introduction of such artificial channels into the coupled equations (28) permits us to exploit the various relations between transition matrices in order to extract the relevant photophysical amplitudes.   Thus using the following relations between the total Green's function G and the transition operator T [32],

$$T = V + V G_o T = V + T G_o V \quad , \tag{35}$$

$$G = G_o + G_o T G_o \quad , \quad G = (E-H)^{-1} \quad , \quad G_o = (E-H_o)^{-1} \quad , \tag{36}$$

one can obtain an expression for the transition amplitude T_{cc1} between an entrance channel $|C1>$ and a real physical continuum (dissociative) channel $|c>$,

$$T_{C1,c} = \exp (i\eta_1) V_{C1,0} G_o^o T_{oc} \quad . \tag{37}$$

G_o^o is the zeroth order (field-molecule) Green's function of the initial bound state $|0>$, η_1 is the elastic phase shift for scattering on the artificial continuum potentials of $|C1>$, and $V_{C1,0}$ is the coupling (weak) between the artificial channel and the bound state.   The numerical solutions of the coupled equations (27-28) including the artificial channel $|C1>$ coupled to the initial state with n photons $|0,n>$ permits us to extract each photodissociation amplitude

$$T_{oc} = T_{C1,c} \exp (- i\eta_1) (V_{C1,0} G_o^o)^{-1} \quad . \tag{38}$$

All quantities on the right hand side of equation (38) can be calculated numerically [24-25]. The above method applies provided the initial state $|0>$ is only weakly perturbed during the multiphoton processes, so that the unperturbed Green's function G_o^o is adequate. This will be the case if the initial state is coupled nonresonantly to resonant processes, as will be shown to occur in the H_2 case (next section). All multiphoton resonant processes and nonadiabatic interactions are calculated exactly in T_{oc}, allowing us to join the weak, perturbative regime $(I < 10^{10}$ W/cm$^2)$ to the strong, nonperturbative regime $(I > 10^{10}$ W/cm$^2)$.

The advantage of the method based on the artificial channel $|C1>$, equation (38), gives direct access to numerical values of the transition photodissociation amplitude T_{oc} from the initial bound state $|0>$ to the final continuum $|c>$ Transition amplitudes are essential quantities to calculate photodissociation angular cross sections or distributions [19]. Thus writing field-molecule states as

$$|j; \Lambda_j J_j M_j> = r^{-1} F_j(J,R)\psi_j(q,R)|\Lambda_j J_j M_j> \ |m_j, n_j> \quad , \qquad (39)$$

where ψ_j corresponds to the electronic eigenfunction of energy $V_j(R)$, q represents the ensemble of electronic and spin coordinates, R is the internuclear separation. The rotational states of the molecule are defined by the normalized symmetric top function $|\Lambda J M>$, where Λ is the electronic angular momentum, J is the total angular momentum and M is the z-component of J [11]. The differential dissociation cross section from some initial state $|0> = |g_j \Lambda_o J_o M_o>$ to the final states $|j> = |j; \Lambda_j J_j M_j>$, is then given by the expression

$$\sigma(J_o M_o; \theta, \varphi) \sim |\sum_j \exp{(iJ_j \pi/2)}(2J+1)^{1/2} D_{M_j \Lambda_j}^{J_j}(\theta,\varphi)T_{oj}|^2 \quad . \qquad (40)$$

4. THE DRESSED MOLECULE.

Having established in the previous section the necessary formalism to treat multiphoton transitions in diatomic molecules beyond perturbation theory, we now expose in detail the method in order to help interpret the recent experimental results of van Linden van den Heuvell [9] and Bucksbaum [10] on the nonlinear photoelectron spectrum of H_2 which exhibits above threshold ionization peaks (ATI), i.e., the ionized electron keeps absorbing photons in the vicinity of the molecular ion core. Each ATI peak now reveals a vibronic structure, as the receding electron remains coupled to the core via coulomb and polarization forces. Furthermore, measurements by Bucksbaum et al. [10] on the proton yield demonstrate unusual yield dependencies on the intensity of the laser.

We limit ourselves in the present work to the experimental laser

wavelength λ = 532 nm. As pointed out by van Linden van den Heuvell
[9], this wavelength allows one to reach the $B^1\Sigma_u^+$ state of H_2 via a
five photon <u>nonresonant</u> transition (see figure 1). A sixth photon
couples radiatively and resonantly the B state to the doubly excited
$2p\sigma_u^2$ electronic state, the so called F state which crosses in a <u>diaba-</u>
<u>tic</u> representation [33] the Rydberg type E electronic state [12]. In
an adiabatic representation, the EF curve forms a double well as does
the GK potential. These two states remain coupled by a nonadiabatic
(non Born-Oppenheimer) coupling. One can however adopt the equivalent
diabatic representation where now the diabatic GF and EK curves cross
(figure 1) and are coupled by a nonradiative <u>nondiabatic</u> coupling due
to the fact that in this representation the molecular electronic
Hamiltonian $\hat{H}_m$ is not diagonal. A residual nondiabatic coupling
$<EK|\hat{H}_{el}|GF>$, the term $\hat{H}_{na}$ in the Hamiltonian (24) is operative. In
fact the diabatic EK and GF electronic potentials were obtained by
deperturbing with a 2 x 2 unitary transformation the spectroscopic
adiabatic electronic states EF and GK (for details see [12], [33]).
This procedure yield a nondiabatic coupling $H_{na}(R)$ which is used in
the coupled equations (28). It is to be emphasized once more that in
the coupled equations formalism, both nondiabatic (we now use this
term in a diabatic representation rather than nonadiabatic which
applies to an adiabatic regime) interstate coupling and radiative
couplings are equivalent from a formal view point. The numerical
procedure presented in the previous section allows for the rigorous
treatment of radiative and nonradiative transitions on an equal
footing, from the perturbative (weak interaction) limit to the nonperturbative (strong interaction) limit.

The sixth photon is thus resonant with the vibrational states of
the GF and EK diabatic electronic potentials which further interact
nondiabatically. A seventh photon now couples radiatively these last
states to the $X^+(^2\Sigma_g^+)$ ground electronic state of H_2^+. In this process,
a free electron is now created so that the electronic transition
moment involves the Rydberg electrons of the E and G states and the
ionizing electron in H_2^+, (assuming that the H_2^+ core is nearly the same
for the E, G and X^+ states (see figure 1)). We emphasize that the F
state, which is doubly excited <u>cannot</u> couple radiatively directly to
the X^+ state; i.e., the electronic transition moment $<2p\sigma_u^2|\vec{r}|1s\sigma_g f_c>$,
where f_c is the ionized electron wavefunction is rigorously zero since
radiative transitions, if one neglects electron correlation, involve
only one electron excitation [12]. We thus have the interesting case
that the B state couples <u>radiatively</u> strongly to the F state, which
then couples <u>nonradiatively</u> to the Rydberg E and G states. It is from
these two Rydberg states that the seventh photon of wavelength 532 nm
can now access resonantly the H_2^+ molecule, leading to ATI when the
ionized electron keep absorbing further photons. This last process

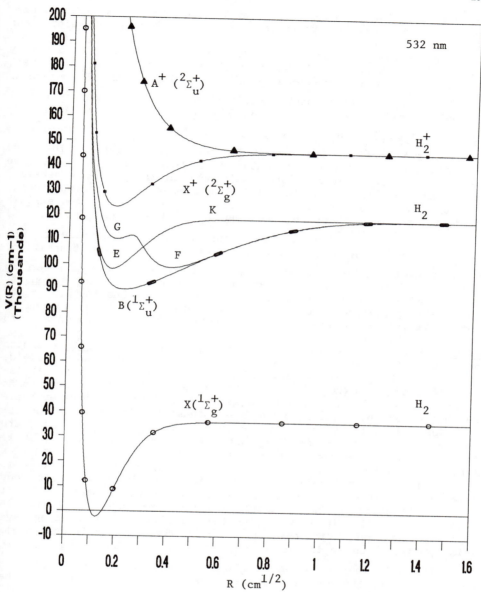

Figure 1.   Ten photon transition at λ = 532 nm leading to ionization and dissociation.

leads to dressing of the electron and various theoretical methods have
been developped over the years to treat this problem [5-6], albeit for
atoms only so far.

What we wish to point out is that in the course of ATI, a purely
electronic process as a first approximation, photons will interact
further with the H_2^+ core leading to a dressing of the H_2^+ molecular
ion. Firstly, a nonresonant three photon transition induces direct
photodissociation from the bound $X^+(^2\Sigma_g^+)$ state to the repulsive, dissociative $A^+(^2\Sigma_u^+)$ of H_2^+. This is seen in figure 1, the standard nonperturbative vertical image of multiphoton transitions. The more
complete nonperturbative representation is that of figure 2 where we
now use the field-molecule states defined in the previous section,
equations (25-26), (i.e., the total wavefunction is linear superposition of products of photon and molecular states). Let us now explain
in detail the meaning of this new representation. The ground $X(^1\Sigma_g^+)$
state with (n+5) photons couples radiatively nonresonantly to the
$B(^1\Sigma_u^+)$ state leaving only n photons after a five photon transition.
Since this transition is nonresonant, it will be weak and can be
treated perturbatively. The remaining transitions, being resonant,
are strong and must be treated nonperturbatively. Thus the B(n) state
is coupled radiatively to two sets of states: the GF (n-1) and GF
(n+1) field-molecule states. The first (n-1) state corresponds to
removal of one photon from the field and is thus ascribed to an absorption. The second (n+1) state is the result of a virtual photon
emission. We remind the reader that the quantized electric field,
equations (8), (21) is explicitly written as the sum of an annihilation (â) and a creation (â⁺) photon operator. The first corresponds
to absorption and the second to emission of photons. We must emphasize that at 532 nm wavelength the B → F tansition is resonant for absorption. Thus the GF (n-1) state crosses resonantly the B(n) state
in the Franck-Condon region for that transition. The B(n) → GF(n+1)
transition is nonresonant and is therefore called a virtual transition
(this transition is responsible for the Lamb shift of electronic
states in vacuum [21-23]). In the field-molecule picture one sees
immediately, figure 2, that this transition is nonresonant. In fact
the GF(n+1) state is $2\hbar\omega$ in energy above the resonant B → F transition. This point help us establish the validity of the rotating wave
approximation, RWA, which neglects all such virtual transitions [5-6].
This approximation is therefore valid only if the Rabi frequency, ω_R
(equation 22), the radiative coupling between the B and F state is
much less than the energy separation between the resonant and the
virtual transition, i.e.

$$\omega_R \ll 2\hbar\omega \quad . \tag{41}$$

This is the main reason why in the X → B five photon transition, only

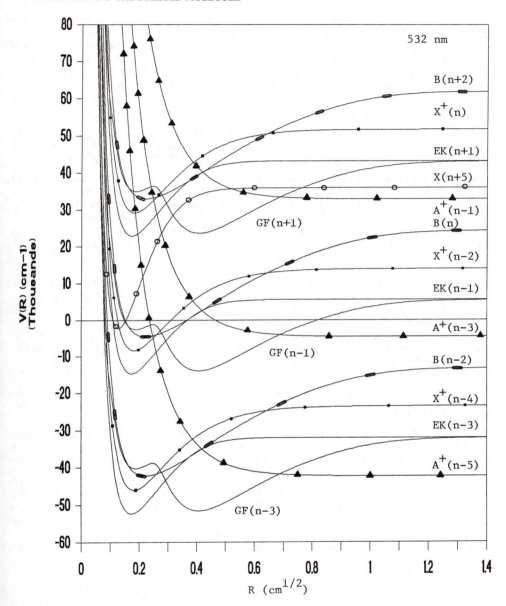

Figure 2. Field-molecule representation of figure 1 including
photon numbers n.

the X(n+5) and B(n) field molecule states are used. The virtual coupling between the X(n+5) and B(n') states, where n' > n can be safely
neglected since the photon absorptions are themselves nonresonant, and
are therefore very weak. In conclusion, every resonant n → n-1 absorption is accompanied by a virtual n → n+1 emission. This explains
therefore the doubling of all electronic states in figure 2.

We now continue to follow the photon paths. The GF states are
coupled nondiabatically (via H_{na}, i.e., $V^m_{en,e'n}$ equation (30)), to the
EF states with the same photon number since this is a <u>nonradiative</u>
transition. Now the Rydberg E and G(n-1) states couple resonantly to
the X^+(n-2) state and virtually to the X^+(n) of H_2^+. The X^+(n-2) state
couples nonresonantly to the A^+(n-3) and virtually to the A^+(n-1)
state. The A^+(n-3) state couples radiatively to X^+(n-4) and X^+(n-2).
The first transition corresponds to the nonresonant absorption of the
<u>ninth</u> photon shown in figure 2. The virtual transition A^+(n-3) →
X^+(n-2) serves to dress the X^+ electronic state, and is depicted in
figure 3. Thus the X^+(n-2) and A^+(n-3) field-molecule states cross at
an energy above the v = 4 vibration of the X^+ ground state of H_2^+. The
symmetric radiative coupling $<X^+|\vec{\mu}|A^+> \vec{\mathcal{E}}$ gives rise to both the ab-
sorption X^+(n-2) → A^+(n-3) and the emission A^+(n-3) → X^+(n-2) processes. Similar crossings occur in the other field-molecule states which
must be added until numerical convergence is achieved. We repeat,
this is due to the fact that the classical coherent electric field $\mathcal{E}$
is a linear superposition of photon states n, equation (15). Finally
we have a transition from the X^+(n-4) to the A^+(n-5) state. This last
state corresponds to the photodissociation-ionization of $H_2(X^1\Sigma_g^+)$ to
$H_2^+(A^2\Sigma_u^+)$ after absorption of ten photons, or the three photon photodissociation of H_2^+.

The figure 2 represents the minimal number of field-molecule
states required for a proper treatment of the nonlinear photoelectron
spectrum of H_2. As the intensity increases, more and more of those
states must be included until numerical convergence is achieved. In
the weak field limit, one recovers of course the direct perturbative
pathway described by figure 1. In the strong field limit, many more
pathways are allowed due to the virtual photon creation processes
which are normally neglected in the RWA regime. Thus the complete
state count as exhibited in figure 2 allows us to bridge the weak and
strong field limits.

The field-molecule representation depicted in figure 2 leads us
to make the following quick predictions. Firstly, crossings of field-
molecular states involving a one photon resonant process become <u>laser-
induced</u> <u>avoided</u> <u>crossings</u> as one increases the field intensity I.
Thus the crossings of the states X^+(n), A^+(n-1); X^+(n-2), A^+(n-3);
X^+(n-4), A^+(n-5) all undergo an avoided crossing as shown in figure 3
for various laser intensities. The new field-molecule states are
obtained by diagonalizing the diabatic 2 x 2 Hamiltonian (in a first

resonant approximation)

$$
\begin{pmatrix}
V_{11}(R) + \hbar\omega & V_{12}(R) \\
V_{21}(R) & V_{22}(R)
\end{pmatrix} , \qquad (42)
$$

giving two new <u>adiabatic</u> states, called the <u>dressed</u> states of the field-molecule system:

$$
V_{\pm}(R) = \frac{V_{11}(R) + V_{22}(R)}{2} \pm 1/2 \left[(V_{11}(R) - V_{22}(R))^2 + 4V_{12}^2(R) \right]^{1/2} , \qquad (43)
$$

where V_{11}, V_{22} are the diabatic (zero-field) molecular electronic potentials (figure 1), V_{12} is the radiative coupling (Rabi frequency, equation (22)). Similar laser-induced avoided crossings occur at the intersections of the B(n), GF(n-1); B(n+2), GF(n+1); B(n-2), GF(n-3) states. These radiative avoided crossings are further perturbed by the nondiabatic interactions with the EK states. These laser induced avoided crossings induce nonperturbative intensity dependent changes in the electronic potential and concomitantly in the vibronic structure of transitions. Such laser induced effects have been considered by various authors [34-37, 15-16]. A detailed study of laser induced resonances, i.e., the nonlinear radiative lifetimes of photodissociating molecular states such as shown in figure 3 has been undertaken by Bandrauk *et al.* [15-16]. In particular, a semiclassical approach used previously in the theory of <u>predissociation</u> of molecules has proven to be very useful in predicting the existence of these new resonances. This is in keeping with the remark made above that in the field-molecule representation, nondiabatic (nonradiative) and radiative interactions are formally equivalent and can be treated simultaneously in a unified formalism. The scattering formalism expounded in the previous section is of course the most convenient method to treat bound and continuum states simultaneously in the presence of large radiative and nonradiative interactions. The experimental observation of a laser intensity dependent vibronic structure of H_2^+ was confirmed recenty [9-10] and was therefore the first report of the laser induced avoided effect illustrated in figure 3.

Figure 2 further demonstrates that at the wavelength of 532 nm five open channels appear, i.e. channels below the initial zero energy. These channels correspond to dissociation of H_2 and H_2^+ into neutral atoms and protons. Thus the $A^+(n-3)$, $X^+(n-4)$ and $A^+(n-5)$ channels will produce H(1S) and H^+ species with kinetic energies corresponding to the difference in energy between the zero line (energy of v = 0, $X^2\Sigma_g^+$ of H_2) (figure 2) and the asymptotic energies of each state. Hence three protons of different kinetic energy are to be expected. The lowest energy proton will emanate from the $A^+(n-3)$ channel as a

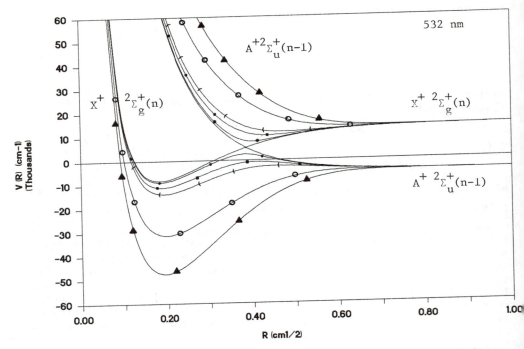

Figure 3. Laser induced avoided crossing between $X^+(n)$ and $A^+(n-1)$ field-molecule states of H_2^+ at various field intensities I (W/cm^2): _____ 0; $\cdots$ 10^{12}; $\square\square\square$ 5×10^{12}; $<<<$ 10^{13}; 000 5×10^{13}; ▲▲▲ 10^{14}.

result of the tunnelling of the vibrational states of $X^+(n-2)$ at the initial zero energy. A second higher energy proton will be produced by the $X^+(n-4)$ channel, which corresponds to the ninth photon process in figure 2, or equivalently the two photon dissociation of H_2^+ via the nonresonant process $X^+(n-2) \to A^+(n-3) \to X^+(n-4)$. Figure 2 tells us immediately that this final channel, $X^+(n-4)$ is coupled radiatively to $A^+(n-5)$, so that a laser-induced avoided crossing will occur between these two channels at the energy 32000 cm^{-1} below the initial zero energy. Thus the yield of the first low energy proton from $A^+(n-3)$ and the second and third higher energy protons from $X^+(n-4)$ and $A^+(n-5)$ are all expected to be nonperturbatively influenced at high intensities by the laser-induced avoided crossings illustrated in figure 3. The high kinetic energy protons from $A^+(n-5)$ are the result of the three-photon nonresonant transition of H_2^+, the last three pho-

tons in figure 1. The field-molecule picture illustrates that this transition will be strongly affected by the two photon transition $X^+(n-2) \to X^+(n-4)$ at high intensities. This transition merits further elaboration.   As indicated above, this is a nonresonant two photon dissociation of the electronic ground state of H_2^+ via the nonresonant repulsive $A^+(n-3)$ state.   Thus the $A^+(n-3)$ serves as a <u>virtual</u> state, i.e., the dissociation products remain on the X^+ potential, but the radiative transition is induced by the A^+ repulsive potential which is nonresonant, and is therefore inaccessible.   This is clearly seen in figure 2 where the $A^+(n-3)$ state is always well above the $X^+(n-4)$ channel.   Thus only resonant processes give rise to crossing potentials, whereas nonresonant processes always have well separated potential surfaces.   This two photon nonresonant transition from bound X^+ nuclear states to continuum X^+ nuclear states has been called by us previously ATPD (above threshold photodissociation) and is the analogue of ATI [27].   The conditions for such processes is large transition dipole moments as in ATI where the ion-electron system gives a dipole moment equal to the distance r between the two.   In ionic molecules such as LiF, the dipole moment of the system is R, i.e. the distance between the ionic moities Li^+ and F^-.   Thus a linearly diverging dipole moment arises, creating a very strong coupling with the radiative field as in the ATI case.   For the $X^+ \rightarrow A^+$ transition, we will show below that the transition dipole moment is R/2, i.e. half the internuclear distance.   Thus in all three cases, similar nonlinear absorption phenomena occur because of the large dipole or transition moments which give rise to very large radiative couplings as the ionized or dissociated species separate.

As in the ATPD of H_2^+ described above, ATPD of H_2 also occurs in the B state of H_2.   Thus from figure 2 one has also an open channel the B(n-2) channel accompanied by the EK,GF(n-3) state, which are coupled radiatively with the B state also with an R/2 transition moment.   Thus neutral H(1S) + H(n=2) atoms are expected to be also created due to ATPD in the neutral B, EF, GK states of H_2.   In the next section we will try to render these predictions quantitative as a result of the numerical efficiency of the coupled equations (27) which enable one to include as many channels as are deemed necessary by convergence criteria.   Furthermore, as emphasized above, one can readily cover the weak field, perturbative regime, I ~ 10^{10} W/cm^2 to the high field nonperturbative limit I ~ 10^{14} W/cm^2, and include simultaneously bound-bound, bound-continuum, radiative and nonradiative transitions in one unified formalism and numerical method.

5.   RESULTS AND DISCUSSION

As discussed in the previous sections, collision theory allows one to obtain, using the artificial channel $|C1>$, transition amplitudes from initial bound states to final bound states or continua.   In our case, we shall be calculating the transition amplitudes from the v = 0 vibrational state of the ground electronic state $X^1\Sigma_g^+$ of H_2 to the various channels that are open according to the field-molecule diagram, figure 2, i.e. all the channels which are below the zero energy line which corresponds to the initial energy.

The input in the coupled equations (27) are the ab-initio poten tials of H_2 and H_2^+ published in the literature [38-39]. These wer interpolated over 3000 points over an internuclear distance of 0.4 t 33 a.u. (in figures 1-3, distance units appear as $cm^{1/2}$ since al energies are reported in cm^{-1}. Thus in H_2, due to the factor of $2m/\hbar$ in the nuclear Schröedinger equation (27), 1 $cm^{1/2}$ = 11 a.u.). In th case of the EF and GK states, since these are calculated adiabaticall [40], these well known double well potentials were deperturbed [12] [33], in order to produce the crossing diabatic potentials GF and EK A gaussian nondiabatic nonradiative interaction V_{12} was found of th

form $V_{12}(R) = 3023 \times \exp [- 38.2 (R - 0.29)^2]$ to give the above cite adiabatic potentials EF and GK when inserted in equation (43).

As to the radiative couplings, two equivalent gauges are possibl [5-6], [26], the electric-field (multipole) gauge or the radiatio field (Coulomb) gauge. The radiative coupling in the first is $e\vec{r} \cdot \vec{\mathcal{E}}$ whereas in the last it is $e\vec{A}/mc \cdot p$. Both gauges will give identica results if a complete set of states is used, since the two gauges ar related by a unitary transformation. The use of one gauge or anothe thus depends on its convenience. In the present problem, the electro nic transitions $B \to F$ and $X^+ \to A^+$ involve excitation to electroni states which have the same asymptotic limits. In fact the $B \to F$ an $X^+ \to A^+$ transitions are both $1s\sigma_g \to 2p\sigma_u$ molecular orbital transition [9]. For these, the transition moment is easily shown to be:

$$\vec{\mu}(R) = < \frac{1s_a + 1s_b}{\sqrt{2}} \ |e\vec{r}| \ \frac{1s_a - 1s_b}{\sqrt{2}} > = eR/2 \quad , \quad (44$$

in the limit of nonoverlapping atomic orbitals $1s_a$ and $1s_b$. Clearl in the asymptotic atomic limit, these moments become infinite implying very strong coupling of the molecule to be electromagneti field upon dissociation. Unfortunately this creates divergent radia tive couplings in the electric field gauge. It is therefore best t use the Coulomb gauge, which as the result of the commutation relatio $p/m = (i/\hbar)[H,r]$, results in the following convergent radiative coupl

$$\frac{A}{mc} P_{ij} = \frac{(E_i(R) - E_j(R))}{\hbar\omega} R/2 \quad . \quad (45$$

since $\vec{\mathcal{E}} = -\frac{1}{c} \frac{\partial \vec{A}}{\partial t}$. Thus as $R \to \infty$, $E_i(R) - E_j(R) \to 0$ and the expressio (45) converges to zero. The Coulomb gauge transition moment (45) wa therefore used in the radiative transition calculations for the tran sitions $B \to F$ and $X^+ \to A^+$. The $X \to B$ five photon transition was simu lated by using an arbitrary weak effective one photon transition, which is of no consequence since this transition is perturbative. Finally the E and G to X^+ transitions being unknown were also given the arbitrary transition moment $\mu = 1$ cm^{-1}. Other transition moments such as B $\rightarrow$ E, G were taken from the literature [41].

Calculations were performed for the transition amplitudes T_{oc}, equation (38) for the open channels below the initial zero energy, corresponding to the v = 0 level of $X^1\Sigma_g^+$ (see figure 2). Since the seventh photon falls in energy just above the v = 3 level of $X^+(^2\Sigma_g^+)$, then the v = 0 to 3 vibrational levels of $X^+(n-2)$ all lie below this zero energy line (figure 2). It is to be emphasized that this figure corresponds to zero kinetic energy of the electron. In actual fact, in the photoionization process $H_2 \rightarrow H_2^+ + e^-$, the electron acquires considerable kinetic energy which is then analyzed, thus exhibiting ATI peaks [9-10]. Since we are interested in the dressing of the molecular ion H_2^+, we can obtain the energies and photodissociation widths of the vibrational levels of H_2^+ in the presence of the laser field by examining the resonance structure of T_{oc}. In fact, these resonances show up in numerical calculation of T_{oc} when one scans as a function of energy the levels of H_2^+ in anyone open channel, from $A^+(n-3)$, $B(n-2)$, $X^+(n-4)$ to $A^+(n-5)$. One takes into account the electronic kinetic energy by shifting up all the H_2^+ channels, i.e. X^+ and A^+. As one calculates T_{oc} as a function of this displacement of the H_2^+ channels (we iterate this corresponds to calculating T_{oc} as a function of electron kinetic energy), one finds that resonances appear corresponding to v = 3, 2, 1, 0 successively from low to high electron kinetic energy. This is as expected, since at low electron kinetic energy, the molecular ion remains in its high vibrational states, whereas at high electron kinetic energy, only low vibrational excitation can remain by conservation of total energy. We illustrate in figure 4 the two resonances attributed to v = 2 and v = 0 levels of $X^+(^2\Sigma_g^+)$ of H_2^+ at the intensity I = 10^{13} W/cm^2. These resonances (normalized to unity by dividing by the maximum value of the resonance) were obtained from T_{oc} where $|c\rangle = |A^+(n-5)\rangle$. They show a Lorentzian behaviour for v = 0 and non-Lorentzian behaviour at v = 2. Thus for the first case, a linewidth can be readily obtained from the half-width at the middle of the resonance curve, whereas in the v = 2 case, the average half-widths at the middle was used as a first order approximation. We emphasize that the same resonances appear in all the open channels, thus lending us to conclude that the energy shifts and widths correspond to all possible photophysical processes included in the calculation as illustrated in the field-molecule representation (figure 2) which goes well beyond the perturbative description depic-

A. D. BANDRAUK

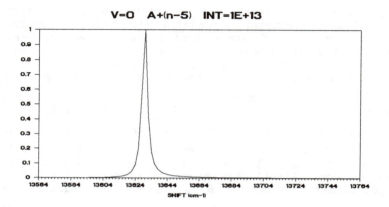

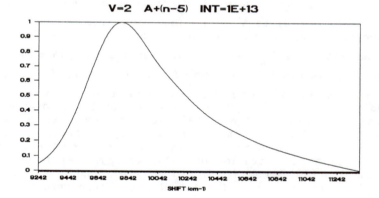

Figure 4. Laser induced resonances attributed as
v = 2, 0 levels of the dressed states, figure 3,
of H_2^+ at I = 10^{13} W/cm^2.

ted in figure 1.

We tabulate in the following table energies corresponding to measured electron kinetic energies and widths of resonances for various intensities as obtained from the numerical calculations described above.

TABLE 1.   Energies E (electron kinetic energy) and widths Γ of vibrational levels v of H_2^+ for various laser intensities I; λ = 532 nm.

I (W/cm^2)	v:	0	1	2	3	4	5	6	7
10^{10}	$E(cm^{-1})$	7116	4926	2864	918				
	$\Gamma(cm^{-1})$	$<10^{-3}$							
10^{11}	E	7204	5018	2958	1016				
	Γ	0.001	0.01	0.005	0.006				
10^{12}	E	8080	5914	3875	1961	177			
	Γ	3.8	2.0	4.3	2.5	3.9			
10^{13}	E	13630	11664	9796	7334	5658	3606	2960	1631
	Γ	4	138	686	980	0.13	502	285	139

At 10^{10} W/cm^2, the radiative interactions are weak so that the spacings of the H_2^+ levels are still that of the unperturbed molecule and linewidths Γ are below 10^{-3} cm^{-1}. Since the lifetime τ = 5 x 10^{12} s/Γ(cm^{-1}), one sees that the radiative (photodissociative) lifetimes of H_2^+ at this intensity are larger than a nanosecond (10^{-9} s). At 10^{12} W/cm^2, one observes now levels with picosecond (10^{-12} s) lifetimes in addition to the appearance of an extra level. This is the effect of the radiative interaction which pushes the dressed adiabatic potential $V_-(R)$ down in energy, equation (43). The large energy shifts are also the result of the laser induced avoided crossing. At 10^{13} W/cm^2, more levels are seen to appear with lifetimes ranging from a femtosecond (10^{-15} s) to a picosecond. Such displacements of the energy levels of H_2^+ have been observed [9-10] and have been attributed to the laser-induced avoided crossing illustrated in figure 3.

We also mentioned the fact that proton yields have been measured and show unusual energy dependences as a function of intensity. In particular it has been noticed by Bucksbaum et al. [10] that at low intensities, high kinetic energy protons predominate, whereas at higher intensities, lower kinetic energy protons predominate. Our calculations show similar behaviour for the proton yield. Thus in the next table we show the maximum intensities of the v = 0 and v = 2 resonances as a function of intensity for the two channels X^+(n-4) and A^+(n-5). The first channel corresponds to two photon ATPD of H_2^+ discussed above and the last, corresponds to the three photon dissociation of H_2^+.

TABLE 2. Maximum intensities of resonances as a function of laser intensity I.

Level: Channels:	$v = 0$		$v = 2$	
	$X^+(n-4)$	$A^+(n-5)$	$X^+(n-4)$	$A^+(n-5)$
$I=10^{11}W/cm^2$	4×10^{-6}	4×10^{-5}	8×10^{-4}	9×10^{-3}
$I=10^{12}W/cm^2$	14	9	3.5	0.2
$I=10^{13}W/cm^2$	10^3	1	65	10^{-1}

One sees clearly from the above table, that at low intensities the $A^+(n-5)$ channel, i.e. the three-photon dissociation channel, is more important than ATPD, the $X^+(n-4)$ channel. As the intensity increases beyond 10^{12} W/cm^2, this trend is reversed. This can be easily rationalized in terms of the field-molecule representation. Thus at low intensities, the radiative interaction between the $X^+(n-4)$ and $A^+(n-5)$ channel at the crossings of these is ineffectual and cannot influence the high kinetic energy protons in the $A^+(n-5)$ channel. From the Landau-Zener probability for a nondiabatic transitions one can predict that high velocity particles at a crossing remain in fact diabatic, i.e., they remain on the same curve [33]. As the intensity increases, the avoided laser-induced crossing predicted from equation (43) creates an upper potential $V_+(R)$ dissociating asymptotically to the $X^+(n-4)$ state. Hence a turnover can be expected eventually in the distribution of dissociating products with lower kinetic energy, i.e. $X^+(n-4)$ products, predominating as a result of the laser-induced avoided crossing of H_2^+. This phenomenon was observed in [10] and was presented as a clear evidence of such an avoided crossing. This laser induced effect creating the new adiabatic potential $V_+(R)$ can also create new stable vibrational states of $V_+(R)$ as we have predicted previously [15-16]. These have yet not been observed experimentally. They should occur in the higher ATI peaks where the electron with high kinetic energy can leave the molecule in the upper potential $V_+(R)$ instead of the lower $V_-(R)$ potential which gives the laser induced resonances enumerated in table 1.

In conclusion, we see that intense lasers modify considerably electronic potentials, altering the vibrational energies and creating laser-induced resonances. An important concept which helps clarify these nonlinear phenomena is the laser-induced avoided crossing, figure 3, in concert with the dressed molecule picture, figure 2. Laser-induced resonances and laser-induced avoided crossings are the result of intense electron electromagnetic field interactions. The electron thus mediates the field-molecule interaction and also serves as a probe for the properties of the dressed molecule.

Acknowledgments - We thank E. Constant, J.M. Gauthier and D. Marchand in the preparation of this manuscript. We also wish to acknowledge support from the Natural Sciences and Engineering Research Council of Canada.

REFERENCES

[1]   Slater, J.C. (1963). Quantum Theory of Molecules and Solids, vol. I, McGraw Hill, N.Y.

[2]   Gallagher, J., Brion, C.E., Samson, J.A.R., and Langhoff, P.W. (1988) "Absolute cross sections for molecular photoabsorption, partial photoionization, and ionic photofragmentation", J. Phys. Chem. Ref. Data 17, 9-153.

[3]   Bawagan, A.D., and Brion, C.E. (1987) "Orbital imaging of the lone pair electrons by electron momentum spectroscopy", Chem. Phys. Lett. 137, 573-7.

[4]   Duke, C.B., Lipari, N.D., and Pietronero, L. (1976) "Electron-vibration interactions in benzene", J. Chem. Phys. 65, 1165-1181.

[5]   Mittleman, M.H. (1982) Introduction to the Theory of Laser - Atom Interactions, Plenum Press, N.Y.

[6]   Faisal, F.H.M. (1987) Theory of Multiphoton Processes, Plenum Press, N.Y.

[7]   Bandrauk, A.D., ed. (1988) Atomic and Molecular Processes with Short Intense Laser Pulses, vol. B-171, NATO ASI Series, Plenum Press, N.Y.

[8]   Cornaggia, C., Normand, D., Morellec, J., Mainfray, G., and Manus, C. "Resonant multiphoton ionization of H_2 via the E, F state", Phys. Rev. A34, 207-215.

[9]   Verschuur, J.W.J., Noordam, L.D., and van Linden van den Heuvell, H.B. (1989) "Anomalies in above-threshold ionization observed in H_2 and its fragments", Phys. Rev. A40, 4383-4391.

[10]   Bucksbaum, P.H., Zavriyev, A., Muller, H.G., and Schumacher, D.W. (1990) "Softening of the H_2^+ molecular bond in intense laser fields", Phys. Rev. Lett. 64, 1883.

[11]   Zarvriyev, A., Bucksbaum, P.H., Muller, H.G., and Schumacher, D.W. (1990) "Ionization and dissociation of H_2 in intense laser fields at 1.064 μm, 532 nm and 355 nm", Phys. Rev. A42, 5500-5513.

[12]   Lefebvre-Brion, H., and Field, R.W. (1986), Perturbations in Spectra of Diatomic Molecules, Academic Press, Orlando, FL.

[13]   Chelkowski, S., and Bandrauk, A.D. (1988) "Coherent pulse propagation in a molecular multilevelmedium", J. Chem. Phys. 89, 3618-3628.

[14]   Corkum, P.B., Burnett, N.H., and Bumel, F. (1989), "Above thres-

hold ionization in the long wave-length limit", Phys. Rev. Lett. 62, 1259-1262.

[15] Bandrauk, A.D., and Sink, M.L. (1981), "Photodissociation in intense laser fields: predissociation analogy", J. Chem. Phys. 74, 1110-1117.

[16] Bandrauk, A.D., and McCann, J. F. (1989), "Semiclassical description of molecular dressed states in intense laser fields", Comments At. Mol. Phys. 22, 325-343.

[17] Su, Q., and Eberly, J.H. (1990), "Stabilization of model atom in super-intense field ionization", J. Opt. Soc. Am. B7, 564-569.

[18] Su, Q., Eberly, J.H., and Javanainen, J. (1990), "Dynamics of atomic ionization suppression and electron localization in an intense high-frequency radiation field", Phys. Rev. Lett. 64, 862-865.

[19] McCann, J.F., and Bandrauk, A.D. (1990), "Two-color photodissociation of the lithium molecule: Anomalous angular distributions of fragments at high laser intensities", Phys. Rev. A42, 2806-2816.

[20] Cohen-Tannoudji, C.C., Dupont-Roc, J., and Grynberg, G. (1989), Photons and Atoms, J. Wiley & Sons, N.Y.

[21] Nguyen-Dang, T.T., and Bandrauk, A.D. (1983), "Molecular dynamics in intense laser field, I - One dimensional systems in infrared radiation, J. Chem. Phys. 79, 3256-3268.

[22] Goldin, E. (1982), Waves and Photons, J. Wiley & Sons, N.Y.

[23] Glauber, R.J. (1963), "Incoherent and coherent states of the radiation field", Phys. Rev. 131, 2766-2788.

[24] Bandrauk, A.D., and Gélinas, N. (1987), "Coupled equations approach to multiphoton molecular processes", J. Comp. Chem. 8, 313-323.

[25] Bandrauk, A.D., and Atabek, O. (1989), "Coupled equations method for multiphoton tansitions in diatomic molecules" in J.O. Hirschfelder, R. Coalson, R.E. Wyatt (eds.), Advances in Chemical Physics, vol. 73, chap. 19, pp. 823-857.

[26] Bandrauk, A.D., Kalman, O., and Nguyen-Dang, T.T. (1986), "Molecular dynamics in intense fields. IV. Beyond the dipole approximation", J. Chem. Phys. 84, 6761-6770.

[27] Bandrauk, A.D., and Gauthier, J.M. (1990), "Above-threshold molecular photodissociation in ionic molecules - a numerical simulation", J. Opt. Soc. Am. B7, 1420-1427.

[28] Bandrauk, A.D., and Turcotte, G. (1983), "Photodissociation angular distribution of diatomics in intense fields", J. Phys. Chem. 87, 5098-5106.

[29] Nguyen-Dang, T.T., Durocher, S., and Atabek, O. (1989), "Direct numerical integration of coupled equations with nonadiabatic interactions", Chem. Phys. 129, 451-462.

[30] Norcross, D.W., and Seaton, M.J. (1973), "Asymptotic solutions of the coupled equations of electron-atom collision theroy for the case of some channels closed", J. Phys. B6, 614-621.

[31] He, X., Atabek, O., and Guisti-Suzor, A. (1988), "Laser induced

resonances in molecular dissociation in intense fields", Phys. Rev. A38, 5586-5594.

[32] Watson, K.M., and Nuttal, J. (1967), Topics in Several Particle Dynamics, Holden Day Publishers, San Francisco.

[33] Bandrauk, A.D., and Child, M.S. (1970), "Scattering theory of predissociation energy shifts and widths", Molec. Phys. 19, 95-111.

[34] Voronin, A.I., and Samokhin, A.A. (1976), "Role of resonances associated with multiphoton transitions in molecules under the influence of an intense light field", JETP (Sov. Phys.) 43, 4-6.

[35] Lau, A.M., and Rhodes, C.K. (1977), "Field induced avoided crossing and new transition channels", Phys. Rev. A16, 2392-2412.

[36] George, T.F., and Yuan, J.M. (1978), "Semiclassical theory of unimolecular dissociation induced by a laser field", J. Chem. Phys. 68, 3040-3052.

[37] Bandrauk, A.D., and Sink, M.L. (1978), "Laser induced preassociation in the presence of natural predissociation", Chem. Phys. Lett. 57, 569.

[38] Kolos, W. (1971), "Ab initio potentials for diatomics", Atomic Data 2, 119-150.

[39] Wolniewicz, L., and Dressler, K. (1985), "The EF, GK and $H\hat{H}$ $^1\Sigma_g^+$ states of hydrogen", J. Chem. Phys. 82, 3292-3299.

[40] Wolniewicz, L., and Dressler, K. (1979), "The $H^1\Sigma_g^+$ - $B^1\Sigma_u^+$ transition of hydrogen", J. Molec. Spectrosc. 77, 286-297.

[41] Wolniewicz, L., "Theoretical investigation of the transition probabilities in the hydrogen molecule", J. Chem. Phys. 51, 5002-5008.

SCATTERING CHAOS IN THE HARMONICALLY DRIVEN MORSE SYSTEM

C. JUNG
Fachbereich Physik
Universität Bremen
2800 Bremen
Germany

ABSTRACT. We study the classical scattering of an electrically charged point particle off the Morse potential under the additional influence of an oscillating electromagnetic field. We find scattering chaos, i.e. a scattering function, which is discontinuous on a fractal subset of its domain. This behavior is explained by the homoclinic/heteroclinic structure of unstable periodic orbits. In the phase space a localized chaotic invariant set Λ (chaotic saddle) is created, whose invariant manifolds reach out into the asymptotic region and influence the generic scattering trajectories. The transition probability to various final energies shows a complicated pattern of singularities.

1. INTRODUCTION

One of the major themes of this conference is the scattering of electrons under the additional influence of a strong electromagnetic field, so called free-free-transitions. Such processes have been observed experimentally for many years [1-7]. The theoretical description of these experiments has been given quantum mechanically within a low frequency approximation [8-11].

In the meantime an interesting new development in scattering theory has emerged in the form of scattering chaos. So far this phenomenon has been investigated for time independent systems mainly, i.e. for systems without external fields (see the reviews [12,13], however, for a preliminary demonstration of chaos in field modified Coulomb scattering see [14]). It is well known that an external oscillating field can drive bound states into chaos and so we expect, that such a field can drive scattering processes into chaos as well. In this article let us try to merge the ideas of free-free-transitions and of scattering chaos. Prior to the investigation of a realistic but complicated system it might be appropriate to investigate a simple model system first in order to demonstrate the basic effects and to try out appropriate methods. A good model for demonstration is the scattering of an electrically charged particle off a local 1-dimensional Morse potential under the simultaneous influence of a harmonically oscillating external field. In periodically driven Morse systems bound chaos has already been found [15,16]. And so

D. Hestenes and A. Weingartshofer (eds.), The Electron, 219–238.

we expect it to be a good candidate for scattering chaos too. To explain
the chaos in this system we essentially use ideas and methods which we
have used before in the treatment of scattering chaos in time independent
systems [17-20]. We modify them in the right way to make them fit to the
time dependent case.

So far chaos is a well defined concept in classical mechanics only.
The concept of quantum chaos has not yet been defined in a general and
satisfactory way. Therefore, in this article the classical side of
scattering chaos will be explained in detail and only in the concluding
section a few remarks on quantum scattering chaos will be made. In this
article we use the methods and the terminology which have become familiar
in nonlinear dynamics and in the mathematically oriented literature on
classical mechanics. Because of a lack of space it is not possible to
explain these issues in detail and we will use these expressions and
methods without comment. The reader, who is not familiar with nonlinear
dynamics, is referred to the textbooks [21-24].

2. SINGULARITIES IN THE SCATTERING FUNCTIONS

The system is given by the Hamiltonian function

$$H(p,q,t) = (p-A\cos(\omega t))^2/2 + \exp(-2q) - 2\exp(-q) \tag{1}$$

q is the 1-dimensional position coordinate, p is the canonically conjugate
momentum, t is the time. ω and A, the frequency and the amplitude of the
external field are free parameters.
The asymptotic limit of scattering is the limit q→∞.
The asymptotic Hamiltonian is

$$H_0(p,q,t) = (p-A\cos(\omega t))^2/2 \tag{2}$$

We introduce the reduced phase

$$\varphi = \omega t - q\omega/p - (A\sin(\omega t))/p \tag{3}$$

p and φ are conserved under the motion generated by H_0. Therefore we can
label asymptotes uniquely by giving p and φ. The set of incoming
asymptotes, denoted by As^{in}, has p<0. The set of outgoing asymptotes,
denoted by As^{out}, has p>0.
To compute a scattering trajectory, we take values of p_{in} and φ_{in}
where p_{in}<0, take some large value of q and compute the corresponding
initial time t_{in} mod $2\pi/\omega$ by inverting (3) and integrate the Hamiltonian
equations of motion generated by H. We stop the trajectory at some
terminal time t_{out} at which q is back in the outgoing asymptotic region
while p>0. The outgoing asymptote is labelled by a pair of values for p_{out}
and φ_{out}. It turns out to be instructive to know the time delay of the
scattering process, which we define by

$$Dt = t_{out} - \frac{q_{out}}{p_{out}} - t_{in} + \frac{q_{in}}{p_{in}} \qquad (4)$$

Now let us make a numerical experiment : We fix values for A and ω, we choose A=0.4, ω=1.4 for reasons which will become clear later. As an example we fix the value of p_{in} at p_{in}=-1.0, scan φ_{in} and plot p_{out} and Dt as function of φ_{in} in Fig.1.

For people not familiar with scattering chaos the result may be surprising : There are intervals of continuity of p_{out} and complicated looking regions in between. In the complicated regions the time delay becomes large. Fig.2 gives a magnification of one of the unresolved complicated regions of Fig.1. In this magnification new substructures become visible. Fig.3 gives a further magnification in which more substructures become evident. We could continue this process and keep on magnifying complicated looking regions and would always find new smaller intervals of continuity interrupted by complicated looking regions which contain still smaller intervals of continuity e.t.c. ad infinitum.

This set of nested structures and substructures on all scales is typical for fractal structures. The set of accumulation points of boundaries of intervals of continuity defines a Cantor set along the φ-axis. Such fractal clusters of discontinuities of the scattering functions have been found in all chaotic scattering systems encountered so far. Therefore, the occurence of this phenomenon has been accepted as defining criterion for classical scattering chaos.

3. TRAJECTORIES IN PHASE SPACE

The first step to explain the behavior displayed in Figs.1-3 consists in looking at phase space trajectories coming from various intervals of continuity. Some examples are shown in Figs.4a-e. In part a φ_{in}=2.0 is chosen, which belongs to the long interval of continuity outside the chaotic region. This trajectory comes in, turns around and goes out again directly. In part b we see a trajectory whose φ_{in} comes from the smaller interval around φ_{in}=3.9 which can be identified clearly in Fig.1. This trajectory comes in, turns around, but does not have enough energy to escape directly. So it returns and comes in a second time. At the second trial it has enough energy to escape. In parts c,d,e we see trajectories, whose φ_{in} comes from still smaller intervals belonging to a deeper level of the fractal hierarchy. They make several turns inside the potential interior.

Figs.1-3 show that smaller intervals of continuity contain trajectories making more revolutions inside the potential interior and leading to longer time delays. In addition, Figs.1-3 indicate that p_{out} goes to zero, if φ_{in} converges to the boundary of an interval. Supposing that the external field is switched off adiabatically at very large distance, then also the kinetic energy of the outgoing particle converges to zero, if φ_{in} converges to the boundary. In total we obtain the following picture : Inside any interval of continuity the scattering trajectories all have the same qualitative structure. At small distances

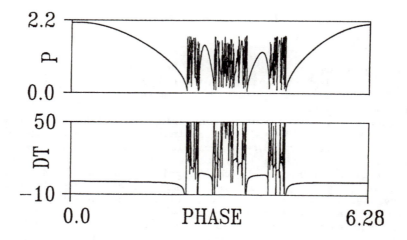

Fig.1: p_{out} (upper frame) and Dt (lower frame) as function of φ_{in} for $p_{in}=-1.0$ and parameter values A=0.4, ω=1.4.

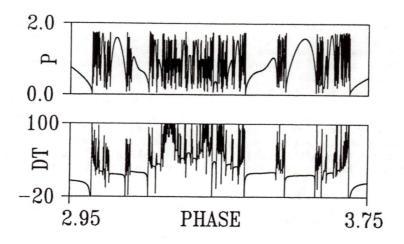

Fig.2: Magnification of Fig.1.

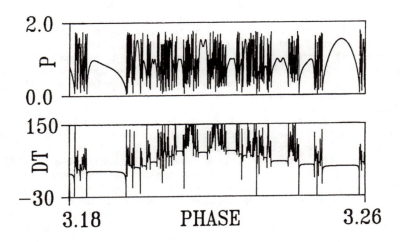

Fig.3: Magnification of Fig.2.

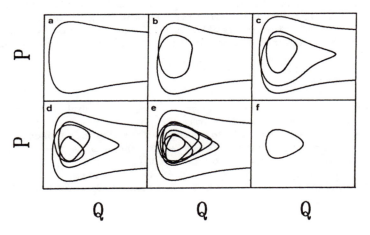

Fig.4: Some phase space trajectories for parameter values A=0.4, ω=1.4.
Parts a-e show scattering trajectories with p_{in}=-1.0 and incoming phases
φ_a=2.0, φ_b=3.8, φ_c=3.55, φ_d=3.46, φ_e=3.19. Part f shows the periodic orbit
γ. Frame boundaries : q∈(-1,3), p∈(-2,2).

beyond the boundary the first part of the scattering trajectories looks similar to the trajectories inside the interval, then however the energy is not sufficient to escape in the same way and the particle returns to the potential interior and makes a few additional turns.

The fractal structure of the intervals is constructed in a hierarchical way : On the 0-th level there is the longest interval $I^{(0)}$ which contains $\varphi_{in}=0$. It contains trajectories coming in, making one single turn and going out again directly (see Fig.4a). There remains one connected unresolved complement $C^{(0)}$ on the φ_{in}-axis (note : because φ is an angle variable, we identify $\varphi=0$ with $\varphi=2\pi$).On the 1st level we cut out of the interior of $C^{(0)}$ the 2 most prominent intervals $I_1^{(1)}$ and $I_2^{(1)}$. As Fig.1 shows, they lie around $\varphi_{in}=2.8$ and $\varphi_{in}=3.9$. These two intervals contain all the trajectories making one additional loop inside the potential interior (see Fig.4b). There remains a complement $C^{(1)}$ which consists of 3 connected components. On the 2nd level we cut out of the interior of $C^{(1)}$ the intervals containing trajectories making two loops in the potential interior (see Fig.4c). We continue this scheme iteratively. In the limit, between the cut out intervals a Cantor set of measure zero remains. It is the set of accumulation points of the boundaries of intervals.

Figs.4a-e give the impression, that scattering trajectories can be attracted by some localized orbits. In the next few sections we shall make this idea more precise. We close this section on phase space trajectories by showing in Fig.4f that localized trajectory which will turn out to be the most simple one and at the same time the most important one. It is a periodic orbit, whose time of revolution is exactly equal to the cycle time $T_c=2\pi/\omega$ of the external field. In the following this trajectory will be denoted by γ.

4. POINCARE SECTION

The best way to get an overview over the periodic orbits consists in studying the appropriate Poincare map M. In the case of periodically driven systems the best type of Poincare map is the stroboscopic map in the (p,q)-plane taken for times t=0 mod $2\pi/\omega$, where ω is the frequency of the external field. This map is constructed like this :

Take an initial point (p_0,q_0) at time t=0, construct the trajectory through this point, follow it until the time $t=T_c$. At this time the trajectory has reached the point (p_1,q_1). Define

$$M(p_0,q_0) = (p_1,q_1) \tag{5}$$

Alternatively we express this in the following way : Let Φ_t be the flow map corresponding to the Hamiltonian (1). Then $M=\Phi_{T_c}$ applied to the plane $t=0 \bmod 2\pi/\omega$ in the 3-dimensional extended phase space.

Any trajectory gives an infinite sequence of points in the Poincare plane by marking its positions in the (p,q)-plane at all times $t_n=n\ T_c$, $n\in Z$. Periodic orbits in the (p,q)-plane, whose time of revolution is an integer multiple of T_c, correspond to periodic points of M.

We shall compare the Poincare plots with the phase portrait of the system in the undisturbed case of A=0, i.e. in the absence of the external field. In Fig.5 a few trajectories of the field free case are plotted in the (p,q)-plane. The broken line marks the separatrix to energy E=0. Trajectories outside the separatrix are scattering trajectories with E>0. Trajectories inside the separatrix are closed bound trajectories with E<0.

The point (0,0) is a fixed point. The time of revolution for trajectories close to this point is $T=2\pi/\sqrt{2}$, i.e. the rotation frequency is $\Omega_0=\sqrt{2}$. Going further away from the origin the time of revolution increases monotonically (the frequency Ω decreases) and towards the separatrix we find $T\to\infty$, $\Omega\to0$. For A=0 and any field frequency ω the Poincare map restricted to the interior of the separatrix is a monotonic twist map, where the winding number of each invariant line is given by $w=\Omega/\omega$.

For weak perturbations, i.e. for small values of A some invariant lines with irrational winding numbers survive. The invariant lines of the field free case having rational winding numbers break into secondary structures around elliptic periodic points and small chaos strips along the invariant manifolds of hyperbolic periodic points. This szenario is standard for generically perturbed twist maps [25,26].

We expect the strongest coupling between field and particle for $\omega\approx\Omega_0$ and for most of our model calculations $\omega=1.4$ is chosen. For $\omega<\Omega_0$, but $|\omega-\Omega_0|$ small, an invariant line of the unperturbed system lying close to the origin is in 1:1 resonance with the external field. For very small amplitude A this line breaks into one elliptic fixed point x_γ and one hyperbolic fixed point x_h. If A is increased a little bit more (approximately at $A\approx0.0015$) x_h collides with x_0, the elliptic fixed point coming from the origin in the field free case. x_h and x_0 destroy each other in a saddle-node bifurcation. Afterwards x_γ becomes the organization center of the complete structure of the Poincare section.

In Fig.6 the case of A=0.04 is displayed. A few initial points (marked by crosses) have been chosen arbitrarily and the next few hundreds of iterates of these initial points are plotted. Starting from the outside of the structure many KAM lines are already broken and some parts of the (p,q)-plane, which are bound in the field free case, merge with the scattering region. With increasing value of A more and more KAM lines are distroyed. Fig.7 shows the situation for A=0.1. Note the different frames of Fig.6 and Fig.7.

x_γ stays elliptic until approximately $A=A_c\approx0.35$, where it changes to inverse hyperbolic. For $A\geq0.4$ no elliptic periodic points and no KAM lines are found. Scattering trajectories can come close to all points of the (p,q)-plane. We find one global chaotic region. However, unstable periodic orbits and aperiodic localized orbits still exist and they play an essential role in the scattering behavior of the system. The fixed

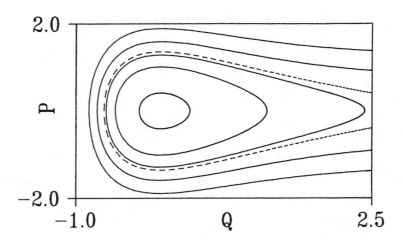

Fig.5: Phase portrait for the field free case A=0. The separatrix between scattering trajectories and bound trajectories is shown as broken line.

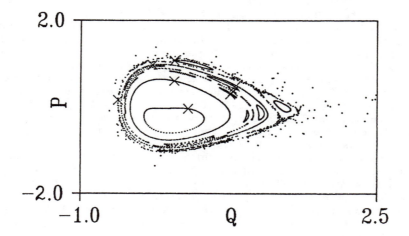

Fig.6: Poincare plot for A=0.04, ω=1.4. A few hundred iterates of some initial points (marked by crosses) under the Poincare map M are plotted.

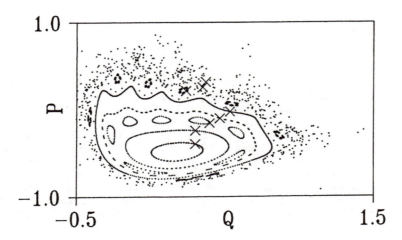

Fig.7: Poincare plot for A=0.1, ω=1.4. A few hundred iterates of some
initial points (marked by crosses) under the Poincare map M are plotted.

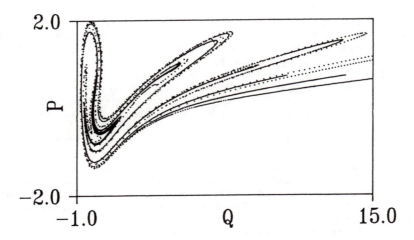

Fig.8: The first few tendrils of the unstable manifold $W^u(\gamma)$ of the fixed
point x_γ for parameter values A=0.4, ω=1.4.

point x_γ of the Poincare map corresponds to the periodic orbit γ shown in Fig.4f.

5. THE CHAOTIC SADDLE

For A=0.4, ω=1.4 the fixed point x_γ in the Poincare plane is inverse hyperbolic. Its eigenvalues are $\mu_1 \approx -2.09$, $\mu_2 = 1/\mu_1$. Now we look at the stable manifold $W^s(\gamma)$ and the unstable manifold $W^u(\gamma)$ of x_γ in the Poincare plane. We take a short segment $S=(x_1,x_2)$ of W^u close to x_γ such that $x_2 = M^2(x_1)$ and plot a few iterates of S in Fig.8. In total, $W^u(\gamma)$ has an infinite number of tendrils reaching out to unlimited distances. And each tendril is a fractal arrangement of an infinite number of lines, going in and out an infinite number of times. In the figures we only can show a few tendrils and only a few branches within each tendril by plotting a finite number of iterates of a finite number of points within the initial segment S (we have chosen 2000 points for Figs.8 and 9).

For q≥10 the Morse potential can not be distinguished from zero within the numerical accuracy. In this outside region the Poincare map acts quite simply :

$$M_{as}(p,q) = (p,q+p{\cdot}T_c) \tag{6}$$

This action of M_{as} in mind, the reader can imagine how the plot would be supplemented under continued addition of further iterates of the points already present. All points of $W^u(\gamma)$ correspond to trajectories which converge towards γ for $t \to -\infty$.

In the same way we take a short segment $R=(x_3,x_4)$ of $W^s(\gamma)$, such that $x_3 = M^2(x_4)$, and plot a few backward iterates of R in Fig.9. All points of $W^s(\gamma)$ correspond to trajectories which converge towards γ for $t \to +\infty$. The comments to Fig.8 apply also to Fig.9, we only have to reverse the direction of time.

Fig.10 shows a smaller neighborhood of x_γ in the Poincare plane and a few branches of $W^s(\gamma)$ and $W^u(\gamma)$ in this region. We see intersections and homoclinic tangencies between W^s and W^u. Most important : If there is any homoclinic point, then all images and preimages of this point under M are homoclinic points too, an infinite number of homoclinic points exist. They correspond to trajectories which converge towards γ for $t \to +\infty$ as well as for $t \to -\infty$. The existence of homoclinic points of M implies the existence of an infinite number of periodic points of arbitrarily high period and an overcountable number of aperiodic localized points of M. We call a point x of the Poincare plane localized, if the q-coordinate of $M^n(x)$ stays limited for $n \to +\infty$ and for $n \to -\infty$.

The localized points form a chaotic invariant set Λ' which is contained in the topological closure of the homoclinic points. Other points, not belonging to Λ', escape to $q \to \infty$ for $t \to +\infty$ or for $t \to -\infty$. Therefore, the chaotic invariant set Λ' in the Poincare plane corresponds to a chaotic invariant set Λ in the extended phase space, a so called chaotic saddle. Λ has measure zero and is a fractal arrangement of an overcountable number of localized trajectories.

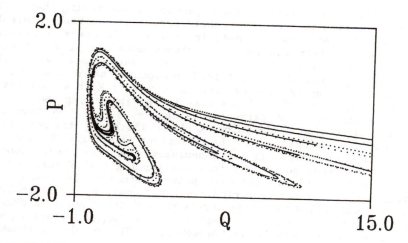

Fig.9: The first few tendrils of the stable manifold $W^s(\gamma)$ of the fixed point x_γ for parameter values A=0.4, ω=1.4.

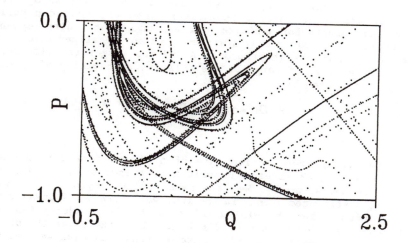

Fig.10: Homoclinic intersections between $W^u(\gamma)$ and $W^s(\gamma)$ in the neighborhood of x_γ for parameter values A=0.4, ω=1.4.

Now we are ready to explain the occurence of scattering chaos : The unstable trajectories in Λ all have their stable and unstable manifolds which reach out into the asymptotic region. Incoming scattering trajectories which start exactly on a stable manifold of a localized orbit, will converge towards this orbit and never escape again. This subset of incoming asymptotes, which lead to capture, has measure zero in the set of all incoming asymptotes. Let us have a look, how this set of exceptional initial conditions looks like in the (p_{in}, φ_{in})-plane.

In out model system at parameter values A=0.4, ω=1.4 we have only one global chaotic region in the Poincare plane. Therefore, the homoclinic structure and the structure of the invariant manifolds of any one particular periodic point is representative of the corresponding structures of any other periodic point. The topological closures of the homoclinic structures of all periodic points coincide. We choose as a representative the most simple periodic point, namely x_γ. We place 10000 points on $W^s(\gamma)$ and follow the trajectories through these points backwards in time to the incoming asymptotic region. In the (p_{in}, φ_{in})-plane we mark the values reached by these 10000 trajectories. The result is shown in Fig.11.

Comparison with Fig.1 shows, that along the line $p_{in}=-1.0$ the intersections with $W^s(\gamma)$ mark exactly the boundaries of chaotic clusters in Fig.1. This is easy to understand : Scattering trajectories become complicated, when they start close to W_s of some localized orbit. And as explained above, the accumulation points of $W^s(\gamma)$ are representative of W^s of any localized orbit and therefore also of W^s for the whole of Λ.

Genuine scattering trajectories have in and out asymptotes, therefore they are never chaotic themselves. However, they can follow localized chaotic orbits for a finite time and trace out finite segments of chaotic motion. The closer they start to W^s of any localized orbit, the longer they stay in the neighborhood of Λ and the larger the time delay Dt becomes. In the limit of large Dt the relative probability of finding a time delay Dt is given by

$$P(Dt) = \exp(-k \cdot Dt) \qquad (7)$$

where k is the average repellation rate of Λ, it is a measure for the instability of Λ. Fig.12 gives a numerical distribution of Dt for the parameter values A=0.4, ω=1.4. We have taken 10^6 trajectories all with $p_{in}=-1.0$ and φ_{in} distributed evenly in the interval $(0, 2\pi)$. The Dt-axis has been cut into boxes of length Δ=0.05 and hits of the 10^6 trajectories into the various boxes have been counted. Fig.12 displays the logarithm of the counts versus the mean delay time of the boxes. From this plot we read off the value k≈6.8.

In total we can characterize the situation of scattering chaos like this : The bundle of incoming scattering trajectories hits Λ. Only a subset of measure zero gets stuck. The other trajectories fly through the gaps of Λ and transport some kind of shadow image of Λ into the outgoing asymptotic region, where we can perform measurements. From the results of these observations we can reconstruct the fractal properties of Λ. With

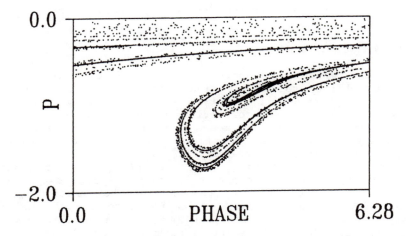

Fig.11: Transport of 10000 points of $W^s(\gamma)$ into the asymptotic (p_{in}, φ_{in})-plane for parameter values A=0.4, ω=1.4.

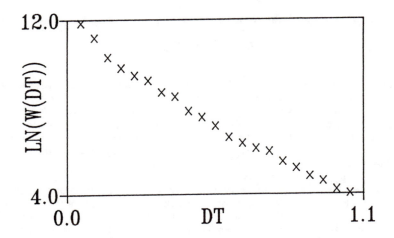

Fig.12: Distributions of time delays. The vertical axis gives $\ln(P(Dt))$ versus Dt on the horizontal axis. The parameter values are A=0.4, ω=1.4.

this picture in mind, we interpret scattering chaos as a version of transient chaos [27,28].

6. TRANSITION PROBABILITIES

The most important measurable quantity in field modified scattering is the transition probability to various final energies of the projectile. Usually in an experiment p_{in} is fixed and the phase φ_{in} is evenly distributed in $(0,2\pi)$. A detector registers outgoing particles with a particular value p_{out} of the final momentum or a particular value E_{out} of the final energy. We imagine that the field is switched off adiabatically in the asymptotic region, such that $E_{out}=p_{out}^2/2$.

To the relative count rates $\sigma(p_{out})$ or $\sigma(E_{out})$ all trajectories contribute, which end with this particular value of p_{out} or E_{out}. These trajectories can be read off from the plot of the function $p_{out}(\varphi_{in})$ plotted in Figs.1-3. In Fig.1 let us imagine a horizontal line at a particular value of p_{out} and register all intersections of this line with the curve $p_{out}(\varphi_{in})$. These intersections mark the initial phases φ_j of those trajectories which contribute to $\sigma(p_{out})$ or $\sigma(E_{out})$. In the case of chaotic scattering there is an infinite number of them. Each one gives the weight c_j to the count rate $\sigma(E_{out})$, where

$$c_j^{-1}(E) = | \frac{dE_{out}}{d\varphi_{in}}(\varphi_j) | \qquad (8)$$

The total transition probability to find E_{out} is given by

$$\sigma(E_{out}) = \sum c_j(E_{out}) \qquad (9)$$

Fig.13 shows the numerical example of $\sigma(E_{out})$ for the parameter values A=0.4, ω=1.4 and the incoming momentum p_{in}=-1.0. 10^6 trajectories have been started with φ_{in} evenly distributed in the interval $(0,2\pi)$. The outgoing energy interval $(0,2.5)$ has been cut into 2500 boxes of length $\Delta E=10^{-3}$ and hits of the 10^6 trajectories into the various boxes have been counted. In Fig.13 the vertical axis gives the logarithm of the count rates versus the final energy on the horizontal axis.

The curve $p_{out}(\varphi_{in})$ displayed in Figs.1-3 has an infinite number of relative extrema, at least one maximum for each interval of continuity. Accordingly, $\sigma(E_{out})$ has an infinite number of singularities of the qualitative shape $c \cdot (E-E_j)^{-1/2}$ for $E<E_j$ and 0 for $E>E_j$. In Fig.13 only a finite number of these singularities is well resolved. The maxima from the various intervals of continuity lie at different values of E_j. On the energy axis they are arranged in a fractal pattern which is a projection of the fractal structure of the chaotic set Λ in the extended phase space (compare the corresponding discussion of angle singularities in the differential cross section given in [19]). Because of the limited accuracy of any measurement of $\sigma(E_{out})$ only a few highest levels of the fractal structure can be resolved.

Here is a brief outline of the explanation, why the pattern of these singularities reflects the fractal structure of Λ : First imagine the transport of $W^u(\gamma)$ into the As^{out} plane. It gives a picture in complete analogy to the transport of $W^s(\gamma)$ into the As^{in} plane shown in Fig.11. Just imagine Fig.11 to be turned upside down. Again $W^u(\gamma)$ is representative for W^u of the whole of Λ. $W^u(\gamma)$ in As^{out} has an infinite number of branches. Each branch j is just one line starting at p=0, going to a maximal value p_j of p_{out} and winding back to p=0. Remember : We identify $\varphi=0$ with $\varphi=2\pi$, therefore As^{out} is a cylinder. The set of all branches of $W^u(\gamma)$ in As^{out} forms a fractal arrangement, reflecting the fractal structure of Λ. Accordingly, also the set of p_j values forms a Cantor set along the p_{out}-axis. The set of outgoing asymptotes which are reached by the trajectories starting inside one particular interval of continuity forms a line in As^{out} of qualitatively the same shape as one branch of $W^u(\gamma)$. The set of accumulation points of the images of all intervals coincides with the accumulation points of $W^u(\gamma)$. Accordingly, the set of maximal p_{out}-values of images of intervals define the same Cantor set as the set of p_j-values coming from maximal values of branches of $W^u(\gamma)$.

7. LOW FREQUENCY LIMIT

So far we have considered the case $\omega \approx \Omega$. What happens for other field frequencies ? For $\omega \gg \Omega$ the external field is so fast, that the particle is not able to follow. The external field is essentially averaged out to zero and no chaotic effects induced by the field are left.

In the other extreme $\omega \ll \Omega$ no chaos occurs either. Let us look at this low frequency limit closer. It is of some interest, since the experiments [1-7] done so far have used infrared lasers whose frequency ω is a lot smaller than atomic energies. Also the theories [8-11] are based on the low frequency limit.

The experiments have been done with incoming energies E_{in} which are large compared to the photon energy and to the interaction energy with the field. No trapping of incoming scattering trajectories occurs. In the picture of Fig.11 the incoming momentum is always far away from p-values which are reaches by W^s of Λ. Therefore, the time which the projectile spends inside the interaction region of the local potential is very small compared to T_c. Then it is not necessary to take into account the variation of the field phase during the projectile-target interaction process. Outside this interaction region the canonical momentum p is a conserved quantity anyway, in the outside region field induced transitions to other momenta never occur. The field phase at which the projectile-target scattering event takes place is just the reduced phase φ given in (3) for the incoming asymptote. Formally we define the phase ψ for any point (p,q,t) of the extended phase space by

$$\psi = \lim [\omega t - q(t) \cdot \omega / p(t) - A \cdot \sin(\omega t) / p(t)] \qquad (10)$$

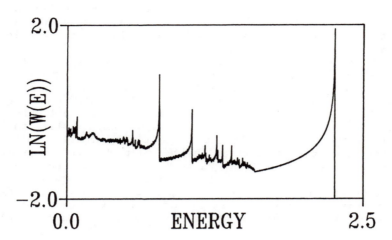

Fig.13: Energy transfer cross section for $p_{in}=-1.0$ and parameter values $A=0.4$, $\omega=1.4$. The vertical axis gives $\ln(\sigma(E_{out}))$ versus E_{out} on the horizontal axis.

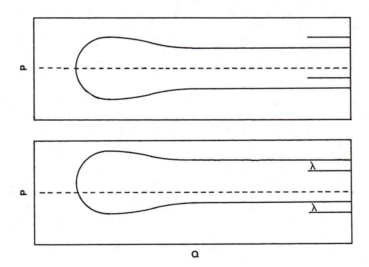

Fig.14: Phase portrait in the low frequency limit. Part a shows the case of $\cos \psi=0$. Part b shows the case of $A \cdot \cos \psi=\lambda$. The q-axis is marked by a broken line.

where the limit t→-∞ is taken along the actual trajectory through the point (p,q,t). ψ is an approximate invariant of the motion in the limit of small ω. The scattering event is governed by the Hamiltonian

$$H = (p - A \cdot \cos(\psi))^2 / 2 + V(q) \qquad (11)$$

In Fig.14 the phase portrait of (11) in the (p,q) plane is shown for 2 values of ψ. Part a displays a line H=const. for the case of $\psi=\pi/2$ mod 2π, which coincides with the field free case. Here the phase portrait curves connect incoming asymptotes with a given value of $p_{in}<0$ with the outgoing asymptote with momentum $p_{out}=-p_{in}$. This is the usual picture for elastic scattering. Part b gives the case of $A \cdot \cos(\psi)=\lambda\neq0$. The line p=0 is marked by a broken line in the plots.

Comparison of a and b shows that in the case of $\cos \psi \neq 0$ the plot looks as if the plot of the field free case would have been shifted by $\lambda=A \cos \psi$. In this case such values of p_{in} and p_{out} are connected by a trajectory, for which $(p_{out}-\lambda)=-(p_{in}-\lambda)$. The difference in energy between the incoming and the outgoing asymptote is

$$\Delta E = \frac{p^2_{out}}{2} - \frac{p^2_{in}}{2} = -2\lambda p_{in} + 2\lambda^2 \qquad (12)$$

Turned the other way around : It is possible, to connect the incoming momentum p_{in} with a particular value of p_{out}, if $||p_{in}| - |p_{out}|| < 2A$. The appropriate initial reduced phase is the one fulfilling

$$\lambda = A \cdot \cos(\psi) = \frac{|p_{out}| - |p_{in}|}{2} = \frac{\Delta E}{\sqrt{2E_{in}} + \sqrt{2(E_{in}+\Delta E)}} \qquad (13)$$

The probability to find an energy transfer of ΔE is equal to the probability to find an appropriate value of λ or ψ. If incoming phases are distributed evenly in $(0,2\pi)$, then this probability to find $\psi= \arccos(\lambda/A)$ is given by

$$R'(\psi) = R(\lambda) = [\pi\sqrt{A^2-\lambda^2}]^{-1} \qquad (14)$$

Compare similar considerations in [29]. The scattering of the projectile off the target is the same as elastic scattering with incoming momentum $p'_{in}=p_{in}-\lambda$ into outgoing momentum $p'_{out}=p_{out}-\lambda$. To obtain a cross section we compare ingoing and outgoing fluxes. Therefore in the inelastic case a momentum ratio is included. In total we obtain

$$\sigma(p_{in}\rightarrow p_{out}) = \frac{p_{out}}{p_{in}} R(\lambda(\Delta E)) \cdot \sigma_{el}(p'_{in}\rightarrow p'_{out}) \qquad (15)$$

where σ is the cross section with field and σ_{el} is the corresponding cross section without field.

In the 3-dimensional case we have a momentum shift only in the direction of the polarization of the field. A and λ become vectors pointing in the direction of the polarization. The quantum counterpart of these transitions only exist for $\Delta E = N\hbar\omega, N \in Z$ and the counterpart of (14) is given by the square of a Bessel function (see discussion in [29])

$$R(\lambda(\Delta E - N\hbar\omega)) = J_N^2(A \cdot (p_{out} - p_{in})/\hbar\omega)$$ (16)

Inserting (16) into (15) explains the result obtained in [8].

Because of the existence of the adiabatic invariant (10) in the extended phase space, the problem has become completely integrable and there is no room left for chaos generated by the interaction with the external field.

8. FINAL REMARKS

In this article some ideas on chaotic scattering coming out of the workshop of nonlinear dynamics have been applied to the harmonically driven Morse system. In the numerical computations the case of $p_{in} = -1.0$ has been demonstrated. What happens for other values of p_{in} ? Fig.11 gives an overview :

There is a critical value $p_c \approx -1.8$, such that for $p_{in} \in (p_c, 0)$ the stable manifolds of Λ will be crossed, when φ is scanned, and chaos occurs. For $p_{in} < p_c$ the incoming asymptotes do not meet W^s of Λ. Accordingly, the trajectories do never come close to Λ and the scattering process is not chaotic. The scattering function $p_{out}(\varphi_{in})$ will be smooth and $Dt(\varphi_{in})$ will be smooth and always finite.

So far we have looked at the case of A=0.4. For increasing A the invariant manifolds of Λ spread over a larger range of p-values and chaotic scattering occurs for a larger interval of incoming p-values, i.e. p_c becomes smaller. In addition, the eigenvalue μ_1 of γ becomes smaller (its absolute value increases) and Λ becomes more unstable. For $A < A_c \approx 0.35$, φ becomes stable and large scale KAM tori exist, which are sticky for incoming scattering trajectories. A small amount of scattering trajectories can stay inside the potential well for a longer time than what is predicted by the exponential law (7). Trajectories inside KAM tori are bound in the potential well for all times in past and future. The field is not strong enough to dissociate these states. If γ is stable and lies inside KAM tori, then the invariant manifolds of Λ can not be represented by the invariant manifolds of γ, which do not exist at all in this case. However, the invariant manifolds of some other unstable periodic orbit, which lies outside of all KAM tori, do the job. For A becoming small, the invariant manifolds of Λ occupy only small strips in the phase space and in the asymptotic region they concentrate in a region near p=0, whose width shrinks to zero in the limit A→0. Otherwise the general szenario is similar to the case of A=0.4 always.

Our whole discussion has been given within classical dynamics. However, scattering investigations are an important source of information in the microworld, where quantum effects are essential. The question of chaos in quantum dynamics is a very delicate one, because chaos in the classical form can never occur in quantum dynamics because of the uncertaincy. The complicated fractal structures which are characteristic of classical chaos, are washed out in the quantum world. If we look at sufficiently small scales, everything becomes smooth in quantum mechanics (for the general problems of quantum chaos see the review [30]).

The best we can do, is to treat classically chaotic systems with semiclassical methods and to look for ways in which features of classical chaos have influence on semiclassical quantities. So far not much has been done in the field of quantum scattering chaos and scattering in the quantum version of system (1) has not yet been investigated. Therefore we only mention a few ideas, which have emerged in the semiclassical treatment of some other chaotic scattering systems. And we suggest that things in system (1) behave in analogy.

The semiclassical scattering amplitude is a sum of terms from the various contributing classical trajectories, which occur in the sum in (9). Accordingly, the semiclassical cross section contains interference oscillations from cross terms between all terms in the amplitude.

A first idea is, to look for chaotic effects in the statistical properties of the fluctuations of the amplitude and the cross section. It has been found, that the quantum cross section of a classically chaotic scattering system has fluctuation properties of the kind we expect for a random matrix system [13,31,32]. Random matrix systems are believed to be generic quantum systems in the same way as chaotic systems are generic classical systems. So the connection between classically chaotic scattering systems and quantum random matrix behavior fits into the general picture.

A second line of thought is to consider the wild fluctuations of the quantum S-matrix as coming from very many poles and to express these poles semiclassically by the unstable periodic orbits of the classical system [33]. This idea runs in parallel to the expression of the poles of the Greens function of bound systems by unstable periodic orbits [34].

A further idea is to Fourier transform the quantum cross section and to interpret the complicated and fractally clustered arrangement of oscillation frequencies as a washed out image of the fractal structure of the classical chaotic saddle Λ [35,36]. Of course, this idea only makes sense, if the system is already in the semiclassical limit, i.e. if the wavelength of the incoming projectile is very small compared to the size of the target. Then very many angular momentum values contribute to the scattering cross section and in the inelastic case very many different final states can be reached from the initial state. In the case of free-free transitions this means, that the photon energy is very small compared to the typical energy change of the electron. Then the discreteness of the possible final energies is not important. In this case we can try to find clusters of singularities in the cross section as a function of final energy in analogy to the classical result of Fig.13.

9. REFERENCES

1. Weingartshofer A., Holmes J., Caudle G., Clarke E. and Krüger H.

(1977) Phys. Rev. Lett. 39, 269
2. Weingartshofer A., Clarke E., Holmes J. and Jung C. (1979) Phys. Rev.
   A 19, 2371
3. Weingartshofer A., Holmes J., Sabbagh J. and Chin S. (1983)
   J. Phys. B: At. Mol. Phys. 16, 1805
4. Wallbank B., Connors V., Holmes J. and Weingartshofer A. (1987)
   J. Phys. B: At. Mol. Phys. 20, L833
5. Wallbank B., Holmes J. and Weingartshofer A. (1987) J. Phys. B:
   At. Mol. Phys. 20, 6121
6. Andrick D. and Langhans L. (1976) J. Phys. B: At. Mol. Phys. 9, L459
7. Andrick D. and Langhans L. (1978) J. Phys. B: At. Mol. Phys. 11,2355
8. Kroll N. and Watson K. (1973) Phys. Rev. A 8, 804
9. Krüger H. and Jung C. (1978) Phys. Rev. A 17, 1706
10. Mittleman M. (1979) Phys. Rev. A 19, 134
11. Rosenberg L. (1979) Phys. Rev. A 20, 275
12. Eckhardt B. (1988) Physica D 33, 89
13. Smilansky U. (1990) Course X of the Les Houches Session LII, ed.
    Giannoni M., Voros A. and Zinn-Justin J., Elsevier, New York
14. Wiesenfeld L. (1990) Phys. Lett. A 144, 467
15. Goggin M. and Milonni P. (1988) Phys. Rev. A 37, 796
16. Heaggy J. and Yuan J. (1990) Phys. Rev. A 41, 571
17. Jung C. and Scholz H-J. (1987) J. Phys. A: Math. Gen. 20, 3607
18. Jung C. and Scholz H-J. (1988) J. Phys. A: Math. Gen. 21, 2301
19. Jung C. and Pott S. (1989) J. Phys. A: Math. Gen. 22, 2925
20. Jung C. and Richter P. (1990) J. Phys. A: Math. Gen. 23, 2847
21. Arnold V. (1978) Mathematical Methods of Classical Mechanics,
    Springer, New York
22. Abraham R. and Marsden J. (1978) Foundations of Mechanics,
    2nd ed., Benjamin Cummings, Reading
23. Lichtenberg A. and Lieberman M. (1983) Regular and Stochastic
    Motion, Springer, New York
24. Guckenheimer J. and Holmes P. (1983) Nonlinear Oscillations, Dynamical
    Systems and Bifurcations of Vector Fields, Springer, New York
25. Chirikov B. (1979) Phys. Reports 52, 263
26. Greene J. (1979) J. Math. Phys. 20, 1183
27. Tel T. (1989) J. Phys. A: Math. Gen. 22, L691
28. Kovacs Z. and Tel T. (1990) Phys. Rev. Lett. 64, 1617
29. Friedland L. (1979) J. Phys. B: At. Mol. Phys. 12, 409
30. Eckhardt B. (1988) Phys. Reports 163, 205
31. Blümel R. and Smilansky U. (1988) Phys. Rev. Lett. 60, 477
32. Blümel R. and Smilansky U. (1990) Phys. Rev. Lett. 64, 241
33. Cvitanovic P. and Eckhardt B. (1989) Phys. Rev. Lett. 63, 823
34. Gutzwiller M. (1971) J. Math. Phys. 12, 343
35. Jung C. (1990) J. Phys. A: Math. Gen. 23, 1217
36. Jung C. and Pott S. (1990) J. Phys. A: Math. Gen. 23, in print

EXPERIMENTS WITH SINGLE ELECTRONS*

Robert S. Van Dyck, Jr., Paul B. Schwinberg, and Hans G. Dehmelt
Department of Physics, FM-15
University of Washington
Seattle, Washington 98195
USA

ABSTRACT: A basic description of past geonium experiments is given and the modifications which allow positron geonium to be formed is described. The use of compensated Penning traps produces a harmonic axial frequency which has a resolution of 10 ppb. By using synchronous detection and a magnetic bottle for coupling, the magnetic resonances become observable. Stability of the radial position in this magnetic bottle is provided by motional (magnetron) sideband cooling. The corresponding magnetic line shapes are primarily determined by the Brownian statistical (axial) motion through this bottle. Finally the beat-note between the nearly degenerate cyclotron and spin precession frequencies defines the anomaly resonance and its value can be determined to ~1 ppb statistical precision by fitting to the noise-modulated Brownian lineshape. Present accuracy has produced the most precise determination of the $\alpha^{-1}(\text{QED})$ and the positron/electron g-factor comparison. Results of new measurements using a phosphor bronze trap are also described which show consistency with previous results using all-molybdenum traps.

1. Introduction

The importance of conducting experiments on this particle (or its anti-particle) resides in our intrinsic interest in understanding the relatively simple system we call "the electron". One of the earliest opportunities in recent history at which it became clear that the electron's structure was not well understood, was provided by the first accurate measurement [1] in 1947 of the hfs in hydrogen and deuterium. A puzzling 0.2% discrepancy occured between the measured and predicted values of this frequency interval. The discrepancy was ultimately traced to the electron's g-factor (the ratio of its magnetic moment and its intrinsic angular momentum) where the prevailing theory by Dirac had predicted this g-factor to be 2.000,000,000,000. The interest to understand the electron's intrinsic structure (in addition to the desire to explain such energy differences as the Lamb shift in hydrogen [2]) helped

---

* Supported by a grant from the National Science Foundation.

D. Hestenes and A. Weingartshofer (eds.), The Electron, 239–293.
© 1991 *Kluwer Academic Publishers. Printed in the Netherlands.*

to inspire the development of the theory of quantum electrodynamics (see Sec. 7 for a very precise g-factor prediction).

The first experiment specifically designed to measure the free electron's g-factor was conducted by Louisell, Pidd, and Crane [3] in 1953, but their accuracy of 1% did not provide an experimental value for the anomalous part "a_e" of the free electron's g-factor defined by

$$g = 2(1 + a_e) \ . \tag{1}$$

Inspired by this pioneering effort, Dehmelt devised a simple experiment in 1958 in which electrons are stored in the field of a positive-ion cloud diffusing slowly in a dense inert gas. This experiment yielded the first *direct* measurement [4] of the anomaly (with a 3% accuracy). When he substituted the quadrupole trap, similar to that shown in Fig. 1, for the optical pumping cell, the illustrious history of the "Penning trap" was initiated at the University of Washington. Much of this early history can be found in a number of previous reviews with various depths of treatment [5-12] and at least one popular review [13] of the experiment is available.

The work initiated by Dehmelt led to the first non-destructive detection in 1968 of electron clouds in the Penning trap at very low pressures ($< 10^{-11}$ Torr), cooled by axial coupling to a resonant tuned circuit [14,15]. Then, noting the obvious limitation on space charge, Dehmelt and colleagues leaped to the logical conclusion that the cloud must be reduced to its irreducible limit, i.e. a single electron, capable of being continuously observed for several days (and months). In 1973, the successfully observed [16] "monoelectron oscillator" made it attractive to complete the development of the second critical aspect of the geonium experiment, that is, the axial-frequency-shift detector [17], also referred to as the "continuous Stern Gerlach effect" [18] (see Sec. 3). This initial fervor of activity culminated in 1977 with the first ever high precision g-factor measurement on a single free electron [19].

The significance of the present series of experiments, using single charged particles, lies in our ability to measure some of its fundamental properties to a very high precision in an environment which is relatively free of unwanted or unaccountable perturbations. In addition, we have demonstrated the ability to isolate both the electron or the positron for an indefinite period of time, making it quite sensitive to mild rf stimulation and detection (subject of Sec. 3). For these experiments, the particle is embedded in a very strong magnetic field which causes it to rotate in a circular orbit within a plane perpendicular to the magnetic field; thus its magnetic moment and charge are made to execute precessional and cycloidal motions in this field (with respective frequencies ω_s and ω_c that are nearly degenerate). The device which we call the Penning trap is used to constrain motion along the magnetic field direction; however, it itself alters the frequencies of these radial motions due to the

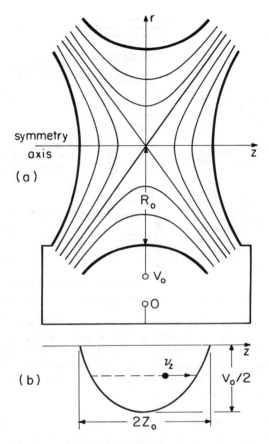

Figure 1. Electric trapping potentials defined by the Penning trap. (a) Electrode surfaces (two endcaps and one ring) are placed upon a given set of equipotentials, amongst the family that satisfies Laplace's equation, by choosing a particular R_0 and Z_0. The z-axis is the rotational axis of symmetry. (b) An axial frequency is obtained by choosing the appropriate bias potential V_0 for the ring relative to the endcaps (from Ref. 7).

electric trapping fields. Because these fields produce the first order perturbations to the observable magnetic frequencies, the trapping device will be described first (see Sec. 2). Because the position of the particle now becomes important due to the specific detection mechanism discussed in Sec. 3, the method used to center the particle is treated next in Sec. 4. The two magnetic modes (cyclotron and precession) are then treated in the following two sections respectively. In all these modes, we discuss the relevant perturbations to the motions in their respective sections. Finally, in Sec. 7, the results of our measurements are presented with a

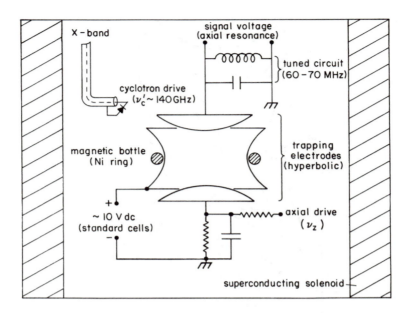

Figure 2. Schematic of the geonium apparatus. The appropriately biased hyperbolic
endcaps and ring electrodes trap the charge axially while coupling the driven harmonic
motion to an external LC circuit tuned to the driven axial frequency. Radial trapping
of the charge is produced by the strong magnetic field obtained from a superconducting
solenoid. Also shown are the microwave multiplier diode and nickel ring for magnetic
coupling to the axial resonance (from Ref. 6).

particular emphasis on the new phosphor bronze trap which was constructed specifically to investigate possible systematics encountered in these experiments.

2. Trapping Device

2.1. BASIC DESCRIPTION

The Penning trap is shown schematically in both Figs. 1 and 2 as a three electrode
trap with cylindrical symmetry. The magnetic field, B_0, (obtained from a stable
superconducting solenoid) is applied along this symmetry axis and is designated
by convention as the z-axis. Using cylindrical coordinates r, z, these electrodes are
machined along hyperboloids of revolution (relative to the z-axis) given by:

$$z^2 = Z_0^2 + r^2/2 \quad (endcaps)$$
$$r^2 = R_0^2 + 2z^2 \quad (ring)$$

(2)

where $2Z_0$ and $2R_0$ represent the minimum endcap separation and the minimum ring diameter respectively. A potential V_0 is applied to the ring electrode relative to the grounded endcaps by means of the standard cells shown in Fig. 2; as a result, the following harmonic potential (shown in Fig. 1(b)) is established:

$$V(r, z) = V_0 \frac{r^2 - 2z^2}{4d^2} \qquad (3)$$

where d is the characteristic trap dimension defined by $4d^2 = 2Z_0^2 + R_0^2$. This voltage distribution then produces the harmonic restoring force that binds the charge axially, assuming the correct sign of the potential has been chosen.

Immersing the entire apparatus in liquid helium produces a low background gas environment in these devices. This low ambient temperature also improves signal-to-noise (S/N) of the detector such that single charged particles can be readily observed. The endcap opposite the LC circuit is used to drive the axial motion which then induces an image current into this external circuit, tuned to the particle's axial frequency. The resulting voltage across the tuned circuit is amplified by the following preamplifier. The magnetic bottle, shown schematically as a nickel wire in Fig. 2 is used to couple the total magnetic moment to the axial resonance. The microwave power which drives the observed cyclotron motion is obtained from an X-band source that is multiplied up to the appropriate frequency by a Schottky diode placed inside the vacuum envelop between an endcap and the ring.

The motion of the trapped electron is described for an ideal Penning trap by the following set of equations:

$$\omega_z^2 = \frac{eV_0}{m_e d^2} \qquad (4a)$$

$$\omega_c = \frac{eB_0}{m_e c} \qquad (4b)$$

$$2\omega_\pm = \omega_c \pm (\omega_c^2 - 2\omega_z^2)^{1/2} . \qquad (4c)$$

The first frequency in Eq. 4 (for ω_z) is associated with the axial harmonic oscillation of the trapped particle and is the subject of Sec. 3. The second is its unperturbed cyclotron frequency (in the absence of the Penning trap). The last equation represents the radial motion; in the ideal case, $\omega_- \equiv \delta_e$ is the frequency of the slow magnetron motion (the subject of Sec. 4) and $\omega_+ \equiv \omega_c'$ is the perturbed cyclotron frequency (the subject of Sec. 5). The composite motion is then a superposition of all three normal mode frequencies, as shown in Fig. 3.

For the ideal trap, it follows that

$$\omega_c' = \omega_c - \delta_e \quad \text{and} \quad \delta_e = \omega_z^2/2\omega_c' . \qquad (5)$$

Three Modes of Motion

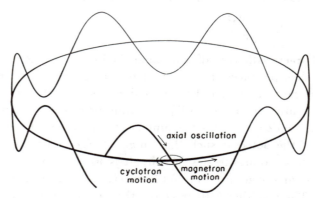

Figure 3. Illustration of the motion of a charge isolated within a Penning trap. The magnetron motion is characterized by an orbit radius (~ 0.014 mm) which is at least 1000 times larger than the typical cyclotron radius. Driven axial amplitudes are also typically less than the magnetron radius since they are usually less than the rms-thermal amplitude (~ 0.018 mm) (adapted from Ref. 13).

However, in real traps, we find that the observed magnetron frequency, ω_m, is slightly larger than δ_e due to trap imperfections and a non-zero angle between $\vec{B}_0$ and $\hat{z}$. But, since $\omega_m \ll \omega_z \ll \omega'_c$, we can ignore the difference between δ_e and ω_m.

The last of the important frequencies is associated with the spin magnetic moment of the electron. It is purely quantum mechanical in nature and is given by

$$\hbar\omega_s = g\mu_B B_0 \tag{6}$$

where μ_B is the Bohr magneton ($\mu_B \equiv e\hbar/2m_e c$). Producing a strong enough oscillating magnetic field at the very high frequency ω_s ($\nu_s \sim 140$ GHz) turns out to be somewhat difficult. Using the near degeneracy of ω_s and ω'_c, it is far more convenient to excite the difference frequency, $\omega'_a = \omega_s - \omega'_c$, which is referred to as the anomaly frequency and is the subject of Sec. 6. Thus, in a Penning trap, the anomaly is obtained from

$$a_e = \frac{\omega'_a - \delta_e}{\omega'_c + \delta_e} \tag{7}$$

which turns out to be quite accurate and insensitive to all leading perturbations such as angle-tilt and electrode eccentricity [20], as well the anharmonicity of the trapping field.

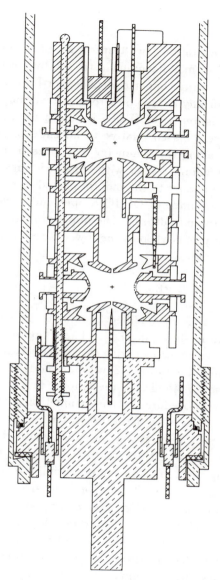

Figure 4. Cross-section of the double trap apparatus used in the electron-positron comparison experiment. An OFHC copper pin base with non-magnetic feedthrus is sealed to an all-metal beryllium copper envelope via a compressed indium O-ring. The lower trap is a well compensated Penning trap used for precision measurements; the upper trap is used only for the storage of positrons which can be transferred into the lower "experiment" trap when needed (from Ref. 6).

2.2. A VACUUM TUBE

The present apparatus containing two traps in series is shown in Fig. 4. Initially both traps consisted of all-molybdenum electrodes, but recently the lower "experiment" trap was rebuilt using copper and phosphor-bronze. The upper "storage" trap was designed specifically to trap positrons and will be described in a following subsection. A field emission point (FEP) is shown in the endcap closest to the pin-base and is negatively biased relative to all endcaps in order to produce an ionizing electron beam along the symmetry axis of each trap. The outer vacuum envelope is made of beryllium copper with threaded rings (of the same material) that are used to apply pressure to the indium melted into each end of the vacuum tube. The pin bases are made of OFHC copper and each contains several cryogenic ceramic-to-constantan feedthrus. The upper pin base (not shown) is used for mounting a sputter-ion pump whereas the lower base locates the experiment trap at the solenoid's field center.

2.3. COMPENSATION FEATURE

One of the main features of our modified Penning traps is the use of compensation electrodes [21] of the form shown in Fig. 4. These "guard" rings protrude into the truncation region between endcap and ring electrodes. (Later versions were also split on one side for "anomaly" excitation.) Several methods are available to tune up the trap. The most common method uses the amplitude asymmetry versus frequency as shown in Fig. 5. The procedure involves symmetrizing the axial resonance using increasingly stronger drives. Until the $6th$-order perturbation enters, symmetry should occur at the same guard setting. A second method involves minimizing the noise in the correction signal of the axial-frequency-shift detector described in the Sec. 3.3. Upon varying the compensation potential, a dc shift in the well depth is also observed by means of this shift detector. The size of this effect was observed to depend exponentially on how close the compensation electrodes are placed in the truncation region relative to trap center and the size of the gap at truncation. In addition, the ability to tune the trap was observed to scale with this shift in well depth for symmetric Penning traps. Later, these observations were verified [22] theoretically by Gabrielse using a numerical relaxation program which computes the exact field in the real Penning trap. For Fig. 5, the normalized $4th$-order coefficient C_4, of the anharmonic potential is negative when the resonance pulls to the low frequency side, thus indicating that the guard potential was set too low.

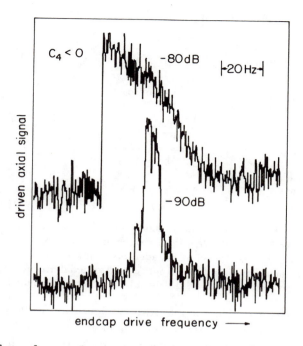

Figure 5. Typical case for an anharmonic axial resonance swept down in frequency. When the normalized coefficient, C_4, of the $4th$-order term in the potential is negative (i.e. the guard potential is set too low), the resonance is pulled to low frequency side. Note that 10 dB less drive nearly restores the completely symmetrical line shape (from Ref. 12).

2.4. ADAPTATIONS FOR POSITRONS

Since the basic geonium apparatus can accept "+" or "−" charges simply by applying the potential of appropriate sign, the full sensitivity of the experiment is available for positrons as well as for electrons with all the systematics that are discussed in the following sections being the same for both. Thus, the principle challenge for this variation of the geonium experiment is the trapping of a single positron which was first demonstrated [23,24] by Schwinberg in 1979. The basic mechanism for losing some of the positron's energy, such that it might fall into a potential well, is by radiation damping of the axial motion due to the resistance of the LC circuit tuned to ω_z. For the radial motion, synchrotron radiation quickly damps the 50–100 keV initial energy in the cyclotron motion within a few seconds down to the thermal range.

Since the requirements imposed on the trapping electrodes for high precision measurements are inconsistent with those imposed in order to achieve an adequate capture rate, it was decided to have two traps, one of which would be optimized

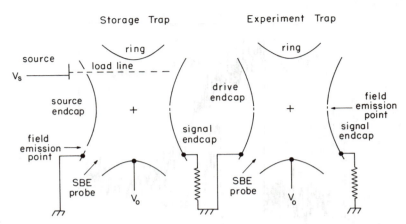

Figure 6. Schematic of the double-trap configuration. The storage trap contains a sealed ^{22}Na positron emitter located at $3R_0/4$. The sideband excitation (SBE) probes are used in the radial centering process for off-axis-loaded positrons and position stabilizing in the experiment trap. Each signal endcap is tuned to the axial frequency via an external inductor for detection of axial motion, rf driven from an opposite endcap (from Ref. 47).

to capture the positrons. This double-trap configuration is shown schematically in Fig. 6. The source of the positrons consists of a 2.6 yr half-life sodium-22 salt in a sealed container and is visible in Fig. 4, located inside the largest endcap (farthest from the pin-base). In order to fit both a field emission point (FEP) and the source within the same endcap at the same radius, as well as to improve the trapping rate, each is mounted at $3R_0/4 \sim 3.55$ mm from the central axis. The FEP is required in order to optimize the centering process using an adequate number of easily obtainable charges. The source hole through this endcap is 1.2 mm in diameter and the line, parallel to the trap axis, which passes through this source hole is referred to as the "load line". The positrons of interest (with 50–100 keV in the radial motion and a cyclotron orbit smaller than the entrance hole) will travel down this load line guided by the axial magnetic field, reflect at the opposite endcap, reflect again at the source cap because the magnetron motion has rotated it away from the load line, and finally lose enough energy after one magnetron period to be permanently trapped. Only a very small fraction of those emitted satisfy all these criteria, typically $\sim 10^{-9}$. Due to the extremely anharmonic nature of this trap, positrons loaded at $3R_0/4$ can not easily be made visible unless there are more than ~ 50 in the cloud. The cloud must first be centered by a strong axial-magnetron sideband drive (see Sec. 4) in order to observe and estimate its size or possibly transfer part of it into the experiment trap.

The actual transfer is accomplished by pulsing the two adjacent endcaps (storage trap signal cap and experiment trap drive cap) to the common ring potential for a

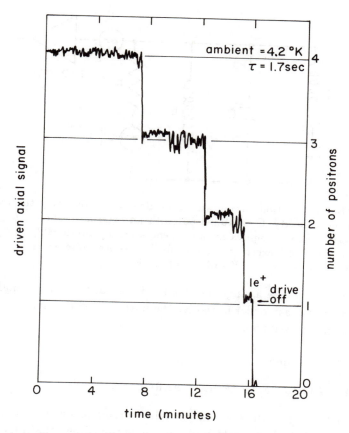

Figure 7. Positron ejection record. Positrons transferred from storage to the experiment trap are detected via strong off-resonance drives (with fixed amplitude and frequency). Then, intense rf pulses at $\omega_z + \omega_m$ are applied to the SBE probe in order to systematically eject one charge at a time until only one remains (from Ref. 45 and 47).

few microseconds. Typical transfer efficiencies between 25 and 50% have been observed; however with pulses much longer than 10 μs, the transfer tends to be very inefficient, possibly due to radial drifting during the passage between traps. Once positrons have been identified to be in the experiment trap, their exact number is estimated (see Sec. 3) and the excess beyond one are systematically ejected (see Fig. 7) by using intense rf pulses, also at the sideband cooling frequency. The rf amplitude of the ejection/cooling pulse is carefully adjusted such that at least 10 consecutive pulses are required in order to eject one positron from the cloud. Note however that empirically the probability for being thrown out increases as the number drops since the individual charges have larger amplitudes as the number decreases (i.e. center-of-mass amplitude scales inversely with the number trapped

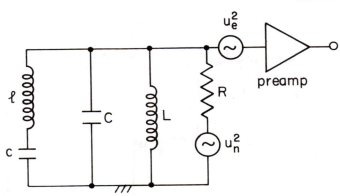

Figure 8. Input detection circuit with equivalent ℓc representing the trapped electron. The trap's net capacity, C, is tuned out with an external inductor, L, yielding a large parallel resistor, R, with equivalent noise generator, u_n^2, presumably at 4 K. The preamp is represented as ideal except for the equivalent series noise generator depicted by u_e^2 which is presumably *not* at 4 K (adapted from Ref. 25).

for fixed drives). Once a single positron is isolated, the drive signal is reduced by orders of magnitude in order to resolve the narrow (4–6 Hz) axial resonance, typical of a well-compensated trap.

3. Axial Resonance

3.1. OBSERVATION OF TRAPPED CHARGE

As indicated in Fig. 2, the trapped particle is driven on one endcap and detected on the other cap. A fairly straight forward calculation shows that driving the trapped electron is equivalent to exciting a series ℓc tuned circuit [25]. For a single electron with mass m_e and charge e, the series inductance becomes

$$\ell_1 = \frac{4m_e Z_0^2}{(C_1 e)^2} \tag{8}$$

where C_1 is a dimensionless constant of order unity which represents the finite geometry of the electrodes. Note that in this expression, $C_1 e$ represents the *effective* charge on the electron at trap center. For an infinite parallel plate arrangement, $C_1 = 1$, but for a Penning trap, $C_1 < 1$ because the effective charge is reduced by the attraction of some lines of force emanating from the electron onto the ring electrode (and is theoretically predicted [26] to be 0.8). The electron oscillator in the Penning trap is therefore replaced by the equivalent capacitor c_1 in series with equivalent inductance ℓ_1 such that $\omega_z^2 = (\ell_1 c_1)^{-1}$. As shown in Fig. 8, the ℓc series

TABLE 1. Summary of experimental parameters for the last two traps.

symbol	description	molybdenum trap	phosphor bronze trap
$\omega_z/2\pi$	axial frequency	64.0 MHz	72.7 MHz
$\omega_m/2\pi$	magnetron frequency	14.5 kHz	18.5 kHz
$\omega_c'/2\pi$	cyclotron frequency	141 GHz	1424 GHz
$\omega_a'/2\pi$	anomaly frequency	164 MHz	165 MHz
$\gamma_z/2\pi$	axial linewidth	6 Hz	6 Hz
$\delta/2\pi$	magnetic bottle step-size	1.3 Hz	± 1.8 Hz
B_0	magnetic field	50.5 kG	50.7 kG
B_2	magnetic bottle	155(4) G/cm^2	$\pm 254(10)$ G/cm^2
V_0	ring-endcap potential	10.2 volts	10.2 volts
$2Z_0$	minimum endcap separation	6.70 mm	6.29 mm
$2R_0$	minimum ring diameter	9.47 mm	7.57 mm
R_m	typical magnetron radius	~ 0.0014 cm	~ 0.0014 cm
R_c	cyclotron radius for $n=0$	114 Å	114 Å
T.C.	trap constant	20.0 MHz/V$^{1/2}$	22.6 MHz/V$^{1/2}$
Q	tuned circuit quality factor	~ 1000	~ 900
C_1	$\vec{E}$-field coupling constant	0.78(4)	~ 0.76
ℓ_1	single lepton inductance	4500 H	3600 H
C	tuned circuit capacitance	15 pf	15 pf
R	tuned circuit resistance	170 kΩ	140 kΩ
C_4'	effective anharmonic coefficient	$\sim 5 \times 10^{-5}$	$\sim 1 \times 10^{-4}$
$B_2'Z_0^2/B_0$	$\vec{B}$-field inhomogeneity	3.4×10^{-4}	$\pm 4.9 \times 10^{-4}$
$eV_0/2m_ec^2$	well depth/rest energy	1.0×10^{-5}	1.0×10^{-5}
kT_z	thermal energy, at $T_z = 4.2$ K	3.6×10^{-4} eV	3.6×10^{-4} eV
$\hbar\omega_c'$	quantum of cyclotron energy	5.8×10^{-4} eV	5.9×10^{-4} eV
$\hbar\omega_z$	quantum of axial energy	2.6×10^{-7} eV	3.0×10^{-7} eV
$\hbar\omega_m$	quantum of magnetron energy	6×10^{-11} eV	8×10^{-11} eV
$-eV_0R_m^2/2R_0^2$	typical magnetron energy	$\sim -5 \times 10^{-5}$ eV	$\sim -5 \times 10^{-5}$ eV

circuit is effectively connected in parallel with the net trap capacitance, C, which is then put in parallel with an external inductor L such that the LC combination is also resonant at ω_z. This external parallel tuned circuit then provides the large series R that damps the axial motion.

Now, when the electron is subjected to an axial rf electric field, the driven motion will induce image currents in the signal endcap or an equivalent current [27]

$$i = \frac{(C_1 e)\dot{z}}{2Z_0} \tag{9}$$

will flow through the ℓc series combination. To be observed, we drop this current across the large resistor R whose value is determined by the Q of the external tuned circuit. As summarized in Table 1 for the recent molybdenum trap, some typical parameters are $C \sim 15 \text{pf}$, $\nu_z \sim 64$ MHz, $Q \sim 1000$, and the equivalent parallel resistance is $R = Q/\omega_z C \sim 170 \text{k}\Omega$. The steady-state axial motion, driven by an electric field at frequency ω with amplitude E_{rf}, has an axial amplitude given by

$$z = \frac{(eE_{\text{rf}}/m_e)}{\{(\omega_z^2 - \omega^2)^2 + \gamma_{z,1}^2 \omega^2\}^{1/2}} \tag{10}$$

where $E_{\text{rf}} = C_1 V_{\text{rf}}/2Z_0$ for a given rf voltage applied to the drive endcap only and $\gamma_{z,1}$ represents the single electron's damping coefficient for the harmonic motion at ω_z. From Eq. 10, it follows that the relative driven on-resonance amplitude Z_d is

$$\frac{Z_d}{Z_0} = \left(\frac{C_1 V_{\text{rf}}}{2V_0}\right)\left(\frac{\omega_z}{\gamma_{z,1}}\right) \approx V_{\text{rf}}/2.5\mu\text{V} \tag{11}$$

where typically $V_{\text{rf}} \sim 10$ nV. From the expression for total energy of a harmonic oscillator $(m_e \omega_z^2 Z_d^2)$, the relative energy for a driven electron in a Penning trap is:

$$\frac{W_k}{eV_0} = \left(\frac{Z_d}{Z_0}\right)^2 = \left(\frac{2V_{\text{rf}}}{C_1 eR\omega_z}\right)^2. \tag{12}$$

It should be noted that the formal definition of Q_z of the axial resonance $(Q_z \equiv \omega_z/\gamma_z)$ is 2π times the average energy stored $(\overline{\ell i^2})$ divided by the energy lost/cycle (power dissipated $\times$ period $= i^2 R/\omega_z$), from which it also follows that the ideal axial linewidth (FWHM) is $\gamma_z = R/\ell$. For a $2\pi(6$ Hz) wide line in the experiment trap shown in Fig. 4, this relation predicts $\ell_1 \sim 4,500$ H for a single electron.

The equivalent current on resonance $(\omega = \omega_z)$ can be found from Eq. 10 by substituting $\dot{z} = \omega z$ into Eq. 9:

$$i = \frac{(C_1 e)(eE_{\text{rf}})}{2\gamma_z m_e Z_0} = \frac{V_{\text{rf}}}{R} \tag{13}$$

where the expression for ℓ_1 has been used. For a moderate drive applied to the endcaps which is effectively -160 dB down from a 13 dBm (or 1 volt rms) drive source, one finds a typical current of $\sim 10^{-13}$ amps for a single electron with ~ 6 Hz axial linewidth. This current also corresponds to a drive energy that is about equal to kT_z when $T_z = 4$ K. As a result, the on-resonance signal amplitude is about 17 nV in this case. If $\Delta\nu_{\text{det}}$ is the detection bandwidth, then the signal voltage can be compared to the rms noise voltage associated with a resistance R given by

$$V_N/(\Delta\nu_{\text{det}})^{1/2} = (4kT_z R)^{1/2} \tag{14}$$

which, for a 170 kΩ resistor in liquid helium, yields 6 nV/$\sqrt{Hz}$. As one can see in Fig. 8, if the trap is ideal and the ℓc circuit is tuned to make it resonate at the parallel tuned circuit frequency, the series circuit would totally short out the resistor R and

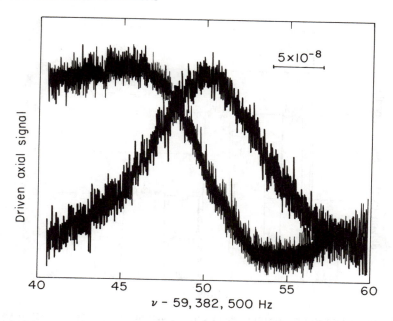

Figure 9. Axial resonance signals at ≈ 60 MHz. The signal-to-noise ratio of this ≈ 8 Hz wide line corresponds to a frequency resolution of 10 ppb. Both absorption and dispersion modes are shown with the latter mode appropriate for the frequency shift detection scheme employed in these geonium experiments (from Ref. 5).

its equivalent noise generator u_n^2. The remaining noise in the experiment would then come from the series equivalent input noise generator u_e^2 associated with the imperfect preamplifier. However, the anharmonic part of the real trapping potential keeps the finite temperature electron from totally shorting the trap. Experimentally, we see the noise reduce by about 50% when on resonance relative to the off-resonance noise. An example of the quadrature components of this axial resonance is shown in Fig. 9. The resolution in this case is $\sim 1 \times 10^{-8}$ or about 0.5 Hz out of 60 MHz, limited by the voltage fluctuations in the standard cells shown in Fig. 2.

3.2. ISOLATING THE SINGLE ELECTRON

When the cloud contains N electrons, the center of mass contains NC_1e effective charge and moves with velocity $\dot{z}_{cm}$, yielding (from Eq. 9) a total current

$$i_{cm} = \frac{NC_1e\dot{z}_{cm}}{2Z_0} . \tag{15}$$

However, inspection of Eq. 13 shows that on resonance and under ideal conditions, the observable voltage signal is simply V_{rf}, the value of the rf potential applied to the drive endcap only. There is no number dependence in this case. Instead, we detune

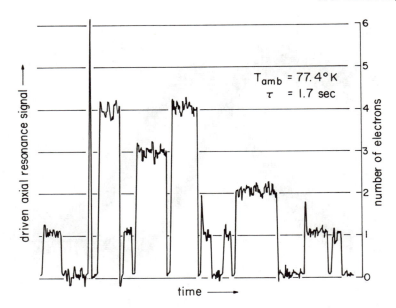

Figure 10. Identification of a single electron. A strong off-resonance signal (with fixed frequency and amplitude) is applied continuously to the drive endcap of the trap. Then, the FEP is turned on for $\sim$ 10 sec with $\sim$ 0.1 nA of current until a nonzero signal is observed. The trap is occasionally dumped and reloaded until the smallest quantized signal is observed (from Ref. 5).

our drive oscillator by an amount $\delta\omega$ such that $\omega_z \gg \delta\omega \gg \gamma_{z,1}$. Again taking $\dot{z} = \omega z$ and evaluating Eq. 10 for $\omega = \omega_z \pm \delta\omega$, we find that the amplitude of the velocity is

$$\dot{z}_{\mathrm{cm}}(\mathrm{off}) = \frac{eE_{\mathrm{rf}}}{2m_e\delta\omega} \tag{16}$$

which is the same for a single or N electrons. From Eq. 15, it follows that

$$V_{\mathrm{sig}}(\mathrm{off}) = N\left[\frac{V_{\mathrm{rf}}R}{2\ell_1\delta\omega}\right] \tag{17}$$

for the off-resonance signal. If the rf amplitude and the detuning are held constant, the signal voltage is proportional to the number of charge quanta in the trap.

To load this charge, a current of $\sim$ 0.1 nA is emitted from the field emission electrode for 10–20 seconds. After each load, the detection system is switched on and the signal observed. Then, the trap is dumped by reversing the sign of the trapping potential. The experiment is repeated and the results are shown in Fig. 10. Note that there is a minimum voltage step, of which all others are integer multiples. After the minimum voltage step is achieved, the drive is reduced to the nominal on-resonance power and the frequency is swept through the resonance,

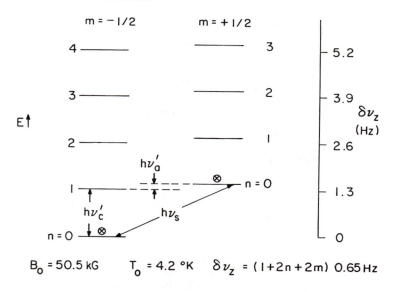

$$B_0 = 50.5 \text{ kG} \qquad T_0 = 4.2 \,^{\circ}\text{K} \qquad \delta \nu_z = (1 + 2n + 2m) \, 0.65 \, \text{Hz}$$

Figure 11. Lowest Rabi-Landau levels for a geonium atom. The axial frequency (shown in the right-hand scale) corresponds to the coupling via the fixed magnetic bottle field for the last molybdenum trap. The lowest state ($n = 0$) which is occupied by the electron or positron 80-90% of the time differs by 1.3 Hz depending on the exact spin state. This is the signature used to indicate that a spin has flipped (from Ref. 6).

obtaining the result shown in Fig. 9. We have also verified that at multiples of this minimum voltage step, the axial resonance linewidth varies as multiples of the R/ℓ_1. The amplitude signal (which is the Lorentzian line shape) is used to monitor the lock signal (see Sec. 3.3.) which is itself derived from the quadrature component (or dispersion shape also shown in Fig. 9).

3.3. MAGNETIC COUPLING

The ultimate goal of this experiment is to observe the magnetic resonances, ω'_c and ω_s (or ω'_a) but observation must come indirectly through the purely electrostatic axial resonance. In order to provide a weak coupling to this frequency, a small magnetic bottle is produced from a ring of nickel (as used in the early compensated Penning traps) or from a superconducting loop of NbTi, either of which is placed in the midplane of the ring electrode. The latter is the secondary of a transformer whose current can be externally adjusted to produce what we refer to as our "variable bottle" [28]. If B_2 represents the quadratic gradient, the net bottle field can be described in cylindrical coordinates, r, z, by

$$B_z = B_0 + B_2(z^2 - r^2/2) \quad \text{and} \quad B_r = -B_2 rz \qquad (18)$$

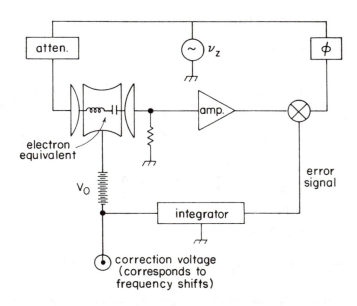

Figure 12. A schematic representation of the superheterodyne system used to detect the electron's motion. The upper half is the standard axial drive and synchronous detection whereas the lower half generates the frequency shift information. The feedback signal is generated by mixing the phase-shifted drive synthesizer with the amplified electron signal to produce an error voltage. This error is integrated and fed back as a correction voltage to the ring bias circuitry, thereby producing a frequency lock (from Ref. 5).

where B_0 is the resultant constant-field term.

The coupling to the axial frequency follows from the orientational potential energy associated with the total magnetic moment $\vec{\mu}$ of the electron relative to the total field $\vec{B}$: $W_m = -\vec{\mu} \cdot \vec{B}$. This term can now be added to the electrostatic potential energy, $-eV(r, z)$, to yield the resultant axial frequency [7]. When combined with the total magnetic moment for the electron, we obtain the quantum-mechanically correct expression (with $g \approx 2$):

$$\omega_z = \omega_{z,0} + (n + m + \frac{1}{2} + \frac{\omega_m}{\omega_c'}q)\delta \tag{19}$$

where n, m and q are cyclotron, spin and magnetron quantum numbers respectively and $\delta = 2\mu_B B_2/(m_e\omega_z) \sim 2\pi(1.3 \text{ Hz})$ for electrons in the last molybdenum trap shown in Fig. 4, before the variable-bottle trap. Figure 11 shows the Landau level scheme for this "geonium atom" and the associated shifts in the axial frequency.

In order to observe these shifts, the axial motion is incorporated into a feedback loop as illustrated in Fig. 12. In this apparatus, the dispersion-shaped axial resonance (shown in Fig. 9) is used as an error signal applied to an integrator, whose output fine-tunes the total ring voltage. If something causes ω_z to shift relative to the synthesizer output, then a nonzero error is integrated until ω_z again coincides with the synthesizer's reference frequency and the error returns to zero. The output of the integrator is the correction signal which will now reflect any shift in the axial frequency due to all sources (i.e. noisy voltage, anharmonic shifts with change in stored energy, tilt in the trap axis, and in particular, cyclotron and/or spin quantum jumps via the magnetic bottle).

3.4. OTHER AXIAL FREQUENCY SHIFTS

The largest perturbation is associated with the non-zero angle, θ, between the magnetic field and the axis of symmetry for the trap. The effect of such a perturbation is to weakly couple all the normal mode frequencies producing small, but observable shifts. In particular, when $\theta \ll 1$, Brown and Gabrielse have predicted [20] that the observable axial frequency $\overline{\omega}_z$ is

$$\overline{\omega}_z^2 \approx \omega_{z,0}^2 \left(1 - \frac{3}{2}\theta^2\right). \tag{20}$$

(In principle, maximizing $\overline{\omega}_z$ will remove the angle θ.) Typically, the relative shift in the axial frequency is $-3\theta^2/4 \sim 10^{-5}$ for $\theta \sim 0.0035$ radians (or $\sim 0.2°$). For small fluctuations, $\delta\theta$, in the average residual angle $\overline{\theta}$, it follows from Eq. 20 that

$$(\delta\omega_z/\omega_z)_\theta = -\frac{3}{2}\overline{\theta}\delta\theta. \tag{21}$$

Because the observed stability of ω_z due to all causes approaches 1×10^{-8}, the average fluctuations must be less than 0.1% of $\overline{\theta}$.

There are also a few small perturbations associated with the non-uniformity of the magnetic field, the non-quadratic terms in the electric potential, and the relativistic corrections. Their effects on all the normal modes have been studied by Brown and Gabrielse. From their analysis [8], one can show that the most significant corrections to ω_z are:

$$\frac{\Delta\omega_z}{\omega_z} = \frac{B_2' Z_0^2}{B_0} \cdot \frac{W_n}{eV_0} + \frac{3C_4'}{2} \cdot \frac{W_k}{eV_0} + \left(6C_4' - \frac{B_2' Z_0^2}{B_0} + \frac{eV_0}{m_e c^2}\right)\frac{W_q}{eV_0} \tag{22}$$

where W_n, W_k, and W_q represent the actual energies associated with the cyclotron, axial, and magnetron motions respectively. The first term is the effective bottle shift associated with cyclotron excitation, where $B_2' = B_2 - (eV_0/2m_e c^2)(B_0/Z_0^2)$ is reduced by the relativistic mass shift. The second term is the effective anharmonic shift of the axial frequency with axial excitation, with reduced coefficient: $C_4' = C_4 -$

$eV_0/4m_e c^2$. In Table 1, we list the relevant dimensionless ratios for our experiment: $B_2' Z_0^2 / B_0$, C_4', and $eV_0/2m_e c^2$; each is less than $\sim 5 \times 10^{-4}$. For the energies, one can obtain W_k from Eq. 12, W_q from $-(eV_0/2)(R_m/R_0)^2$ and W_n from $(n + \frac{1}{2})\hbar\omega_c'$ with $n < 10$. Using some parameters from Table 1, we find that all energy ratios should be less than 5×10^{-4}, yielding the upper limit on the relative shifts in ω_z, due to these perturbations, at 3×10^{-7}.

Finally however, there is one large shift in axial frequency which is used for calibration purposes. When very strong anomaly drives are applied to one endcap or to the split guard rings (see Sec. 6), the axial frequency is found to shift proportional to the drive power. For anomaly power applied to the drive endcap only, as in the very earliest geonium experiments, the shift was a direct measure of the amplitude of the electric field used as an off-resonance axial drive. When applied to the guard rings on the other hand, the circulating current in a second LC circuit, tuned to ω_a', develops a differential voltage across the loop inductance of the guards which is then capacitively coupled over to the endcaps. Since the drive cap typically has a very low impedance, unlike the signal cap, the rf drive is again asymmetrically applied. The shift in axial frequency itself is due to the production of a pseudo-potential associated with the rf trapping field as routinely observed in Paul traps [29].

4. Magnetron Motion

4.1. METASTABLE RESONANCE

The next normal mode which is relatively easy to observe is the slow guiding-center motion in the radial plane at the magnetron frequency, ω_m. The primary interest in determining this frequency is that it allows us to determine the angle between $\vec{B}$ and $\hat{z}$. This is quite important in the double trap configuration shown in Fig. 4 because positrons can not transfer into the experiment trap (see Sec. 2.4.) if the angle is much greater than $0.25°$. Actually, this motion should properly be called an anti-resonance since it has a radial potential hill, not a binding potential well that would naturally restore any displacement from equilibrium. The motion, however, is indeed metastable since the decay out of the trap by all dissipative forces is very slow, as experimentally verified to be $\gg 10^5$ sec.

In order to observe this frequency, an additional drive at ω' is applied to a guard probe that protrudes slightly into the gap between endcap and main ring electrode in order to produce the sideband excitation (SBE). This drive field is not uniform at the center of the trap, and will contain an axial component of the form $\vec{E}' = Ay\hat{z}$ where coordinate axes are preferentially chosen such that the probe lies in the yz-plane. Thus if $A = A_0 \sin(\omega_z + \omega_m)t$ and $y = R_m \sin\omega_m t$ for the instantaneous position of the electron, then, the sideband drive becomes:

$$\begin{aligned}
\vec{E}' &= [A_0 \sin(\omega_z + \omega_m)t][R_m \sin\omega_m t]\hat{z} \\
&= \frac{A_0 R_m}{2} \cos[\omega_z t]\,\hat{z} - \frac{A_0 R_m}{2} \cos[(\omega_z + 2\omega_m)t]\,\hat{z}\,.
\end{aligned} \tag{23}$$

Thus, the first term appears to be a constant axial drive, similar to the one used to lock up the axial resonance. Note that the strength of the drive field is proportional to R_m. Now suppose this auxiliary drive is not exactly resonant, i.e. $\omega' = \omega_z + \omega_m + \delta$; then, a sharp resonant "perturbation feature" exists δ in frequency away from ω_z. This feature includes a beating of the free magnetron motion with the driven motion and when this SBE drive is swept, such that $\delta \to 0$, the beat note is faithfully reproduced in the correction signal shown in Fig. 13 (for the phosphor bronze trap), assuming the loop time constant is not too long. By sweeping both directions, the onset of perturbation-lock allows a determination of ν_m to better than 0.01 Hz or 1 ppm.

4.2. CHANGING THE MAGNETRON ORBIT

An uncontrolled parameter could exist when electrons are loaded into the trap, due to the random initial magnetron orbit radius. Since the amount of radial wander with time out of the trap depends on the initial orbit radius and since the magnetic field is purposely made inhomogeneous (see Sec. 3.3.), the electron's cyclotron frequency would vary in time in some non-linear way. Therefore, by necessity, this means shrinking the magnetron orbit to as close to zero as possible.

The nature of the centering process is best visualized by considering conservation of energy. When a photon of energy, $\hbar(\omega_z + \omega_m)$, is absorbed from the SBE probe drive, $\hbar\omega_m$ is absorbed by the magnetron motion while the remaining part, $\hbar\omega_z$, is harmlessly added to the axial motion which quickly damps away because of the strongly coupled tuned circuit, held at some fixed temperature T_z. Now, the magnetron energy is strongly dominated by the negative radial potential hill given by $-m_e\omega_z^2 R_m^2/4$, compared to the very small kinetic energy $-m_e\omega_m^2 R_m^2/2$. Thus, adding a positive amount of energy $\hbar\omega_m$ to this negative hill will reduce the magnetron radius, R_m. This we designate as a "cooling" or "centering" drive, in

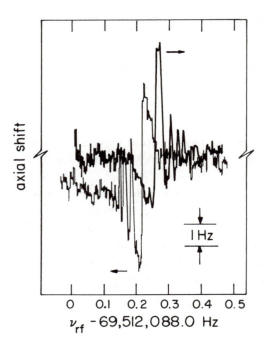

Figure 13. The $\omega_z - \omega_m$ cooling resonance using SBE probes in the phosphor bronze trap. The magnetron sideband of the axial resonance is observed as a pulling of the locked axial resonance and a subsequent beating between the excited magnetron motion and the applied magnetron drive. Such resonances allow ω_m to be determined to 1 ppm and the magnetron radius to be reduced because of the absorbed energy $\hbar\omega_m$.

contrast to the application of a sideband at $\omega_z - \omega_m$ which drives the electron radially out of the trap. This technique [5] is also referred to as "motional sideband cooling".

In principle, cooling will continue to occur by application of this upper sideband drive until the occupation quantum numbers (k for axial, q for magnetron) in the two separate motions become equal [8], i.e. $q = k$. In terms of the thermodynamic temperature T_z of the axial motion, the average axial energy can be written $\hbar\omega_z(k + \frac{1}{2}) = k_B T_z$. If the driven energy, used to lock the axial frequency, is much less than $k_B T_z$, then T_z is essentially that of the thermal reservoir produced by the strongly coupled tuned circuit and the preamplifier. The minimum magnetron energy is determined by the condition:

$$-m_e\omega_z^2 R_m^2/4 = -\hbar\omega_m(q + \frac{1}{2})\,\big|_{q=k} \tag{24}$$

and the resulting theoretical minimum orbit, $R_{\min}$, can then be written

$$R_{\min} = \left(\frac{2ck_B T_z}{e\omega_z B_0}\right)^{1/2} \tag{25}$$

which explicitly shows the dependence on magnetic field. It is also insensitive to well depth varying as the inverse $4th$ root. For typical conditions listed in Table 1, $R_{\min} \sim 6 \times 10^{-5}$ cm. However, we are unable to obtain this theoretical limit due to some unknown heating mechanism since we observe [5] ~ 20 times this value. One possible cause could be the finite angle between $\vec{B}$ and $\hat{z}$ which could couple the modes, thus allowing the axial drive to be a direct source of residual heating.

4.3. FREQUENCY SHIFTS

As it happens, the magnetron frequency is very sensitive to the angle of tilt, θ, which occurs between $\vec{B}$ and $\hat{z}$ as well as another parameter, ϵ, which is associated with the asymmetry of the electrodes (or simply their average ellipticity). From the 1982 analysis of Brown and Gabrielse [20], the observed magnetron frequency $\bar{\omega}_m$ can be approximated by the following formula (for $\epsilon \ll 1$ and $\theta \ll 1$):

$$\bar{\omega}_m \approx \bar{\delta}_e \left[1 - \frac{\epsilon^2}{2} + \frac{9}{4}\theta^2\right] \tag{26}$$

where $\bar{\delta}_e$ is defined in terms of the observed axial and cyclotron frequencies: $\bar{\delta}_e = \bar{\omega}_z^2 / 2\bar{\omega}_c$. The condition $\epsilon \ll 0.01$ is expected to hold for these high precision Penning traps (by construction) and thus may be ignored in most cases in comparison to the effect of the tilt angle θ. All large compensated Penning traps used in electron g-2 work have always had $\bar{\omega}_m > \bar{\delta}_e$ which is consistent with the dominance of θ.

Again, there are a few minor perturbations associated with the non-uniformity of the magnetic field, the non-quadratic terms in the electric potential, and the relativistic corrections. From the 1986 analysis by Brown and Gabrielse [8], we find that the most significant of these corrections to ω_m are:

$$\frac{\Delta\omega_m}{\omega_m} = \left\{\frac{2B_2 Z_0^2}{B_0}\right\}\frac{W_n}{eV_0} + \left\{6C_4 - \frac{B_2 Z_0^2}{B_0}\right\}\frac{W_k}{eV_0} + \left\{6C_4 - \frac{2B_2 Z_0^2}{B_0}\right\}\frac{W_q}{eV_0} \tag{27}$$

where all (neglected) relativistic terms are less significant by at least $(\omega_z/\omega_c')^2 < 3 \times 10^{-7}$. Each term in this shift is at least an order of magnitude smaller than the one obtained from Eq. 26 and is also below our level of sensitivity except for excitation energies > 0.01 eV.

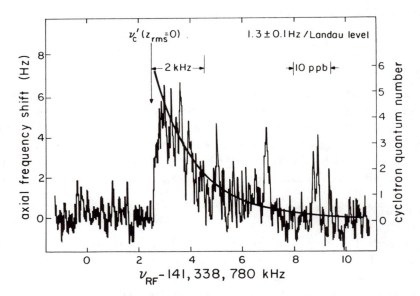

Figure 14. Single electron cyclotron resonance for the 1.3 Hz bottle in the last molybdenum trap. The characteristic magnetic line shape has an exponential tail that reflects the Boltzmann distribution of axial states coupled via the z^2 term in the magnetic field. The sharp edge feature corresponds to the field at the trap center and the solid line represents a fit to the data with a width corresponding to a 5 K axial temperature (from Ref. 6).

5. Cyclotron Resonance

5.1. LINE SHAPE

The shape of the cyclotron resonance is almost exclusively determined by the magnetic bottle and the thermal Boltzmann distribution of axial states. In addition, the relativistic pulling effect has some influence on the line shape, causing the resonance to appear somewhat broader than expected [30]. It is beneficial, when possible, not to have the axial locking drive applied if the true cyclotron line shape is to be determined. This follows from the z^2 term in the magnetic bottle which will contribute a cross term of the form $Z_{th} Z_d$ that will also broaden the resonance (Z_{th} is the thermal amplitude of the axial motion). Necessarily then, the detection and excitation for this resonance are alternated and the observed shifts in ω_z signal the onset of the cyclotron resonance, assuming that the frequency is swept up in magnitude for a positive magnetic bottle. Figure 14 shows a typical cyclotron resonance taken with this scheme for the fixed positive bottle in the last molybdenum trap shown in Fig. 4. A large negative bottle would tend to yield a resonance which is nearly the mirror image of this positive bottle line shape, except for the effect of the relativistic mass shift on the leading edge.

The main requirement of the alternating scheme is that the alternating rate not be too slow compared to $1/\tau_c$ where τ_c is the classical cyclotron decay time predicted by radiation damping. The alternating scheme allows time for axial energy to decay away by injecting a delay $\gtrsim 100$ ms before onset of cyclotron excitation. In addition, the axial damping time ($\tau_z \sim 50$ ms) needs to be short compared to τ_c in order that some signal is registered in the detection part of the cycle. It is worth noting however, that if the axial locking drive is scaled down significantly such that lock is just barely possible, a continuous cyclotron resonance can be swept out whose low-frequency edge agrees very well with the alternating low-frequency edge (shown in Fig. 14), but without quite as good resolution. Alternating conditions were quite favorable in the previous molybdenum traps since τ_c was measured to be ~ 1 sec [6]; in this case, it was possible to achieve a 1-ppb short-term resolution of the field at the center of the trap. However, in the case of the phosphor bronze trap, we now find that $\tau_c \sim 0.1$ sec and, therefore, we must use the low-axial drive method for deducing ω_c'; because of the lower S/N of this method, resolution was about a factor of 2 worse.

The high frequency tail of the cyclotron resonance shown in Fig. 14 represents the Boltzmann distribution of axial states, and the $1/e$ linewidth corresponds to

$$\Delta\omega_c' = \frac{eB_2 k_B T_z}{m_e^2 c \omega_z^2} = \frac{k_B T_z}{\hbar \omega_z}\delta \tag{28}$$

where δ is defined for Eq. 19. If this expression is used to experimentally determine the axial temperature, we often find that $T_z \sim 10$ K. This could be due to the equilibration to an amplifier whose temperature is greater than ambient, but some of the discrepancy is believed to be due to the relativistic pulling effect [30]. Here, the onset of excitation of the edge for positive magnetic bottles immediately pulls the true resonance edge to a lower frequency, thus artificially reducing the apparent peak response. Since this reduction will be more significant (in terms of absolute signal) near the peak than on the tail of the resonance, the shift has the effect of broadening the $1/e$ resonance linewidth.

5.2. PERTURBATIONS TO THE CYCLOTRON FREQUENCY

The largest shifts in ω_c' are due to the non-uniformity of the magnetic field and the relativistic mass shift. Again using the 1986 analysis of Brown and Gabrielse [8], it can be shown that

$$\frac{\Delta\omega_c'}{\omega_c'} = \left\{\frac{-eV_0}{m_e c^2}\right\}\frac{W_n}{eV_0} + \left\{\frac{B_2' Z_0^2}{B_0}\right\}\frac{W_k}{eV_0} + \left\{\frac{2B_2 Z_0^2}{B_0}\right\}\frac{W_q}{eV_0} \tag{29}$$

where the conspicuous absence of the C_4 term occurs because its contribution is reduced by the factor $(\omega_z/\omega'_c)^2 < 3 \times 10^{-7}$. Again using some of the parameters listed in Table 1, this equation becomes

$$\frac{\Delta\omega'_c}{\omega'_c} = \{-0.2W_n + 3.4W_k + 6.8W_q\} \times 10^{-5} \tag{30}$$

where all energies are measured in eV. The first term represents the relativistic correction and corresponds to ~ 1.2 ppb per integer change in the cyclotron quantum number. For a positive bottle, swept up in frequency, it should not affect the edge resolution but only the height of the excitation, unless the noise pedestal of the $14th$ harmonic of the X-band source from the Schottky diode heats up the cyclotron temperature and allows the resonance to be prematurely pulled into excitation. Normally, the average occupation level in a 50 kG magnetic field is 0.2 at 4 K. The second term is responsible for the shape of the cyclotron resonance and was the subject of the last section. The last term generates the most concern for us in our g-2 experiment since a change in the metastable magnetron orbit yields a corresponding change in ω'_c and ω_s. It illustrates the importance of cooling the magnetron motion. Assume that R_m has a typical value [5] of ~ 0.0014 cm; then $W_q \sim -5 \times 10^{-5}$ eV. Thus, a -3 ppb relative shift occurs in the cyclotron frequency which must be *stable* over time if a 1 ppb accuracy is to be maintained. Ideally, if one could reach the cooling limit, the relative shift could be kept below 0.02 ppb. The use of the variable bottle represents our first attempt to reduce or eliminate the effect of the magnetic bottle in these measurements.

Another possible source of shifts in ω'_c that was considered in earlier sections was associated with the angle of tilt θ between $\vec{B}$ and $\hat{z}$, as well as possible ellipticity ϵ in the trapping electrodes. In terms of $\bar{\delta}_e$ defined in Sec. 4.3., the observed cyclotron frequency $\bar{\omega}'_c$ can be written in the form

$$\bar{\omega}'_c \approx \omega_c - \bar{\delta}_e - \frac{\bar{\delta}_e}{\bar{\omega}'_c}(\bar{\omega}_m - \bar{\delta}_e) \tag{31}$$

which differs from Eq. 5 only by the discrepancy between measured and calculated magnetron frequencies, scaled by the ratio of magnetron and cyclotron frequencies. The relative shift is given by

$$\frac{\Delta\omega'_c}{\bar{\omega}'_c} = \frac{1}{4}\left(\frac{\omega_z}{\omega'_c}\right)^4\left(\frac{\bar{\delta}_e - \bar{\omega}_m}{\bar{\delta}_e}\right) \tag{32}$$

where the quantity $(\omega_z/\omega'_c)^4 \approx 10^{-13}$ for parameters chosen for our experiments.

(see Table 1) and the last factor is usually no larger than 10^{-4}. Thus, this shift is totally negligible for precision in the foreseeable future.

However, a much more serious problem arises for the phosphor bronze trap which contains the variable bottle. This latter feature consists of a closed superconducting loop embedded in the midplane of the ring electrode. Since total magnetic flux through this loop must be fixed for all time (i.e. $B\cos\theta = $ constant), a change in the direction of the axis of the loop corresponds to a change in area of the loop normal to the magnetic field. Currents are then established in the superconducting wire which permanently alters the magnetic field at trap center according to $\delta B/B = \bar{\theta}\delta\theta$ where $\bar{\theta} \gg \delta\theta$ now represents the average residual angle between the axis of the superconducting loop (which should be the z-axis by construction) and the magnetic field. From Eq. 21, variations in the cyclotron frequency can be related to fluctuations in the axial frequency due to variations in θ:

$$\left(\frac{\delta\omega_c'}{\omega_c'}\right)_{loop} = \frac{2}{3}\left(\frac{\delta\omega_z}{\omega_z}\right)_\theta. \tag{33}$$

The limitation that the observed axial stability is $\sim$ 1-Hz out of 72 MHz for all causes in the variable-bottle trap leads to 9×10^{-9} for the upper limit to fluctuations in the magnetic field. As it happens, we observe on the order of 10 ppb wander in the field for this trap. The impulses to the magnetic axis are believed to be due to the movement of a hydraulic elevator which is about 10 meters from the magnet. We observe $\sim$ 1 mG field changes at the distance of 10 meters from this elevator as it moves from ground to the fourth floor of the physics building (as required for day-time classes). Clearly, the use of the variable bottle puts a severe constraint on the allowed variations in θ or else requires that $\bar{\theta}$ be reduced by an order of magnitude (which is indeed possible).

5.3. CAVITY EFFECTS

One observation that was particularly suspicious in our early work was the determination that the optimum alternation rate for observing cyclotron resonances in the molybdenum trap shown in Fig. 4 (at 50.5 kG) was about 0.5 Hz (i.e. 1 second on and 1 second off). It was apparent to one of us (H.G.D.) that the excitation to higher cyclotron quantum states was not decaying as quickly as expected from the classical free-space damping time of 0.1 sec at this field. This took the form of a warning from Dehmelt [9] as early as 1981 to guard against shifts in ω_c' "due to accidental resonances with standing wave or 'cavity' modes" inside the trap structure.

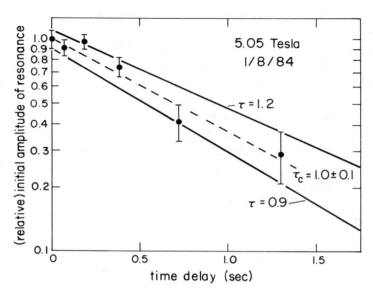

Figure 15. Cyclotron damping time for a single electron in a microwave cavity (i.e. last molybdenum trap). By delaying onset of axial detection drive in the normal alternation scheme, the initial resonance amplitude is found to decay exponentially with a time constant which is ten times longer than predicted for free space radiation (from Ref. 6).

To observe the inhibition of spontaneous emission, the initial amplitude of the cyclotron edge is measured for various fixed time delays after turning off microwave power and turning on the axial detection drive. The response time for the axial motion was less than 50 ms and thus did not affect these measurements. The relative initial amplitude is then plotted on a semi-log graph versus the fixed delay time as shown in Fig. 15 in order to determine [6] that the classical decay time was 1.0 ± 0.1 s, or about 10 times longer than τ_c predicted by free space radiation damping. It was quite fortuitous that these early experiments exhibited such long decay times. It meant that much less microwave power was required to see a sharp $Z_{rms} = 0$ frequency edge, and therefore put less demand on the strength of the carrier relative to the multiplied-up noise pedestal of the Klystron used in the microwave frequency chain. This type of inhibition in a Penning trap was first directly measured and reported [31] by Gabrielse and Dehmelt for a trap which did not have a magnetic bottle, but did have a much more open structure. As a result, they found a decay time τ_c which ranged from 86 to 347 ms for B_0 near 60 kG. In the phosphor bronze trap, some effort was made to reduce the Q of the microwave cavity and subsequent measurements of τ_c [32] indicated that the trap had a cyclotron decay time in excellent agreement with the classical free-space value at ~ 50 kG.

The presence of shifts in the cyclotron frequency could only be detected by changing the magnetic field but even with this change, the true unshifted frequency would be difficult to deduce without some theoretical model as a guide. Since the g-factor is related to the ratio of spin precession frequency divided by the cyclotron frequency, a shift of several parts in 10^{12} in ω_c without a corresponding shift in ω_s would produce a comparable systematic error in the measured g-factor (and 10^3 times larger shift in a_e). Since present experimental and theoretical precision for the electron g-factor is on the order of several parts in 10^{12}, this systematic effect becomes very important and represents the principle motivation for our repeating the single electron g-factor measurements in the new phosphor bronze trap with a somewhat more open cavity structure.

The first complete theoretical treatment [33] of the effect of a microwave cavity on the cyclotron motion of the trapped electron was based on a cylindrical model which was chosen at the time because of the ease of theoretical treatment that the geometry afforded. It predicted that for a cavity with a $Q \approx 1000$, shifts in the cyclotron frequency could exceed 70 parts in 10^{12}. However, a closer look [34] at the model suggested that for this same $Q \approx 1000$ with $\tau_c(\text{cavity}) \approx 10\tau_c(\text{vacuum})$, the maximum and probable shifts would be on the order of 8 and 4 parts in 10^{12} respectively. The latter uncertainty thus represents the limitation to the accuracy of the last precision measurement [35] of the electron's g-factor inside a fixed-bottle, molybdenum trap. To see if a $Q \sim 1000$ was reasonable, an experimental technique was devised to directly observe the cavity modes [36], based on the well-established bolometric technique [14,25] for clouds of several hundred electrons located at the center of the Penning trap. The results of such studies using the molybdenum trap shown in Fig. 4 was an indication that this trap has a lower than expected Q (probably less than 500).

It should also be noted that this work had stimulated further theoretical interest in pursuing a numerical calculation of the mode structure in an ideal Penning trap [37]. Unfortunately, subtle variations in slits and holes in the structure as well as deviations from the perfect family of hyperboloids do not allow for any reasonable agreement between the observed and the predicted mode structure for this particular g-2 Penning trap. Future efforts to account for the mode structure will either concentrate on using very low Q cavities that can not exhibit significant frequency pulling or else on measuring the few nearest modes on each side of our operating frequency in hopes of predicting [34,37,38] the true cyclotron frequency. The latter approach requires much more experimental effort, but does have the advantage of putting less dependence on an absolutely clean microwave source and may be necessary for the future relativistic mass shift method of detecting spin flips.

In addition, there is also a more subtle electrostatic perturbation due to the intrinsic constraint of a real charged particle surrounded entirely by conducting surfaces. The simple model of a point charge located inside a grounded sphere of radius a has been used to predict [39] the electrostatic shifts due to the image charge located outside the sphere. Using the experimental condition that $\overline{\omega}_z$ is held constant by the feedback loop (described in Sec. 3.4.), the observed shift in $\omega'_{c,0}$ for a single electron at the center of the trap becomes

$$\overline{\omega}_{c,1} = \omega'_{c,0} - 3\Delta/2\overline{\omega}_{c,1} \tag{34}$$

where $\omega'_{c,0}$ represents the cyclotron frequency in the limit of zero trapped electrons and $\Delta = e^2/m_e a^3$. Since it was experimentally determined that a is approximately R_0 in a 3 times smaller quadring Penning trap [40], we predict that for a single electron in a g-2 trap that

$$\left(\frac{\Delta\omega'_c}{\overline{\omega}'_c}\right)_{image} = \frac{3m_e c^2}{R_0^3 B_0^2} \approx 5 \times 10^{-15} . \tag{35}$$

This is indeed negligible and may be ignored for all the g-2 experiments.

5.4. OBSERVED SHIFTS IN PHOSPHOR BRONZE TRAP

5.4.1. *Variation with Microwave Power.* An example of a shift in ω'_c with microwave power was published earlier [35] and appeared to vary quadratically. The source of this shift was a mystery whose explanation is still not available, though we know much more about it now. In particular, with the implementation of the variable bottle in the phosphor bronze trap, we have found that this shift depends on the applied bottle strength. In Fig. 16, we have plotted this shift versus relative X-band power (with arbitrary reference) for the two extremes of the variable bottle. The relative power levels are obtained by calibrating the rectified dc current through the multiplier diode versus a known amount of microwave attenuation. The zero shift limit near 4.3 mW corresponds approximately to the apparent zero for the 14th harmonic, below which the cyclotron resonance abruptly disappears (i.e. the $n = 1$ level is not excited). The difference between this shift and that which was reported earlier could be due to either the different multiplier diode used or the lack of direct X-band power radiating onto the electron since a high pass filter is used in the phosphor bronze trap. A direct determination of B'_2 is possible from the magnitude of the single quantum step associated with the spin flip according to Eq. 19. For the positive bottle extreme, $B'_2 = 245(10)$ G/cm^2 and for the negative extreme, $B'_2 = -262(10)$G/cm^2. These values would correspond to an intrinsic

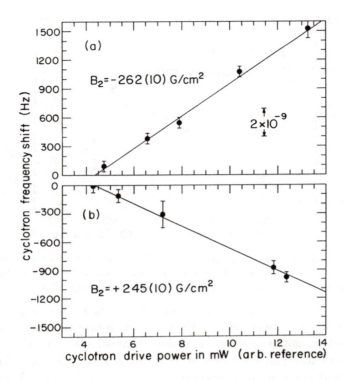

Figure 16. Residual systematic shift of the cyclotron frequency versus applied X-band microwave power for a single electron in the phosphor bronze trap. The solid curves represents a linear fit to these data with the constraint that the "zero" occurs at fixed power level which can no longer excite the resonance.

bottle of $-8.5(14)$ G/cm^2, which is consistent with -6 G/cm^2 predicted by the relativistic mass correction. However, the data does not show symmetry about zero effective magnetic bottle. The slope for maximum negative bottle is about 40% larger than the magnitude of the slope for the maximum positive bottle; but if the shift is proportional to the applied B_2 quadratic gradient, there should be complete symmetry since only the sign of the current in the magnetic bottle was reversed.

Another characteristic of this shift is that it appears to be associated only with the measurement of the cyclotron edge and to not be present in anomaly resonances if the simultaneously-applied microwave power is non-resonant. Part of the motivation for using such non-resonant drives was to eliminate this type of shift by causing the same shift to occur in the anomaly resonance (which it did not do in this case).

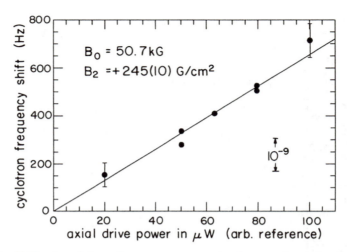

Figure 17. Shift of cyclotron frequency versus simultaneously applied weak axial drive for a single electron in the phosphor bronze trap. Adequate data was not available for the maximum negative bottle, but the sign of the shift is consistent with the sign of B_2.

In our discussion of Eq. 29, we also did not include a term proportional to the cyclotron energy which depended on the magnetic bottle strength, because it was many orders of magnitude smaller than the relativistic term as well as the size of the shifts observed. Upon reviewing other terms in this theoretical shift equation, only the magnetron energy term is consistent with both the dependence on B_2 and the sign of the shift (recall W_q is negative and observed cyclotron shifts are indeed negative for positive bottles). The size of the required magnetron orbit which would give such shifts is also quite believable:

$$R_m^2 = R_{m,0}^2 - \frac{2B_0}{B_2}\left(\frac{\delta\nu_c'}{\nu_c'}\right). \tag{36}$$

Thus assuming that $R_{m,0}$ is close to the observed cooling limit of 0.0014 cm, then a 1-ppb shift in the cyclotron frequency would corresponds to a 10% increase in magnetron radius for the maximum bottle strength used.

5.4.2. *Variation with Axial Drive Power.* On occasion, we find that it is necessary (or convenient) to measure ν_c' simultaneously with axial drive applied, i.e. without alternating excitation and detection periods. Subsequently, we observe shifts in the cyclotron frequency proportional to the power of the axial drive. Figure 17 illustrates the typical shifts associated with axial drive, measured in the new phosphor bronze trap. The (arbitrary) reference power, at unity, is taken to be the same for both runs that were combined to yield the graph. We have also observed that the cyclotron shift with axial power reverses sign when B_2 is reversed. Thus, the shift

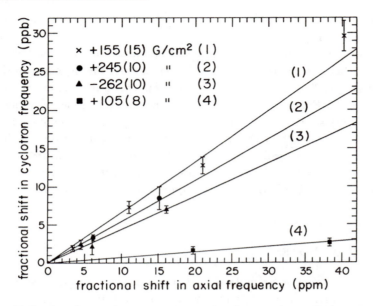

Figure 18. Shift of cyclotron frequency versus simultaneously applied anomaly drives for a single electron in the phosphor bronze trap. The strength of that drive is measured as a fractional shift in the axial frequency according to its effect on well-depth for an rf-trapping field. From the available data, these positive shifts appear to vary quadratically with B_2.

is expected to vary linearly with B_2 and should be described by the second term in Eq. 29. This equation predicts that a shift of 1 ppb should correspond to an axial energy of 0.00003 eV. From Eq. 12, this energy corresponds to $V_{rf} = 4$ nV which agrees quite well with our estimate for the single endcap drive actually applied.

5.4.3. *Variation with Anomaly Drive Power.* The one shift which we can find no plausible source is the shift in cyclotron frequency associated with applied anomaly drive. In the past, we attributed this shift to an off-resonant axial drive whose non-zero average position in the magnetic bottle caused a true shift in the effective magnetic field. As a field shift, this effect was expected to affect w'_a in the same way and thus not affect the determination of a_e to first order. Figure 18 summarizes all available data on this effect at $B_0 = 50$ kG. The data for the old molybdenum trap (i.e. $B_2 = 155(15)$ G/cm^2) have been properly scaled to account for the different axial and anomaly frequencies relative to those of the phosphor bronze trap. Note that this older data does not agree with any data from the newer trap, presumably because of different capacitive effects in the two traps. The curious observation is that the shift is always positive, independent of the sign of B_2 and, from the three values of the bottle shown for the phosphor bronze trap, the shift appears to vary quadratically with the magnitude of B_2. Clearly, this shift is not characterized by the second term in Eq. 29.

These observations also rule out many other possible shift mechanisms. For instance, the relativistic mass shift is even smaller than the B_2 contribution and it has the wrong sign. The anharmonic potential shifts are still smaller yet and have no dependence on B_2. Even a relativistic $\vec{v}_c \times \vec{E}_r$ magnetic field has been investigated, but is likewise very small. Measurements of ν_m during application of anomaly power do not find absolute shifts in the magnetron frequency comparable to those in ν'_c. Obviously, more data is needed in order to discover the true nature of this effect. Also we must continue to test a_e versus anomaly power in all future experiments to see if any part of the shift in ω'_c does not appear in ω'_a for a_e determinations.

6. Anomaly Resonance

6.1. CLEAN "BEAT-NOTE"

The anomaly resonance is actually a beat-note between the precession rate ω_s of the magnetic moment about $\vec{B}_0$ and the cyclotron rotation at $\omega_c \approx \omega_s$ about the same axis. This beating was dramatically illustrated by the work on electrons and positrons completed in 1971 by Rich et al. [41], and the work on positive and negative muons completed in 1977 by the CERN collaboration [42]. Excitation of the resonance is produced by a simultaneous two photon process, in which the spin is flipped and the cyclotron state changes by one unit (see Fig. 11). This yields an advantage of 3 orders of magnitude in precision for g-factor measurements over any method which independently measures ω_s and ω_c since major perturbations are reduced accordingly by the simultaneous transitions.

To give a particular example, the relative shifts in ω_s due to the bottle field are (but not including terms of order $\omega_z^2/\omega_c'^2$ [8]):

$$\frac{\Delta\omega_s}{\omega_s} = +\left\{\frac{B_2 Z_0^2}{B_0}\right\}\frac{W_k}{eV_0} + \left\{\frac{2B_2 Z_0^2}{B_0}\right\}\frac{W_q}{eV_0} . \tag{37}$$

As with the cyclotron frequency, terms associated with the anharmonic potential do not appear at this order of precision. Upon comparing this equation with corresponding terms for the relative cyclotron perturbation, given in Eq. 29, the difference frequency, $\omega_s - \omega'_c$, then has the same functional form shown in Eq. 37 but now for $\Delta\omega'_a/\omega'_a$.

The shift in ω'_a associated with the relativistic mass increase has been considered in our previous review [7] and there we obtained

$$\omega'_a = \omega_{a,0}\left(1 - \frac{W_k}{2m_e c^2}\right) \tag{38}$$

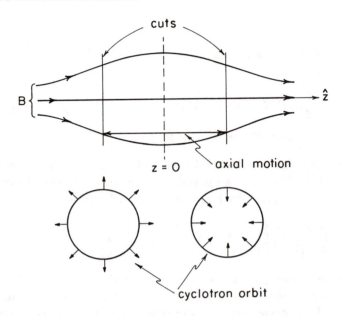

Figure 19. Mechanism for inducing spin flips by the electron's axial motion in the magnetic bottle. From the electron's frame of reference, a magnetic field is seen to be rotating at ω_c'. This field is then modulated by the axial motion at ω_a', thus yielding sidebands at $\omega_c' \pm \omega_a'$ where $\omega_s = \omega_c' + \omega_a'$ (from Refs. 5 and 18).

where a large 1-ppb dependence on cyclotron energy is now absent. Thus, if W_k is coherent axial energy comparable to kT_z, then, $\Delta\omega_a'/\omega_a' \sim -0.35$ ppb.

.2. AXIAL EXCITATION TECHNIQUE

An added benefit of using a magnetic bottle for detection is that it can be used fortuitously to generate the appropriate perpendicular spin flipping field, readily associated with standard NMR experiments. The technique is based on axial amplitude modulation, at the anomaly frequency, of the non-homogeneous magnetic field seen by the electron, whose component in the radial plane has the form given by Eq. 18: $b_r = -B_2 r z$. Figure 19 illustrates the motion through this bottle at the two extremes of the axial motion, where for this discussion the magnetron motion can be safely ignored. From the frame of reference of the electron, going around its cyclotron orbit, it sees a magnetic field rotating about the z-axis, which points in opposite directions at the two extremes of the motion as shown. Thus, from this rotating frame, the radial magnetic field is given by:

$$b_r = -B_2 R_c Z_a \sin(\omega'_c t) \sin(\omega'_a t) = b_r(1) + b_r(2) \tag{39}$$

where

$$b_r(1) = \frac{1}{2} B_2 R_c Z_a \cos(\omega'_c + \omega'_a)t$$
$$b_r(2) = -\frac{1}{2} B_2 R_c Z_a \cos(\omega'_c - \omega'_a)t \ . \tag{40}$$

The cyclotron radius, R_c, in a 50.5 kG field and for the $n = 0$ ground state cor
responds to ≈ 114 Å. The axial amplitude Z_a, produced by the electric field E_r
applied to one endcap, is given by Eq. 10 when far off resonance:

$$\frac{Z_a}{Z_0} = \left(\frac{Z_0 E_{rf}}{V_0} \right) \left[\left(\frac{\omega'_a}{\omega_z} \right)^2 - 1 \right]^{-1} \approx \left(\frac{V_{rf}}{10 V_0} \right) . \tag{41}$$

For our typical conditions (see Table 1), this gives $Z_a \sim 0.003$ cm for a 1 volt r
drive yielding an amplitude $B_2 R_c Z_a = 0.53 V_{rf} (\text{in}\,\mu G)$ assuming V_{rf} is in volts. The
radial component $b_r(1)$ corresponds to the appropriate perpendicular spin flipping
field at frequency $\omega'_c + \omega'_a = \omega_c + \omega_a = \omega_s$, whereas the second component will
correspond to an insignificant non-resonant perturbation. From standard NMR
theory, the $b_r(1)$ component will then produce spin flips with a Rabi frequency

$$\omega_{Ra} = \frac{2\mu_B}{\hbar} b_r(1) \tag{42}$$

which is $\sim 2\pi(0.7 \text{ Hz})$ for a 1 volt rf endcap drive for the last molybdenum trap
shown in Fig. 4. The average time between flips is then given [7] by

$$\langle T \rangle = 2\pi \frac{\Delta \omega'_a}{\omega_{Ra}^2} \tag{43}$$

which, on resonance, is about 6 sec for a 3 Hz wide anomaly line.

Unfortunately, applying a volt of anomaly power would also produce a significan
amount of liquid helium evaporation from a 50Ω terminated drive line. Such time
varying heat loads then upsets the equilibrium of the axial frequency lock loop and
reduces the effective S/N. However, a reduction in anomaly power would ther
require much longer data-taking periods. This problem was alleviated somewhat by
applying a simultaneous microwave drive in the tail of the cyclotron resonance in
order to heat the cyclotron motion into higher excited states on the average with
correspondingly larger values of R_c and therefore larger $b_r(1)$.

6.3. GUARD-RING EXCITATION TECHNIQUE

Every major systematic shift encountered so far appears to be related to the strength of the detection magnetic bottle. To get around these shifts, the bottle must be reduced which means some other mechanism is needed to generate the required spin-flip field. As a first step, a direct current loop is made possible by splitting the guard rings on one side only. The injection of current is arranged to yield currents flowing in opposite directions for each guard (i.e making a pair of counter-rotating current loops) and is therefore expected to produce a constant radial magnetic gradient (near trap center) of the form [6]

$$B_1 \approx 1.3 I_{rf} f_s \text{ Gauss/cm} \tag{44}$$

where I_{rf} is in amps and f_s is the rf electrode shielding factor. The measured spin flipping rate indicates that $f_s \sim 0.1$. A tuned circuit attached to the guard rings also enhances the production of rf current. The radial field then becomes:

$$b_r = B_1 R_c \sin(\omega_a' t) \sin(\omega_c' t) = b_r(1) + b_r(2) \,. \tag{45}$$

One immediately notes the similarity to the two components described in Eq. 39, in which the amplitude $B_2 R_c Z_a$ is replaced by $B_1 R_c$. If R_c is again associated with $n = 0$, this amplitude becomes $\approx 0.74 I_{rf} f_s (\text{in } \mu G)$, assuming I_{rf} is in amps.

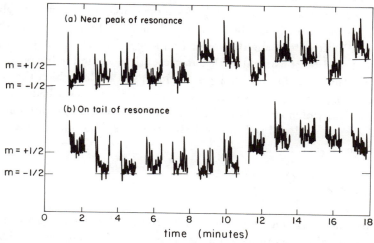

Figure 20. Example of spin-flip data. The alternation between detection drive and anomaly excitation is much slower than for cyclotron excitation in order to minimize the power in the anomaly field. The excitation frequency is kept constant for each 12-20 alternation cycles while spin-flips are being detected in the correction signal. Note the increased rate of flipping near the peak of the resonance shown in (a) compared to the rate on the exponential tail shown in (b) (from Ref. 6).

6.4. RESONANCE LINE SHAPE

The excitation cycle for anomaly resonances involves turning off the axial drive at ω_z, and turning on the anomaly drive at ω_a' with the microwave drive also applied but tuned into the tail of the exponential line shape of the cyclotron resonance The subsequent detection cycle then involves turning off both anomaly and cyclotron drives while the ω_z drive registers the floor of the lock-loop correction signal Examples of the type of anomaly data obtainable with this method are shown in Fig. 20. The anomaly drive was kept fixed in frequency for some specified number of excitation/detection cycles and then the number of successful flips is plotted versus the applied drive frequency. An example of such early resonances is shown in Fig. 21(a) for run 109. The dotted line is added only to aid in recognizing the resonance shape. Before a theoretical line shape was available with which to fit this data, an estimate had to be made as to where the leading edge would be for $Z_{\text{rms}} = 0$, typically taken at one-third of peak height. The lack of a sharp edge is due to the noise modulation (or statistical fluctuation) of the anomaly resonance by the axial motion within the inhomogeneous magnetic field. The estimation process produced a serious "apparent" systematic shift of the anomaly with anomaly power This shift is related to the tendency for this type of resonance to saturate (or to be power broadened). A maximum spin flipping rate is 0.5 (with equal probability of being left in spin-up or spin-down state). As more anomaly power is applied, the resonance edge is broadened and the "apparent" anomaly frequency (and therefore a_e) is artificially shifted to a slightly lower frequency. However, the extrapolation to zero power should be consistent with our presently accepted value of a_e.

The basis for this unusual line shape is found in Eq. 18 for which the magnetic field depends weakly on the instantaneous value of z^2. As described in Sec. 5.1. the magnetic width, which represents the $1/e$ linewidth, will be proportional to the average axial energy, and will reflect the Boltzmann distribution of axial states at some temperature T_z:

$$\Delta\omega_a' = \omega_a' \frac{B_2 Z_0^2}{B_0} \cdot \frac{k_B T_z}{eV_0} . \tag{46}$$

In other words, the weak magnetic coupling to the axial frequency also causes the random thermal axial fluctuations to appear in all magnetic resonances. This effect was accurately described [43] by Brown in 1984 by writing the stochastic average of the axial motion as a functional integral which transcribes it into a soluable one-dimensional, quantum-mechanical barrier penetration problem. The resulting line shape profile $\chi(\omega)$ as a function of frequency is given for very weak rf drives by

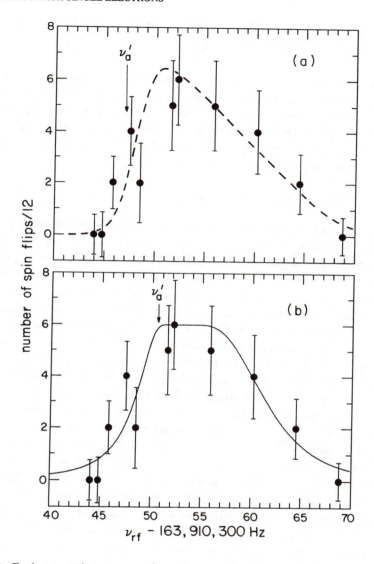

Figure 21. Early anomaly resonance (run 109 taken in the last molybdenum trap) showing strong saturation on the peak. These early runs, obtained by plotting the number of flips observed out of a fixed number of tries vs anomaly frequency, were found to show a systematic shift towards lower $\nu_a'(Z_{rms} = 0)$ with increasing applied power. As shown in (a), this arose from the artificial choice of determining the resonance edge at $\approx 1/3rd$ of peak height. As the new line shape fitting shows in (b), the correct $Z_{rms} = 0$ is near the top of the peak (adapted from Ref. 6).

$$\chi(\omega) = \frac{4}{\pi}\mathrm{Re}\left\{\frac{\gamma'\gamma_z}{(\gamma'+\gamma_z)^2}\sum_{n=0}^{\infty}\frac{(\gamma'-\gamma_z)^{2n}(\gamma'+\gamma_z)^{-2n}}{(n+\frac{1}{2})\gamma'-\gamma_z/2+\gamma_c/2-i(\omega-\omega_0)}\right\} \tag{47}$$

where Re denotes the real part, $\gamma' \equiv (\gamma_z^2 + 4i\gamma_z\Delta\omega_a')^{1/2}$, γ_z is the corresponding axial linewidth, and γ_c is the natural cyclotron damping linewidth in the absence of the magnetic bottle, However, moderately hard anomaly drives are typically used such that near the peak, the probability for a spin flip approaches $\frac{1}{2}$ as indicated earlier. If t_e represents the period of time that the excitation is on, then the net probability for observing a spin flip is given by

$$\wp(t_e) = \frac{1}{2}\left\{1 - \exp[-\pi\omega_{\mathrm{Ra}}^2 t_e\chi(\omega)]\right\} \tag{48}$$

where the strength of the drive is contained in the Rabi frequency ω_{Ra}. Note that when ω_{Ra} is sufficiently large, $\wp \to \frac{1}{2}$ (i.e. saturated). If run 109, shown in Fig. 21(a), is fitted to Eq. 48, one sees in Fig. 21(b) an excellent example of this saturation effect where the $Z_{rms} = 0$ position is near the peak instead. Using fitted data, the apparent systematic shift of a_e with anomaly power totally disappears.

Recent work in the new variable-bottle phosphor-bronze trap has required that we modify our excitation method due to the 10 times shorter cyclotron decay time (now comparable to the axial damping time). The result of the 10 times less dwell time in the $n = 1$ state is a strong asymmetry between spin-up and spin-down transition rates. As a result, we now prepare each spin-flip event by first forcing the electron into the spin-up state. This is accomplished by strong (saturating) anomaly drives with some cyclotron assisting to further enhance the transition probability. However, once prepared in the up-state, the anomaly power is reduced to low nominal levels, corresponding to (rf-trapping) axial shifts of 200-300 Hz, and the cyclotron-assist is turned off. Figure 22 shows two examples of resonances obtained with this method at the extremes of the variable magnetic bottle. Figure 22(a) shows the anomaly shape for the maximum negative bottle case when $\gamma_z/2\pi \approx 1.0$ Hz The solid curve represents the theoretical fit which yields such fitted parameters as $\Delta\omega_a/2\pi \approx -3.4$ Hz and $\omega_a/2\pi = 164,992,769.36$ Hz. Figure 22(b) then gives the usual anomaly resonance associated with the maximum positive bottle, for which $\gamma_z/2\pi \approx 0.5$ Hz. Again, the solid curve is a fit which yields $\Delta\omega_a/2\pi \approx 3.9$ Hz and $\omega_a/2\pi = 164,791,544.80$ Hz. With this particular excitation method, the fits as shown in Fig. 22 are obtained by letting the theoretical number of observations per data point be twice the actual experimental number. This is done in order to use the same fitting program and at the same time to avoid saturation at $\frac{1}{2}$ since saturation now occurs at unity.

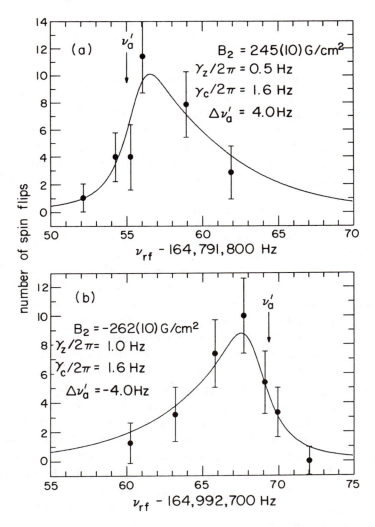

Figure 22. Anomaly data taken at the two extremes of the variable bottle in the phosphor bronze trap. The natural cyclotron linewidth, $\gamma_c = 2\pi(1.6 \text{ Hz})$, is characterized by the free-space damping at $\tau_c = 0.10$ sec. The axial linewidth, γ_z, was reduced in each case by shifting the axial resonance off the tuned circuit during the anomaly excitation, whereas $\Delta\nu'_a$ represents the fitted bottle width for the actual anomaly resonance. The line shape now clearly shows the expected exponential tail, but with the low-frequency edge still smeared by the axial noise modulation. Error bars are derived from the binomial distribution, and are computed with the least-squares-fitted curve as the true parent distribution.

7. Results, Conclusions, and Goals

7.1. THE g-FACTOR RESULTS

From the time of our first demonstration in 1976 [19] that these geonium experiments could indeed determine the g-factor of the free electron to very high precision (i.e almost 20 times better than any previous method at this first demonstration), there have been four major experimentation periods in which significant improvements have been made. The first of these is the work of 1979 [9,44,45] which is concerned with the measurement of the g-factor at three different magnetic fields, using our first molybdenum compensated Penning trap in a glass envelope:

$$a_e(1979) = 1\,159\,652\,200(40) \times 10^{-12} \,. \tag{49}$$

The use of three different fields was done at the time as a very simple expedience for finding an accurate g-factor since systematics were expected to depend on relative field homogeneity. The next period that ended in 1984 [6] noted the use of a new (molybdenum) double-trap configuration (shown in Fig. 4) and several improvements in technique, most notably the use of guard-ring drive and line shape fitting to eliminate the major systematic error. The work of the next to the last period in 1987 [35] used the same double trap to uncovered the presence of systematic dependencies on power, the prediction of the cavity shift and yielded the precision comparison of the electron with the positron. Finally, the work that ended in 1989 [46] utilized our new phosphor bronze trap which features a low Q cavity in order to test for cavity shifts. It also features the variable magnetic bottle which allows us to compare g-factors with different values for B_2. Thus, another very real division of our geonium experiments can be characterized by the use of either a traditional (asymptotically symmetric) molybdenum Penning trap with a fixed magnetic bottle in which $R_0^2/2Z_0^2 = 1.0$ or the special (nearly orthogonal) phosphor bronze trap with a variable bottle in which $R_0^2/2Z_0^2 = 0.72$. A true orthogonal trap [22] would have $R_0^2/2Z_0^2 \approx 0.674$ and is characterized by the non-interaction of guard compensation potential and the absolute well depth, V_0.

7.1.1. Traditional Molybdenum traps. Much of this early work concentrated primarily on developing hardware and techniques, and a little on systematics. For example, the development of a new all-metal vacuum envelope became necessary because of the unreliability of the glass envelopes. The introduction of the cyclotron-assist for the spin-flip cycle allowed us to use less anomaly power and thus reduce the disruption to both the magnetic field stability and the detection S/N from the increased liquid helium evaporation. For the same reason, we also introduced the

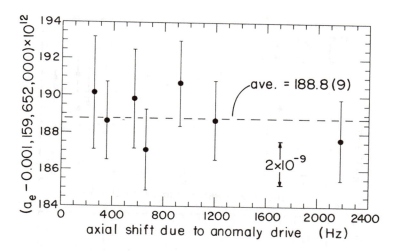

Figure 23. Summary of 1983-84 anomaly runs using guard-ring excitation and line shape fitting. The relative power of the anomaly drive is again measured as an axial frequency shift due to the rf effect. The residual anomaly-power dependence is < 1-ppb over the range of drives used, but there was an undetected (at that time) −4 ppb error due to the residual systematic shift of the cyclotron frequency (from Ref. 6).

guard-ring excitation for producing the observable spin flips. To narrow the anomaly resonances, we established an alternating detection/excitation scheme which also eliminates the perturbation of the detection axial drive on the anomaly frequency.

By far the most significant improvement in technique was the utilization of the line shape fitting program to predict the "true" location of the $Z_{rms} = 0$ anomaly edge. By taking power broadening into account, the fitting process successfully eliminated the systematic shift in past data that had an adequate amount of the line shape with which to fit to the theoretical shape. This effect was discussed in Sec. 6.4. The apparent systematic which was reported earlier [44,45,47] is clearly not present in the runs shown in Fig. 23 which comprises the 1984 data.

Another improvement in technique was the addition of a small shift in trapping potential during the ω_c' and ω_a' excitation phases. This shift is done to reduce the axial damping linewidth, γ_z. The Brownian-motion line shape theory [43] has confirmed that the anomaly resonance will have a sharper [9] $Z_{rms} = 0$ edge if this axial linewidth is reduced well below the magnetic resonance width.

At this point, our precision had made it possible to observe a residual shift of the cyclotron frequency with cyclotron drive power (as described in Sec. 5.4.). After correcting the cyclotron resonances for this effect, the 1987 data is displayed in a plot of δa_e versus microwave power (in Fig. 24(b)) which is directly related to the rectified current through the multiplier diode and similarly versus rf anomaly power

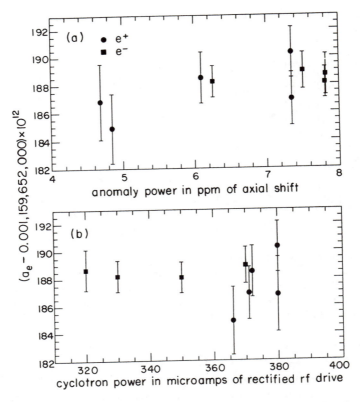

Figure 24. Summary of most recent molybdenum trap data used in the electron/positron comparison. In (a), the data is plotted vs anomaly power measured in ppm of the axial shift when anomaly power is applied. In (b), the same data is plotted vs cyclotron power measured in microamps of rectified rf drive. Any residual systematic shift is below the 1-ppb level of uncertainty (data obtained from Ref. 35).

(in Fig. 24(a)) which is related directly to the axial frequency shift produced by the anomaly drive as discussed in Sec. 3.4. Because the *same* microwave diode, microwave generator, and trap were also used for the 1984 data, we are therefore justified in correcting each run of the earlier data series for the shift in cyclotron frequency associated with the strength of the microwave field. The data shown in Fig. 23 contain this correction. As a result,

$$a_e(1984)_{\text{corr.}} = 1,159,652,189(4) \times 10^{-12} \tag{50}$$

which is 4 ppb lower than the published result [6]. The uncertainty was limited by apparent magnetic field fluctuations (from background gas collisions inside a leaky vacuum tube), but was at least an order of magnitude less than for the 1979 data.

The statistical error was about half of the reported uncertainty.

Finally, our improved precision has brought considerable theoretical interest in possible cavity shifts due to the mode-pulling effects. Using the model of a cylindrical cavity as developed by Brown et al. (1985) [33], and the constraint that the classical decay time is inhibited by a factor of 10 (shown in Fig. 15) in the neighborhood of mode-structure (believed to exist for a comparable sized cylindrical cavity at 50 kG), it was estimated that a probable shift in a_e of no more than 4×10^{-12} would exist; the following g-factor anomaly was thus quoted [35]:

$$a_e(1987) = 1\ 159\ 652\ 188.4(4.3) \times 10^{-12} \tag{51}$$

where a statistical error of 0.62×10^{-12} in the weighted average of all electron runs was combined in quadrature with our estimate of 1.3×10^{-12} for the uncertainty in the residual microwave power shift and the 4×10^{-12} potential cavity-mode shift discussed above. Also mentioned at the beginning of this section, similar runs were taken at this same time using positrons instead. These runs are included in Fig. 24 and give essentially the same result within their uncertainties [35]:

$$g(e^-)/g(e^+) = 1 + (0.5 \pm 2.1) \times 10^{-12} \tag{52}$$

where the common cavity-shift estimate is not included in the final uncertainty.

Phosphor Bronze Trap. This most recent work features a trap which is about 10% smaller than our previous traps, has a phosphor bronze ring electrode and compensation guards, and OFHC copper endcaps. The structure was purposely left more open than the traditional molybdenum traps in order to spoil the Q of the microwave cavity. At high field, it indeed appears to be low since the classical decay time is measured to be the free space value in the immediate vicinity (0.2%) of 50.7 kG. However, at 36 kG, the trap looks less open because of the now longer wavelength. Thus, over a 2% range, the cyclotron decay time was measured to vary from 2 to 4.5 times the free space value. Again using the cylindrical model as our guide, we estimate that the cavity Q for this trap could be on the order of 100. In fact, according to Brown *et al.* [37], cavity shifts in hyperbolic cavities are expected to produce shifts only half as large as those of comparable cylindrical cavities. Thus, this trap should produce results relatively free of cavity shifts in comparison with earlier molybdenum traps, assuming that one can make these measurements at fields where the cyclotron decay time is near the free space value.

The primary implication of a 10 times shorter classical decay time is that less uninterrupted dwell time exists in the excited Landau levels (see Fig. 11). As a result, much more anomaly power is needed to flip the spin from the required $n = 1$ level, if the spin is initially in the down state where it predominantly resides in the

TABLE 2. Summary of runs taken in the phosphor bronze trap using the variable-bottle feature. Power shifts associated with axial, cyclotron, and anomaly drives are summarized as a net shift in the third column and corrected anomalies have statistical errors included. All cyclotron calibrations are made relative to "spin-down" state.

Run	Magnetic bottle (G/cm^2)	Relative a_e shift (ppb)	[(Corrected a_e)− 1 159 652 000] ×10^{-12}
161	+245(10)	−0.7(7)	182.0(2.8)
162	+245(10)	−0.7(7)	189.0(2.6)
164	+245(10)	+0.5(9)	189.4(3.3)
165	+245(10)	+0.7(8)	191.3(2.3)
168	−262(10)	+2.0(8)	183.7(2.4)
169	−262(10)	+1.0(8)	182.8(2.7)
170	−262(10)	+1.8(9)	177.7(2.3)
171	+245(10)	+6.9(9)	181.2(3.6)
172	+245(10)	+1.5(8)	189.0(1.9)
173	+245(10)	+2.4(7)	187.9(2.0)
175	−262(10)	−3.4(8)	188.4(2.2)
176	−262(10)	−1.5(7)	182.5(2.3)
177	−262(10)	+2.9(8)	184.1(1.8)
179	−262(10)	+2.3(8)	188.6(2.6)

$n = 0$ level. The observed detrimental effect of high anomaly power is reduced axial S/N due to the boiling of liquid helium. Thus, we first prepare the electron by forcing it into the spin-up state as described in Sec. 6.4. As indicated, when analyzing this data, we report to the fitting program twice as many attempts to flip the spin as the actual number used. The program expects a saturation to occur with a relative probability of 0.5 for the spin to flip on each try. However, when we start with the spin-up state, then we can approach a probability near unity for flipping to the spin-down state for moderately hard drives because of the low probability of leaving the lower energy state. Clearly, this simple approximation fails in the limit of very high power since again, the probability of flipping the spin approaches 0.5 irrespective of initial state. Thus, we have been very careful to keep well away from even the apparent saturation at unity.

The second implication of a 10 times shorter classical decay time is associated with observing the cyclotron resonance line shape as described in Sec. 5.1. Our axial S/N is not adequate for the fast alternating sequence required ($\sim$ 5 Hz) for taking the cyclotron resonance in comparison with the optimal rate of 0.5 Hz used in the old molybdenum traps when $\tau_c \approx 1$ sec. In addition, higher excited Landau levels now decay away even faster, thus requiring much more microwave power, even when using continuous axial detection. This requirement put a severe constraint on the microwave source, which had to be greatly improved in order to get a clean source that did not artificially broaden the resonances.

Because of the use of continuous detection, we are now required to *correct* the cyclotron frequency for the measured shifts due to axial drive as shown in Fig. 17 in addition to those due to the applied microwave field as shown in Fig. 16. For the same reason, we must also *add* the shifts shown in Fig. 18 due to the anomaly power which is applied during the observation of the anomaly resonance, since again axial S/N was not adequate to allow for the simultaneous application of off-resonance anomaly power during the observation of the cyclotron resonance. Table 2 lists this series of runs taken in the phosphor bronze trap, with half taken at each extreme of the variable bottle. The net corrections are also listed, and only in one case is this correction not comparable or less than the final uncertainty in the measured anomaly. In Fig. 25, we show the anomaly plotted versus W_k, W_c, and W_a for these runs, after corrections have been applied.

Finally, as indicated in Sec. 5.2., a magnetic field wander of ~ 10 ppb is observed in this trap which contains a closed superconducting loop. As a result, the anomalies measured in the runs listed in Table 2 are spread out in a somewhat non-statistical distribution between two limits, about 12 ppb apart. This effect is clearly evident in Fig. 26 which is a histogram of all runs taken in the phosphor bronze trap at the two limits of the variable magnetic bottle. There is a possible residual B_2 dependence in these results since the average of the runs using $B_2 = 245 \text{G}/\text{cm}^2$ exceeds the average of the runs using $B_2 = -262 \text{ G}/\text{cm}^2$ by about 3.1×10^{-12}. Therefore, because of the nature of this distribution, it seems only prudent to report the simple average with the given standard deviation as an estimate for the uncertainty of the mean:

$$a_e(1990) = 1\,159\,652\,185.5(4.0) \times 10^{-12} \,. \tag{53}$$

As evident from Table 2, the uncertainty in Eq. 53 is about twice the typical statistical errors of the individual runs (which is most of the error listed in the fourth column). This result clearly agrees with the measurements shown in Eq. 49-52. Because of the possibly 10 times lower cavity Q, we are justified in not including any uncertainty associated with unknown cavity shifts at this level of confidence.

7.2. "a_e" AND QED

The intrinsic interest in this very precise measurement of the electron's g-factor resides in the ability of theorists to calculate the same quantity to roughly the

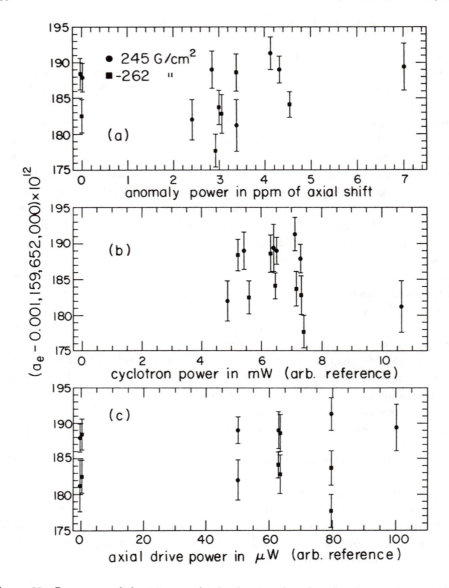

Figure 25. Summary of the 14 runs obtained using the phosphor bronze trap at the two extremes of the variable bottle. In each case, the computed anomaly is plotted versus the amount of (a) anomaly, (b) microwave, and (c) axial drive power applied during the excitation part of each cycle. At the given statistical level of uncertainty, no residual shift remains.

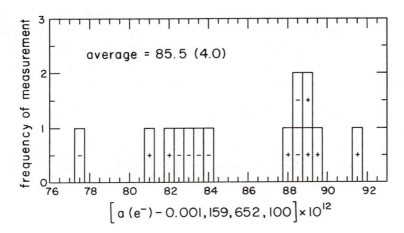

Figure 26. Histogram of the 14 runs obtained using the phosphor bronze trap. The "+" or "−" sign represents the sign of the variable bottle used for each indicated run. The distribution is believed to be a representation of the wandering magnetic field which occasional would make relatively fast fast jumps to a new field value between cyclotron calibrations.

same accuracy via the powerful theoretical technique of Quantum Electrodynamics (QED) which is based on the processes of virtual emission and absorption of photons and the polarization of the vacuum by electron/positron pairs. These radiative corrections can be arranged into an infinite power series of the quantity α/π where α is the fine-structure constant, and C_i are the corresponding coefficients of the power series. This particular lepton system is also fortunately blessed with understandable and calculable effects due to the strong and weak interactions such that they do not limit the present accuracy of the g-factor prediction.

The leading term has one Feynman diagram due to a single virtual photon exchange and its coefficient C_1 was first shown [48] by Schwinger to be exactly 0.5. The second term allows for two virtual photon exchanges or a single vacuum polarization bubble containing an e^+/e^- pair, yielding a total of 7 Fynman diagrams. The C_2 coefficient in front of this term is known analytically to be $-0.328,478,965,\dots$ (most recently predicted [49] by Sommerfeld and Petermann). The next order of complexity contains 72 Fynman diagrams and the corresponding C_3 coefficient has been evaluated in part analytically [50] and in part numerically [51]: $C_3 = 1.175,62(56)$. As if this was not difficult enough, Kinoshita undertook the mammoth job of evaluating [52] the 891 Feynman diagrams associated with the $8th$ order term, yielding $C_4 = -1.434(138)$. The remaining contributions due to the other-lepton loops, hadronic terms, and the electroweak effect have all been collected [53] into the term: $\delta a'_e = 4.46 \times 10^{-12}$.

In order to make connection between theory and experiment, one needs to have an accurate measurement of the fine-structure constant, α. Unfortunately, the tremendous improvement in precision of the g-2 experiment and theory have both exceeded the present improvements in the accuracy of α. At present, the most accurate independent measure of α is by the quantum Hall effect [54]:

$$\alpha^{-1}(\text{QHE}) = 137.035\ 997\ 9(3\ 2) \tag{54}$$

which can be compared to the fine-structure constant computed from the ac Josephson value of $2e/h$ combined with the most recent low field determination [55] of the gyromagnetic ratio γ_p' (for protons in a water sample):

$$\alpha^{-1}(\text{acJ}\&\gamma_p') = 137.035\ 984\ 0(5\ 1)\ . \tag{55}$$

These two values do not even agree within their combined uncertainties. Using the first of these (because it is more accurate), the theoretical value for a_e [52] is

$$a(\text{theory}) = 1\ 159\ 652\ 140(28) \times 10^{-12} \tag{56}$$

where the error is almost exclusively due to the measurement error of α. Because of this limitation, it is preferable to determine α from the g-2 experiment [33] and QED theory [52]:

$$\alpha^{-1}(\text{QED}) = 137.035\ 992\ 22(94) \tag{57}$$

where the uncertainty is still primarily determined by the theoretical error. However, it is anticipated that one day there will be comparisons of all three of these major ingredients [α, a_e(expt.), and a_e(theory)] at the one part per billion level of precision. This will provide a far more stringent test of QED than is presently possible.

7.3. POTENTIAL FUTURE EFFORTS

The best method for handling the cavity problem for the future has not yet been determined. It has been suggested by Brown, et al [37] and Dehmelt [34,38] that the characteristics of several neighboring modes can be investigated and the theoretical model can then be used to predict the region in frequency space which is free of shifts. Such regions do exist, but considerable effort may be required to find them. However, it is always preferable, if a choice exists, to design the experiment such that all regions of space are relatively shift free. More than likely, both low cavity

Q traps will be used and the nearest modes will be studied by actually measuring g-2 at the maximum shift positions.

The next task will be the improvement of axial sensitivity in order to allow still smaller axial frequency shifts to be observed, with the intent of reducing the size of the magnetic bottle. The clear evidence is that all major systematic effects, other than cavity shifts, depend directly on the strength of this bottle in some way. A more stable voltage source (possibly from a Josephson junction array) would translate immediately into improved S/N, since voltage fluctuations in the standard cells are believed to produce the present 10 ppb resolution. A higher Q tuned circuit would also generate a larger signal for the same oscillation amplitude in the trap.

However, since the real goal is to achieve an almost perfectly uniform magnetic field (with some means of detecting magnetic moments), two possible methods have been suggested for which this condition is approached by at least an order of magnitude. One method involves utilizing the existing relativistic bottle in some way. The primary impediment to this approach is axial sensitivity which presently requires 10–20 times larger bottles just to see a spin flip. Several elegant methods have been proposed [30,56-59] which might enhance the effect in order to yield the required S/N, but no technique has yet demonstrated that it will work. The second method involves another application of the variable magnetic bottle. Here, the B_2-term can be modulated [60] at some slow rate and a PSD can be used on the axial-frequency-shift detector to pick out the corresponding modulated magnetic moment (see Eq. 19). This technique has been attempted, but has so far achieved only partial success. In all cases, the primary objective is improvement in precision (though improvement in accuracy is also desirable) when the test to be performed is an ultra high precision comparison of electron and positron g-factors. Systematics such as cavity shifts or B_2-related shifts will be common to both. Thus, a statistical improvement is all that is required for the next improved comparison of matter and antimatter.

References

1. Nafe, J.E., Nelson, E.B., and Rabi, I.I. (1947) 'The Hyperfine Structure of Atomic Hydrogen and Deuterium', Phys. Rev. **71**, 914-915.
2. Lamb, W.E. and Retherford, R.C. (1947) 'Fine Structure of the Hydrogen Atom by a Microwave Method', Phys. Rev. **72**, 241-243; also Retherford, R.C. and Lamb, W.E. (1949) 'Shift of the $2^2S_{\frac{1}{2}}$ State in Hydrogen and Deuterium", Phys. Rev. **75**, 1325.
3. Louisell, W.H., Pidd, R.W., and Crane, H.R. (1953) 'An Experimental Measurement of the Gyromagnetic Ratio of the Free Electron', Phys. Rev. **91**, 475 and (1954) Phys. Rev. **94**, 7-16.

4. Dehmelt, H.G., (1956) 'Spin Resonance of Free Electrons Polarized by Exchange Col-
    lisions', Phys. Rev. **109**, 381-385.
5. Van Dyck, Jr., R.S., Schwinberg, P.B., and Dehmelt, H.G. (1978) 'Electron Magnetic
    Moment from Geonium Spectra', in B. Kursunoglu, A. Permutter, and L.F. Scott
    (eds.), *New Frontiers in High-Energy Physics*, Plenum Publishing, N.Y. pp. 159-
    181.
6. Van Dyck, Jr, R.S., Schwinberg, P.B., and Dehmelt, H.G., (1984) 'The Electron and
    Positron Geonium Experiments', in R.S. Van Dyck, Jr. and E.N. Fortson (eds.),
    Atomic Physics 9, World Scientific, Singapore, pp. 53-74.
7. Van Dyck, Jr., R.S., Schwinberg, P.B., and Dehmelt, H.G. (1986) 'Electron Magnetic
    Moment from Geonium Spectra: Early Experiments and Background Concepts', Phys.
    Rev. D **34**, 722-736.
8. Brown, L.S. and Gabrielse, G. (1986) 'Geonium Theory: Physics of an Electron or Ion
    in a Penning Trap', Rev. Mod. Phys. **58**, 233-311.
9. Dehmelt, H.G. (1981) 'Invariant Frequency Ratios in Electron and Positron Geonium
    Spectra Yield Refined Data on Electron Structure', in D. Kleppner and F.M. Pipkin
    (eds.), *Atomic Physics 7*, Plenum Publishing, N.Y. pp. 337-372.
10. Wineland, D.J., Itano, W.M., and Van Dyck, Jr., R.S. (1983) 'High-Resolution Spec-
    troscopy of Stored Ions', in B. Bederson and D. Bates (eds.), *Advances in Atomic
    and Molecular Physics*, Vol 19, Academic Press, N.Y. pp. 135-186.
11. Dehmelt, H.G. (1988) 'A Single Atomic Particle Forever Floating at Rest in Free Space:
    New Value for Electron Radius', Physica Scripta **T22**, 102-110.
12. Van Dyck, Jr., R.S. (1990) 'Anomalous Magnetic Moment of Single Electrons and
    Positrons: Experiment', in T. Kinoshita (ed.), *Directions in High Energy Physics*,
    Vol. on "Quantum Electrodynamics," World Scientific Publishers, Singapore.
13. Ekstrom, P. and Wineland, D. (1980) 'The Isolated Electron', Sci. Am. **243**, 104-121.
14. Dehmelt H.G. and Walls, F.L. (1968) 'Bolometric Technique for the rf Spectroscopy
    of Stored Ions', Phys. Rev. Lett. **21**, 127-131.
15. Walls, F.L. (1970) 'Determination of the Anomalous Magnetic Moment of the Free
    Electron from Measurements Made on an Electron Gas at 80°K Using a Bolometric
    Technique', Ph.D. thesis, University of Washington; also Walls, F.L., and Stein, T.S.
    (1973) 'Observation of the g-2 Resonance of a Stored Electron Gas using a Bolometric
    Technique', Phys. Rev. Lett. **31**, 975-979.
16. Wineland, D.J., Ekstrom, P., and Dehmelt, H. (1973) 'Monoelectron Oscillator', Phys.
    Rev. Lett. **31** 1279-1282.
17. Dehmelt, H., Ekstrom, P., Wineland, D., Van Dyck, R. (1974) 'Landau Level Depen-
    dent ν_z-Shifts in the Monoelectron Oscillator', Bull. Am. Phys. Soc. **19**, 572.
18. Dehmelt, H. (1986) 'Continuous Stern-Gerlach Effect: Principle and Idealized Appa-
    ratus', Proc. Natl. Acad. Sci. USA **83**, 2291-2294.
19. Van Dyck, Jr, R.S., Schwinberg, P.B., and Dehmelt, H.G. (1977) 'Precise Measure-
    ments of Axial, Magnetron, Cyclotron, and Spin-Cyclotron-Beat Frequencies on an
    Isolated 1-meV Electron', Phys. Rev. Lett **38**, 310-314.
20. Brown, L.S. and Gabrielse, G. (1982) 'Precision Spectroscopy of a Charged Particle in
    an Imperfect Penning Trap', Phys. Rev. A **25**, 2423-2425.
21. Van Dyck, R.S., Jr., Wineland, D.J., Ekstrom, P.A., and Dehmelt, H.G. (1976) 'High
    Mass Resolution with a New Variable Anharmonicity Penning Trap', Appl. Phys.

Lett. **28**, 446-448.
22. Gabrielse, G. (1983) 'Relaxation Calculation of the Electrostatic Properties of Compensated Penning Traps with Hyperbolic Electrodes', Phys. Rev. A **27**, 2277-2290.
23. Schwinberg, P.B. (1979) 'A Technique for Catching Positrons in a Penning Trap via Radiation Damping', Ph.D. thesis, University of Washington [available from University Microfilms International, Ann Arbor, MI.].
24. Schwinberg, P.B., Van Dyck, Jr, R.S., and Dehmelt, H.G. (1981) 'Trapping and Thermalization of Positrons for Geonium Spectroscopy', Phys. Lett. **81A**, 119-120.
25. Wineland, D.J. and Dehmelt, H.G. (1975) 'Principles of the Stored Ion Calorimeter', J. Appl. Phys. **46**, 919-930.
26. Gabrielse, G. (1984) 'Detection, Damping, and Translating the Center of the Axial Oscillation of a Charged Particle in a Penning Trap with Hyperbolic Electrodes', Phys. Rev. A **29**, 462-469.
27. Shockley, W. (1938) 'Currents to Conductors Induced by a Moving Point Charge', J. Appl. Phys. **9**, 635-636; also, Sirkis, M. and Holonyak, N. (1966) 'Currents Induced by Moving Charges', Am. J. Phys. **34**, 943-946.
28. Van Dyck, Jr., R.S., Moore, F.L., Farnham, D.L., and Schwinberg, P.B. (1986) 'Variable Magnetic Bottle for Precision Geonium Experiments', Rev. Sci. Instrum. **57**, 593-597.
29. McLachlan, N.W. (1947) *Theory and Applications of Mathieu Functions*, Oxford University Press, N.Y., p. 20.
30. Dehmelt, H., Mittleman, R., and Liu, Y. (1988) 'Relativistic Cyclotron Resonance Shape in Magnetic Bottle Geonium', Proc. Natl. Acad. Sci. USA **85**, 7041-7043.
31. Gabrielse, G. and Dehmelt, H. (1985) 'Observation of Inhibited Spontaneous Emission', Phys. Rev. Lett. **55**, 67-70.
32. Van Dyck, Jr, R.S., Schwinberg, P.B., and Dehmelt, H.G. (1988) 'Damping Time Measured in a Low Q Penning Trap', Bull. Am. Phys. Soc. **30**, 2349.
33. Brown, L.S., Gabrielse, G., Helmerson, K., and Tan, J. (1985) 'Cyclotron Motion in a Microwave Cavity: Lifetime and Frequency Shifts', Phys. Rev A **32**, 3204-3218; also (1985) 'Cyclotron Motion in a Microwave Cavity: Possible Shifts of the Measured Electron g Factor', Phys. Rev. Lett. **55**, 44-47.
34. Dehmelt, H., et al. 'Practical Zero-Shift Tuning in Geonium', (unpublished).
35. Van Dyck, Jr, R.S, Schwinberg, P.B., and Dehmelt, H.G. (1987) 'New High-Precision Comparison of Electron and Positron g Factors', Phys. Rev. Lett. **59**, 26-29.
36. Van Dyck, Jr., R., Moore, F., Farnham, D., Schwinberg, P., and Dehmelt, H. (1987) 'Microwave-Cavity Modes directly Observed in a Penning Trap', Phys. Rev. A (Brief Reports) **36**, 3455-3456.
37. Brown, L.S., Gabrielse, G., Tan, J., and Chan, K.C.D. (1988) 'Cyclotron Motion in a Penning Trap Microwave Cavity', Phys. Rev. A **37**, 4163-4171.
38. Dehmelt, H. (1987) 'Single Atomic Particle at Rest in Free Space: Shift-Free Suppression of the Natural Line Width?', in W. Persson and S. Svanberg (eds.), *Laser Spectroscopy VIII*, Springer-Verlag, New York, pp. 39-42.
39. Van Dyck, Jr., R.S., Moore, F.L., Farnham, D.L., and Schwinberg, P.B. (1989) 'Number Dependency in the Compensated Penning Trap', Phys. Rev. A. **40**, 6308-6313.
40. Van Dyck, Jr., R., Schwinberg, P., and Bailey, S. (1980) 'High Resolution Penning Trap as a Precision Mass-Ratio Spectrometer', in J.A. Nolen, Jr. and W. Benenson (eds.),

Atomic Masses and Fundamental Constants 6, Plenum, New York, pp. 173-182.

41. Wesley, J.C. and Rich, A. (1971) 'High-Field Electron g-2 Measurement', Phys. Rev. A **4**, 1341-1363; and Gilleland, J.R. and Rich, A. (1972) 'Precision Measurement of the *g* Factor of the Free Positron', Phys. Rev. A **5**, 38-49.

42. Bailey, J., *et al.* (1979) 'Final Report on the CERN Muon Storage Ring Including the Anomalous Magnetic Moment and Electric Dipole Moment of the Muon, and a Direct Test of Relativistic Time Dilation', Nuc. Phys. B **150**, 1-75.

43. Brown, L.S. (1984) 'Geonium Lineshape', Ann. Phys. (NY) **159**, 62-98; also Brown, L.S. (1984) 'Line Shape for a Precise Measurement of the Electron's Magnetic Moment', Phys. Rev. Lett. **52**, 2013-2015.

44. Van Dyck, Jr., R.S., Schwinberg, P.B., and Dehmelt, H.G. (1979) 'Progress of the Electron Spin Anomaly Experiment', Bull. Am. Phys. Soc. **24**, 758.

45. Schwinberg, P.B., Van Dyck, Jr, R.S., and Dehmelt, H.G. (1984) 'Preliminary Comparison of the Positron and Electron Spin Anomalies', in B.N. Taylor and W.D. Phillips (eds.), *Precision Measurement and Fundamental Constants II*, Natl. Bur. Stand. (U.S.) Spec. Publ. 617, pp. 215-218.

46. Van Dyck, Jr., R.S., Schwinberg, P.B., and Dehmelt, H.G. (1990) 'Consistency of the Electron *g*-factor in a Penning Trap', submitted to Book of Abstracts for the Twelfth Int. Conf. Atomic Physics (ICAP-12), and this publication.

47. Schwinberg, P.B., Van Dyck, Jr, R.S., and Dehmelt, H.G. (1981) 'New Comparison of the Positron and Electron g Factors', Phys. Rev. Lett. **47**, 1679-1682.

48. Schwinger, J. (1948) 'On Quantum-Electrodynamics and the Magnetic Moment of the Electron', Phys. Rev. **73**, 416-417.

49. Sommerfield, C. (1957) 'Magnetic Dipole Moment of the Electron', Phys. Rev. **107**, 328-329; Petermann, A. (1957) 'Fourth Order Magnetic Moment of the Electron', Helv. Phys. Acta **30**, 407-408.

50. Levine, M.J., Park, H.Y., and Roskies, R.Z. (1982) 'High-Precision Evaluation of Contributions to g-2 of the Electron in Sixth Order', Phys. Rev. D **25**, 2205-2207.

51. Cvitanovic, P. and Kinoshita, T. (1974) 'Sixth-Order Magnetic Moment of the Electron', Phys. Rev. D **10**, 4007-4031; also Kinoshita, T. and Lindquist, W.B. (1977) 'Improving the Theoretical Prediction of the Electron Anomalous Magnetic Moment', Cornell preprint CLNS-374.

52. Kinoshita, T. and Lindquist, W.B. (1990) 'Eighth-order Magnetic Moment of the Electron V. Diagrams Containing No Vacuum-polarization Loop', Phys. Rev. D **42**, 636-655; see also Kinoshita, T. (1988) 'Fine-Structure Constant Derived from Quantum Electrodynamics', Metrologia <u>25</u>, 233-237; and Kinoshita, T. (1989) 'Accuracy of the Fine-Structure Constant', IEEE Trans. Instrum. Meas. <u>38</u>, 172-174.

53. See, for instance, Kinoshita, T. (1978) 'What Can One Learn from Very Accurate Measurements of the Lepton Magnetic Moments?', in B. Kursunoglu, A. Permutter, and L.F. Scott (eds.), *New Frontiers in High Energy Physics,*, Plenum Publishing, New York, pp. 127-143; also Kinoshita, T. and Lindquist, W.B. (1981) 'Eighth-Order Anomalous Magnetic Moment of the Electron', Phys. Rev. Lett. **47**, 1573-1576.

54. Cage, M.E., *et al.* (1989) 'NBS Determination of the Fine-Structure Constant, and of the Quantized Hall Resistance and Josephson Frequency to Voltage Quotient in SI Units', IEEE Trans. Instrum. Meas. **38**, 284-289.

55. Williams, E.R., *et al.* (1989) 'A Low Field Determination of the Proton Gyromagnetic

Ratio in Water', IEEE Trans. Instrum. Meas. **38**, 233-237.

56. Dehmelt, H., Van Dyck, R., and Schwinberg, P. (1979) 'Proposal for Detection of Geonium Spectra via Radial Displacement', Bull. Am. Phys. Soc. **24**, 491.

57. Dehmelt, H., Van Dyck, R., Schwinberg, P., and Gabrielse, G. (1979) 'Proposal to Detect Spin Flips in Geonium via Linked Axial Excitation', Bull. Am. Phys. Soc. **24**, 675.

58. Dehmelt, H. and Gabrielse, G. (1981) 'Faster, Simpler Schemes to Distinguish $n = 0, 1$ in Geonium', Bull. Am. Phys. Soc. **26**, 797; (1984) 'Comb Excitation Scheme for Resolving the Cyclotron Spectrum of Geonium', Bull. Am. Phys. Soc. **29**, 44; (1984) 'Quasi-Thermal, Multi-Step Excitation Scheme for Geonium Cyclotron Spectroscopy', Bull. Am. Phys. Soc. **29**, 926.

59. Gabrielse, G., Dehmelt, H., and Kells, W. (1985) 'Observation of Relativistic, Bistable Hysteresis in the Cyclotron Motion of a Single Electron', Phys. Rev. Lett. **54**, 537-539.

60. Schwinberg, P.B. and Van Dyck, Jr., R.S. (1981) 'Geonium Spectroscopy Using a Modulated Magnetic Bottle', Bull. Am. Phys. Soc. **26**, 598.

EXPERIMENTS ON THE INTERACTION OF INTENSE FIELDS WITH ELECTRONS

A. Weingartshofer
Laser-Electron Interactions Laboratory
Department of Physics
St. Francis Xavier University
Antigonish, Nova Scotia
Canada   B2G 1C0

ABSTRACT.   This is an attempt to present an overview of currently performed experiments that investigate some fundamental aspects of the interaction of electrons with intense electromagnetic fields (laser and microwave).   The electrons are free or in a continuum state of the atom or ion.

1.   THE ROLE OF A GOOD EXPERIMENT

At the moment, we have the situation where the art of experimentation in intense laser-atomic physics - which provides much information on laser-electron interactions - is ahead of theoretical and computational methods.   Many laboratories have not only provided some astounding new observations but, more important, they have also presented them to the physics community in a quantitative form.   The results are impressive but the systems that have been investigated are complex and in order to explain them, theorists have to make approximations and find it difficult to go beyond qualitative agreement with experimental data.   To try to develop a theoretical model that can incorporate "all" the subtle aspects of multiphoton ionization and also the quantitative observations of the generation of "multiple harmonics", is surely a challenge.

Experimentalists are not only discovering new phenomena but they are also exploring them in quantitative terms and as a function of fundamental parameters like intensity or polarization.   One of their major problems is to provide absolute values of laser intensities, ionization cross sections or reaction rates.   They are essential to differentiate among the theories that have been proposed to explain the same phenomena.   The laser laboratories at C.E.N. de Saclay, France are making great efforts in this direction [1].

Theoretical approaches may succeed in reproducing qualitatively, most of the observed features and, in principle, this would be sufficient if the theory can also establish their dependence on specific atomic structure or single out the possible situations that involve primarily an explanation in terms of "free electron dynamics in intense lasers", i.e. "the simpleman's theory".

D. Hestenes and A. Weingartshofer (eds.), The Electron, 295–310.
© 1991 *Kluwer Academic Publishers. Printed in the Netherlands.*

Experiments in multiphoton ionization (MPI) and above-threshold ionization (ATI) are very complex phenomena because of the large number of parameters that can enter the process. Of less complexity are experiments in simultaneous electron-multiphoton excitation of atoms (SEMPE) because we have control over important parameters that determine these off-energy shell processes and where ionization can be avoided. Of still lesser complexity is the "elastic" version of the previous experiments: elastically scattered electrons in the presence of an intense field, also known as laser-induced free-free (FF) transitions. They all are important in applied problems.

From the point of view of theory, they all are complex. There is also a positive aspect if one considers that there is a common link between them that can be explored to increase our understanding of intense laser-atomic physics. Processes that can be controlled experimentally can provide a testing ground for theoretical approximations and new ideas before they are applied somewhere else, i.e. the approximations and ideas can now be modified or extended with confidence. It is an obligation of the experimentalist to devote time and effort in designing experiments to test the limitation of widely used approximations and establish their regime of application. We have pursued this fundamental philosophy in our Laser-Electron Interactions Laboratory.

2. GENERAL OVERVIEW

In dealing with laser-electron interactions it is convenient to distinguish five basic experimental situations that describe the state of the electron with respect to an atom and the intense radiation (laser or microwave).

a) a free electron: since free electrons cannot absorb photons, for the interaction to take place effectively a "laser-stimulated mechanism" must be invoked which is known in the literature as "stimulated Thomson scattering" (elastic) and "stimulated Compton scattering" (inelastic). This gives rise to light-induced forces that change electron energies and momenta and are referred to as "ponderomotive" effects. Multiples of photons can only be exchanged with the radiation field in the presence of a third body, i.e. a potential, and therefore can only occur in the case of a bound electron or an electron colliding with an atom, i.e. a scattering process, as described below.

b) a bound electron: bound in the ground state of the atom or "quasi-free", bound in a high Rydberg state close to the ionization limit. The electron can also form a negative ion, i.e H$^-$ which has an electron affinity of -0.754 eV. Current experiments of interest investigate the ionization of atoms when illuminated with intense electromagnetic radiation (laser or microwave). The ion yield as a function of field intensity can be examined but principally they rely on the interpretation of the photoelectron spectra of the ejected electrons recorded with an electron spectrometer. The experiments fall into four categories: multiphoton ionization (MPI), above -threshold ionization (ATI), multiphoton detachment (MPD), and tunnelling if the intensity reaches a critical value.

c)   electrons scattering from an atom in the presence of an intense laser field are referred to as laser-induced free-free (FF) transitions, although the initial and final electron scattering states are NOT completely free states, only asymptotically free states (even though the time of collision may only last 10^{-15} s). It is also possible to experiment with scattering resonances which occur at well-defined electron impact energies, and the electron attaches itself to the target atom forming a temporary negative ion, nevertheless a truly bound state that may live 10^4 times longer than the duration of a normal collision.

d)   electron-multiphoton-atom a simultaneous interaction, carried out in a controlled three-beam experiment.

e)   electrons produced in new sources; i.e. Penning electrons.

Free electrons in (a) originate in experiments with bound electrons described in (b). As a matter of fact, they play a crucial role in the overall processes of MPI, MPD and the more intriguing phenomenon of ATI where the electron absorbs a much greater number of photons than necessary for the atom to be ionized. ATI experiments have greatly increased our understanding of the interaction of intense fields with free electrons. Intense laser atom physics relies on these electrons for information on the behaviour of atomic structure when exposed to high intensities (10^{14} W/cm^2). Recent developments of femtosecond laser pulses has provided a tool to investigate these properties [2]. See also article by A. Bandrauk in this Book. The field has made remarkable progress and has reached a very interesting state but there are also many challenges, especially in the theoretical interpretation of the complexity of processes that occur. Since there is a necessity to understand laser-matter interaction (for applied reasons) it has become one of the most active areas of research and is well-documented [3,4]. The more elementary laser-electron interaction represents one part of the overall program and the intention in this presentation is to discuss some relevant aspects of this interaction and establish some common links with the more complex processes. First, we want to summarize some general notions.

Experiments with microwave fields occupy a very unique place in this area because of their special techniques and they also represent the limit of the low-frequency regime. The contribution of Tom F. Gallagher to this book illustrates all this in a magnificent way in an article by the title "Ionization in Linearly and Circularly Polarized Microwave fields". Microwave ionization of Rydberg atoms can investigate high order multiphoton ATI (~10^5 microwave photons) processes and show in which way they are fundamentally different from ATI at optical frequencies.

It was mentioned above that intense laser-atom physics relies on the ejected electrons for information of atomic structure. Another important source of information is the emitted radiation in an MPI process which contains very high harmonics of the laser field. This radiation is now being examined quantitatively and systematically in C.E.N. de Saclay,

France. The essential aspects of this phenomenon is reviewed in this book in an article by the title: "Multiple Harmonic Generation in Rare Gases at High Laser Intensity" prepared by Anne L'Huillier, Louis-André Lompre and Gérard Mainfray.

Two very recent areas are now demonstrating their worth as new powerful tools to probe both, laser-electron interactions as well as particular aspects of intense laser-atom physics. Although multiphoton processes are being observed, all the complexities of MPI are avoided by working at low laser intensities, 10^8 W/cm^2. These new areas are described in this book by Barry Wallbank in "Absorption and Emission of Radiation During Electron Excitation of Atoms" and by Harald Morgner in "Penning Ionization in Intense Laser Fields". This area also illustrates some possible mechanisms that can lead to laser-controlled chemistry.

The process of multiphoton ionization of an atom under the influence of an intense oscillating electromagnetic fields is a fundamental question. Leo Moorman addresses the problem from a new point of view in an article entitled "Microwave Ionization of H Atoms: Experiments in Classical and Quantal Dynamics", Moorman discusses the theory of microwave ionization based on experiments performed in the laboratory at Stony Brook, NY, which reveals that the process may be divided into different regions of distinct behavior depending on the precise experimental situation. Some regions may be described using classical atomic dynamics while others require quantum atomic dynamics. The author discusses the level of accuracy of this division. Microwave ionization of atoms has become one of the experimental and theoretical prototypes for studying quantum chaos. It is of interest to observe that also in the last paper of the theory section, Christof Jung addresses a similar problem by proposing a theory to merge the ideas of free-free transitions and scattering chaos.

Finally in the theory section of this book, André Bandrauk discusses the theory of intense laser interaction with molecules in a very timely article by the title "The Electron and the Dressed Molecule". Molecules present a system of higher complexity and therefore is the restricted domain of only a small group of investigators. His predicted laser-induced effects have been observed experimentally and the results illustrate magnificently that this area is important for fundamental as well as applied reasons.

Our current understanding of MPI and ATI is, that, somehow we can think of it as a four stage event. First, the atom is deformed by a dynamical Stark shift. Two, the deformed, or dressed, atom absorbs photons and forms an ion and an electron in a continuum state in the combined fields of the laser and ion (long-range potential). Three, laser-induced electron transitions between continua may take place and even laser-stimulated recombination. Four and final stage, the electron is released and it has to escape from the laser field as a free electron and ultimately find its way to the detector.

When does the electron come free, and when can we assume free electron laser interaction? There is no clear answer to this, but, at this point, one distinguishes two aspects: the "quiver" and the "drift" motion of the electron. The former is referred to as the "ponderomotive" aspect of the laser-electron interaction and this is now

fairly well understood, although some confusion existed in the past [5].
The "drift" aspect of this interaction which can impart large amounts of
kinetic energy to the ejected electron, as has been demonstrated in
recent experiments [6,7], has raised some interesting questions. These
are typical ATI experiments where the energy comes from the absorption of
photons in excess of the minimum number required for ionization.
However, this raises the question: where does the momentum come from
since photons carry very little momentum [8]? Some further reference to
this will be made in Section 4.

The oft-made analogy between electron scattering and the ejection of
an electron during the ionization process is a very valuable concept that
has been explored repeatedly, e.g. Collisions and Half-Collisions with
Lasers is the title of a workshop organized by N. K. Rahman and
C. Guidotti [9]. It has been pointed out [8] that the analogy is far
from complete. In the low frequency regime one essential difference is
that, although both the collision duration and atomic orbital period are
short compared to the cycle time, in a scattering process the electron
passes by the core only once, while in an ionization process the electron
orbits the core many times.   Rosenberg [10] has suggested that
multiphoton ionization (MPI) may be thought of as the second half of an
induced resonance reaction. It is not difficult to realize that there is
a link between microwave excitation and ionization of highly excited
Rydberg atoms and "resonant electron scattering" in a laser field via the
formation of a temporary bound state. In this context it is illuminating
to refer the reader to page 192 of Bandrauk's article on the physics of
the multiphoton ionization of the H_2-molecule, where he describes the
photoionized electron as continuing to absorb photons that create above-
threshold ionization peaks which reflect the vibronic structure of the H_2^+-
ion "dressed by the intense field".

It is appropriate at this point to recall one important result that
we have learned in scattering experiments in the presence of a radiation
field [11,12]. The measurements clearly demonstrate the transfer of
multiples of the photon energy of a CO_2 - laser into the kinetic energy
of electrons while being scattered on an atomic target. This is
basically a confirmation of the Kroll and Watson equation [13] which
plays a fundamental role in laser-electron interactions and has occupied
the minds of many theorists and presents challenges to the
experimentalists. Further reference to this type of quantified "inverse
bremsstrahlung" or laser-induced free-free (FF) transitions of electrons
will be made in Sections 3 and 4.

Negative ions represent a unique system for the investigation of
MPI and ATI. They have usually one bound state and electron correlations
play a primary role in determining binding energies. Negative hydrogen
is a typical example, the ion has only one bound state (the ground state)
and, in contrast to ionization of atoms, the detached electron [14,15]
experiences a short-range ($1/r^4$) potential as it leaves the ion rather
then the usual long-range Coulomb potential seen in MPI and ATI. The
four- stage event of the process is considerably simplified. Experiments
of MPD of electrons from negative hydrogen, H^-, have recently been
performed [16] where a beam of ions (produced externally) was intersected
by a focused CO_2 TEA laser. We want to remark that negative ions also

play an important role in electron scattering, i.e. electronically excited "Feshbach resonances" that form a system of bound "temporary negative ions" [17]. These resonant states have also been investigated in the presence of a strong radiation field [18,19]. The negative ions are produced (internally) and since the photon energy (hω) is much larger than the resonance width (Δ), the field changes sign many times during the resonance lifetime so that decay by tunnelling is not very likely. Further reference to this will be made in Sect. 4.

3. LASER-STIMULATED PROCESSES

Once the electron is free (see stage four above) it can no longer absorb photons but it will continue exchanging momentum and energy with the laser field in a very significant way. This can only be explained in the photon language of quantum electrodynamics, i.e. high-intensity coherent light scatters from a free electron in a stimulated manner known as "stimulated" Thomson scattering (elastic: stimulated scattering among different momentum eigenmodes) or "stimulated" Compton scattering (inelastic: stimulated exchange of photons between different frequency modes) [20,21,22].

These theoretical notions originated with the early development of high intensity laser beams and were summarized by Kibble in 1966 [22] in a paper with the title "Mutual Refraction of Electrons and Photons" where the complete analogy between the processes of refraction of light by electrons and of electrons by light is emphasized. Kibble goes on to show that the important effects in most cases of practical interest (huge number of photons) can all be understood in purely classical terms. Solving Newton's equation for an electron subjected to a Lorentz force that is turned on during the time scale of one radiation cycle will set the electron into a "quiver" motion. Thomson scattering arises because the electron is accelerated by the external electric field, and thereby absorbs and reradiates energy [21]. Dealing with the fine details of the interaction of the electron (mass $\underline{m}$ and charge $\underline{e}$) with the spatially varying $\underline{E}$ and $\underline{B}$ fields of the radiation field (i.e. the focused laser) is referred to by the word "ponderomotive" and we speak of ponderomotive forces (i.e. a field gradient force) and a ponderomotive potential, which is the time-averaged kinetic energy of the oscillating electron

$$Wp = \frac{e^2Eo^2}{4m\omega^2}, \qquad Wp = 9.34x10^{-14}\lambda^2I \ eV$$

where Eo is the electric field amplitude, I the intensity in W/cm^2, ω and λ the angular frequency and wavelength of the radiation. This energy depends inversely on the square of the laser frequency and therefore ponderomotive effects are very important for infrared and microwave radiation.

Experiments suggested in the paper of Kibble [22] have been attempted [23], but it is only very recently that confirmation of these effects has been carried out in a series of very elegant experiments conducted at AT&T Bell Laboratories [24,25,26]. These results clarified the role of the ejected free electrons in MPI and ATI processes and have been of fundamental assistance in the interpretation of the observed photoelectron energy spectra.

Ponderomotive forces are dramatic examples of the predominance of stimulated processes in free electron-laser interactions. In this regard, it is illuminating to consider another example that was mentioned before, FF transitions or electron scattering from an atom under the influence of an external field [12]. During the scattering event, the electron is accelerated and, on the basis of classical electrodynamics, an accelerated charged particle will emit electromagnetic radiation, i.e. bremsstrahlung. For an electron with an incident energy, $E_i = 10$ eV, the "spontaneous" emission of a photon with a perceptible energy has only negligible probability. However, if the electron scattering event takes place inside a laser field, then the induced analog of bremsstrahlung can occur and we will observe a laser "stimulated" phenomenon where absorption as well as emission will occur with equal probability. Notice, however, that in contrast to stimulated Thomson scattering, during the electron-atom collision, multiples of laser photons ($\hbar\omega$) are exchanged i.e. the scattered electrons can only increase or decrease their kinetic energy in units of laser photons, and this is now observed routinely [11,27,28].

It is of interest to remark that the emitted photons in FF transitions have not been investigated. This project is in the planning stage. It has been predicted [29] that there is a high probability that these emitted photons contain a high proportion of multiple frequencies of the driving wave frequency, i.e. the most elementary example imaginable of "multiple harmonic generation".

Multiple harmonic generation in the MPI of the rare gases at high laser intensity has been investigated in C.E.N. Saclay, France and some exciting experimental results and their interpretation are discussed in this book in an article prepared by Anne L'Huillier, Louis-André Lompre and Gérard Mainfray. The phenomenon has been known for over 20 years [30] but it is only recently that it is being investigated systematically and quantitatively.

Both laser-stimulated Thomson scattering and FF transitions take place in intense radiation fields where a very large number of coherent photons are being scattered and therefore we are, in principle, dealing with a classical radiation field. FF transitions become a semi-classical problem because the interaction of the electron beam with the target atom has to be treated quantum mechanically [11,12]. On the other hand, stimulated Thomson scattering assumes a totally classical physical picture as was described above, i.e. a classical electron in an oscillating electromagnetic field [21,22].

4. INTERESTING SITUATIONS THAT RAISE SOME QUESTIONS

When an atom is ionized, the free electron is born with an amount of ponderomotive (i.e. quiver) energy Wp and its motion is the superposition of a "drift" motion and a "quiver" motion, which as it travels through the beam waist (finite size, 20 μm) is reconverted into translational motion by the field gradient force (ponderomotive force), thereby leaving the electron energy unchanged, owing to the conservative nature of the potential. This reconversion of energy was predicted 25 years ago by Kibble [22] and has recently been observed experimentally by Bucksbaum et al [24,26] who demonstrated in a very elegant way the equivalence of the electron's leaving the focus of the field to leaving the region of a potential energy Wp.

This reconversion cannot be consummated in the case of ultrashort pulses that terminate before the electrons have time to traverse the dimensions of the focused pulse. A substantial fraction of the quiver energy cannot be recovered and the energy is effectively transferred back to the laser field. Two recent experiments on ATI processes were performed under similar conditions by Gallagher [6] and Corkum et al [7] with circularly polarized radiation in the long wavelength limit: CO_2 - laser and microwaves. The results were remarkable, the electrons picked up an unusually high drift momentum, while it was just remarked above that for ultrashort pulses the energy is returned to the wave and does not contribute to ATI so, where does the drift motion come from?

The interpretation of Gallagher [6] and Corkum et al [7] is that these are free electron phenomena and the observations are primarily the result of the interaction of a newly freed electron with the laser field. The electrons are removed from the primary influence of the atom almost instantaneously by tunnelling or dc field ionization (fraction of cycle of the radiation) and therefore are released into a well defined phase of the electric field. The quantitative discussion is based on classical physics. The fact remains, however, that these observations bring a new insight to ATI of a laser-electron interaction that will generate much discussion in the physics community.

Two theorists, Cooke and Shakeshaft [8,14] have written a very interesting article on these observations which they describe as a "puzzle" with the following words "the electron is released into the field with a small mechanical momentum, and so one may wonder how the electron can later acquire a large momentum ..... while the radiation field provides the photoelectron's energy in a photoionization experiment it cannot provide the photoelectron's momentum". They propose a non-radiative transfer mechanism that underlines the ideas of electron scattering in their theory. The drift energy originates in the absorption of photons followed by an electron-core scattering that is elastic so the drift momentum really originates in the electrostatic force which initially binds the electron to the atomic core, they call it a "soft scattering of the electron from the atomic core to which it is bound". The theory is illustrated with very vivid physical images and it will surely stimulate many minds to try to improve on the model. However, a new theoretical interpretation is not excluded.

Although the analogy between scattering and ionization in a low-frequency field is far from complete, it has been used very successfully as a basis for a theory that explains several of the characteristic structures observed in ATI spectra. This is clearly illustrated in an article by Kupersztych [31] with the title "Inverse Half-Bremsstrhlung in MPI of Atoms in Intense Light Beams". In the first part of the problem

Kupersztych takes basically the same approach as Gallagher [6] and Corkum et al [7], i.e. the electron is removed from the ground state into the continuum in a time shorter than an optical period and then comes strongly under the influence of the intense electromagnetic field but still acted on by the long-range Coulomb field of the ion. The electron can absorb photons and momentum while receding from the ion. The ionized atom and the collision centre are the same and, therefore, the name half-collision and the similarity with free-free (FF) transitions as a mechanism to absorb additional photons. The only difference is that in FF transitions we scatter electrons from a neutral atom while in the ATI process the scattering is from an ion (long-range Coulomb field). This approach has the merit that it represents an attempt to investigate the physical relationship that may exist between ATI and FF processes. Is there a clear connection, and is it possible to detect the influence of the long-range Coulomb field of the ion on FF processes? Mittleman [32] has written a very comprehensive article on the Coulomb extension of the Kroll and Watson theory [13] which shows special peculiarities that we don't fully understand.

The FF process that has been discussed so far is basically an elastic collision (the target atom changes no internal energy) and the duration of the collision (approximately $\hbar/E_i$, where E_i is the incident energy of the electron) is much shorter than the laser period ($1/\omega$) in the low-frequency regime. Therefore, photon absorption will occur primarily immediately before or after the collision but is very unlikely to occur during the collision proper. This is the essence of the low-frequency approximation and the condition is that $\hbar/E_i \ll 1/\omega$, or $\hbar\omega \ll E_i$, i.e. also the name "soft-photon" approximation. The consequences are that the atom takes on the role of "passive" observer and simply supplies a potential to consummate the FF process. This was first recognized by Kroll and Watson [13] in their formulation of FF processes, i.e. the FF cross section is simply the radiationless elastic cross section multiplied by a factor that reflects only the properties of the laser and the electron but not the atom. This result has been confirmed, so far, by experiments but attempts are being pursued intensely to determine the circumstances under which the breakdown of the "soft photon" approximation can be observed [11,27,28,33].

One possible approach is to scatter electrons resonantly by making $E_i = E_r$, the resonance energy, so that the electron is captured temporarily forming an electron-core scattering resonance with a narrow width Γ and a lifetime $\hbar/\Gamma$ which is 10^4 times longer than $\hbar/E_r$, i.e. a true bound state is formed in the presence of the radiation field. One might expect some significant influence on the cross section of the FF process under the new circumstances. The expected changes were examined [19a,19b] by measuring the FF cross sections of the p-wave neon and argon resonances around 16.1 and 11.1 eV for the incident electron energy. Significant abnormalities were observed that cannot be predicted by the soft-photon approximation [34]. These results are, of course, very promising and are now being investigated in our laboratory at higher laser intensities.

5. ELECTRON-MULTIPHOTON-ATOM SIMULTANEOUS EXCITATION

In FF processes the atom remains in the ground state and, therefore, we
are examining elastic collision processes. In contrast, in this series
of experiments the structure of the atom will be modified, first by the
laser field and then excitation under the simultaneous action of one
electron and several photons i.e. a three-body multiphoton collision
performed in a controlled three-beam experiment. The process can be
monitored by detecting the excited state directly (if a metastable state)
or indirectly by observing the emitted UV radiation, the energy and
angular distribution of the scattered electrons, or the ion yield if one
is probing the ionization threshold in order to determine the ionization
cross section as modified by the laser field or investigating laser-
stimulated electron-ion recombination.
        So far, we have used a TEA CO_2 - laser with a peak power of 10^8 W/cm^2
operated in the multimode optical configuration and linearly polarized.
These experiments present a new approach to examine both photon-electron
interactions and laser-electron-atom interactions. One may want to
investigate how electrons and photons can couple to perform a specific
function or explore atomic structure under the influence of an intense
field by monitoring electron-multi-photon excitation processes, i.e. an
alternative way to understand intense laser-atom physics for a specific
system that is readily accessible. This condition poses limitations but
one has to remember that the investigation has the advantage of being
performed in a controlled manner and can produce quantitative
information, i.e. cross sections and their dependency on several
parameters like intensity, polarization, etc.
        This technique is only at its beginning and so far it has been used
to measure cross sections for the following.

    i)   Simultaneous off-energy shell excitation of He 2^{3}S by an
          electron and one to four photons. Notice that the process
          starts with a singlet state He 1^{1}S and therefore we are also
          observing an interesting electro exchange mechanism [35].

    ii)   Simultaneous electron-photon excitation of metastable He (2^{1}S),
          and the 2^{3}P states of neon and argon [36].

    iii) Quantitative experimental investigations of the effects of
          laser intensity and polarisation on the formation of He 2^{3}S by
          simultaneous electron-photon excitation [37].

    iv)   Absorption and emission of radiation during electron excitation
          of the 2^{1}S and 2^{1}P states of helium. We report electron spectra
          resulting from the inelastic scattering of 45-eV electrons from
          He atoms in the presence of an intense CO_2 - laser [38].

        One can foresee a wealth of new experiments of this nature as a
testing ground of laser-electron interactions and atomic physics in the
presence of intense lasers. As a way of illustration I would like to
make a few specific comments about the type of experiments described

under (i) above. Barry Wallbank will discuss the example of case (iv) in an article in this book.

5.1 COMMENTS ON THE OFF-ENERGY SHELL EXCITATION He 1^1S → He 2^3S

Geltman and Maquet [39] have examined these results and their calculated cross sections are in good qualitative agreement with them. This is interesting since they considered a simple extension of the Kroll and Watson formula [13] for laser-assisted collisions which would demonstrate that the "soft-photon" approximation has still some validity under these circumstances. Further comments on this will be made below.

Trombetta [40] is intrigued by the electronic exchange that has been observed in the excitation "singlet → triplet" state of helium and has calculated cross sections for the electronic exchange in laser-assisted elastic electron-hydrogen collisions. He has reached the remarkable conclusion that the exchange mechanism may strongly dominate the differential cross section. This, is of course, a very significant result since this would be a clear example of the use of a laser to control specific excitation mechanisms that may have practical applications, i.e. laser-assisted chemistry.

In this inelastic process, all three pairs of collisional subsystems violate energy conservation. We have, then, the possibility of observing, for example, inelastic electron-atom differential cross-sections or total excitation cross-sections at electron kinetic energies which are off the "energy shell" from the point of view of the electron-atom subsystem. Sundaram and Armstrong, Jr. [41] have taken the viewpoint that ATI is a highly complex process and make an interesting analysis of the participating mechanisms which involve both excited bound states and continuum states, <u>all invoke intermediate, off-energy shell transitions</u>. It is our hope that in the future our experiments may contribute to this analysis.

Finally, it is illuminating to interpret the results from the point of view of the "simpleman's theory" [42] which in this case is really a qualitative discussion of the Kroll and Watson [13] approximation discussed in section 4. The basic idea, as illustrated in the diagram,

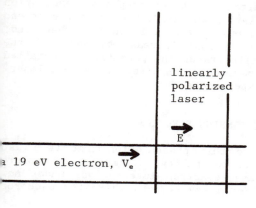

is that we are viewing an "instantaneous" collision between an electron (with velocity V_e) and an atom inside a laser field ($E = E_o \cos wt$). The electron will acquire an oscilatory motion (i.e. "quiver") and therefore a quiver velocity $V_q(t) = (eE_o/m\omega)$ E sin ωt. The duration of the collision is "ultrafast" compared to the period of the laser $(2\pi/\omega)$, and therefore the electron will collide with the atom at a given time t_o with an instantaneous velocity $V = V_e + V_q(t_o)$ and a

linearly polarized laser

E

a 19 eV electron, V_e

kinetic energy $\frac{1}{2}mV_2^2 = \frac{1}{2}m [V_e + V_q(t_o)]^2$ or $\frac{1}{2}m [V_e^2 + V_q^2 + 2V_e \cdot V_q(t_o)]$. It is easy to see that the middle term is negligible compared to the third term which is the one of interest in this discussion. We wish to evaluate the term $2V_e \cdot V_q(t_o)$ and demonstrate that under optimum conditions (i.e. $\sin \omega t_o \sim 1$) can acquire an energy equivalent to FOUR CO_2 - laser Photons, so that the combined energy of the incident electron (19.34 eV) plus the FOUR PHOTONS (4 x 0.120) add up to 19.820 eV or the excitation energy for the transition He $1^1S \rightarrow$ He 2^3S.

And indeed, this can be demonstrated for the following experimental conditions: The incident electron has an energy of 19.3 eV with a corresponding velocity $V_e = 2.6 \times 10^8$ cm/s and we are using a pulsed TEA CO_2 - laser of intensity I = 10^8 W/cm^2 and angular frequency $\omega = 2 \times 10^{14}$ s^{-1} and a wavelength = 10.6 μm. Assuming $\sin wt \sim 1$, substitution will result in

$$2 V_e (eE_o/m\omega) \times 1 = 2 \times 2.6 \times 10^8 \times 25.6 \times \sqrt{10^8} \times 10.6$$

$$= 4.5 \times 10^{-13} \text{ ergs} = 0.48 \text{ eV (FOUR PHOTONS)}$$

The "simpleman's theory" provides a physical picture consistent with energy conservation involving FOUR-PHOTONS, but it takes more sophistication to reproduce the "quantized " photon effects observed in the experiment [37] which indeed are qualitatively reproduced in the calculations of Geltman and Maquet [39] mentioned above.

It is difficult, however, to visualize a mechanism to explain the TEN-PHOTON absorption/emission that has been measured in FF processes in the backward electron scattering direction and the same laser intensity. In this geometry one would expect a cancellation of effects.

Changing the plane of polarization of the laser reduces the effect, also in accord with the "simpleman's theory", however when V_e and E are perpendicular to each other the cross section should totally vanish but the experimental results show that this is not the case. We observe a very significant signal that has not been explained.

The absorption of photons implies not only energy but linear and angular momentum as well. The problem of linear momentum is not obvious and was discussed in Sect. 4. Here we want to make reference to the angular momentum. In the simultaneous electron-photon excitation experiment, linearly polarized radiation was used, i.e. each photon can change the angular momentum quantum number ℓ by ± 1, on the other hand, with circularly polarized radiation ℓ must increase by one with the absorption of each photon. This illustrates the potential of detailed examination that these types of experiments can offer to explore laser-electron interactions for future experiments.

6. NEW SOURCES OF ELECTRONS: THE PENNING ELECTRON

It was said in the introduction to this presentation that it is an obligation of the experimentalist to design new experiments to explore intense laser-atom physics and laser-electron interactions under new conditions. So far we have considered experiments where the electron originates in a MPI process or in a controlled electron beam. In contrast, the ionization energy to produce Penning electrons is provided from an excited rare gas atom in a collision process where the partners simply have to come in "contact". It could be interpreted as autoionization of this collision complex which is a very fast process (10^{-15}s). This represents new initial conditions which offer various experimental possibilities. The important point that we want to make here is that "field-free" Penning ionization is well understood and the observed electron-energy spectra of many systems are very sensitive to external perturbations. In principle, we have here a new system that can provide information on both, laser-electron interactions or the "perturbed atomic structure" by an intense laser field.

The Penning technique uses controlled two beam experiments and examines the emitted Penning electrons according to their energy and their angular distribution. To monitor the experimental parameters the standard procedure is to compare the Penning electron spectra with those produced by a conventional VUV photoionization source (He resonance line, $\hbar\omega$ = 21.2175 eV), i.e. one photon. Two remarks have to be emphasized: one, the experiments are performed under good controlled conditions second, the VUV photon ionization source can also offer some interesting experimental possibilities with intense lasers physics.

These "new idea" experiments are discussed in this book in an article by Harald Morgner.

7. REFERENCES

1. Morellec, J., Normand, D. and Petite, G. (1982) Nonresonant multiphoton ionization of atoms, Adv. At. Mol. Phys. 18, 97-164.
2 (a) Agostini, P. and Petite, G. (1989) Shifts in atomic understanding, Physics World, September, 47-50.
   (b) Crance, M. (1990) Nonperturbative ac Stark shifts in hydrogen atoms, J. Opt. Soc. Am. B7, 449-455.
3. (a) Cooke, W. E. and McIlrath, T. J., eds. (1987) Feature on multielectron excitation in atoms, J. Opt. Soc. Am. B4, 705-862.
   (b) Kulander, K. and L'Huillier, A. eds. (1990) Feature on high-order processes in intense laser fields, J. Opt. Soc. Am. B7, 403-685.
   (c) Bandrauk, A. D. ed (1988) Atomic and molecular processes with short intense laser pulses, Plenum Press.
4. (a) Mittleman, M. H. (1982) Introduction to the theory of laser-atom interactions, Plenum Press, New York.
   (b) Chin, S. L. and Lambropoulos, P. eds. (1984) Multiproton ionization of Atoms, Academic Press.
   (c) Faisal, F. H. M. (1987) Theory of Multiphoton Processes, Plenum Press, New York.
   (d) Agostini, P. and Petite, G. (1988) Photoelectric effect under strong irradiation, Contemp. Phys. 1, 57-77.
5. Miloni, P. W. and Ackerhalt, J. R. (1989) Keldysh approximation, A^2, and strong-field ionization, Phys. Rev. A 39, 1139-1148.

6. (a) Gallagher, T. F. (1988) Above-threshold ionization in low-frequency limit, Phys. Rev. Lett. 61, 2304-07.
    (b) Fu, P., Scholz, T. J., Hettema, J. M. and Gallagher, T. F. (1990) Ionization of Rydberg atoms by a circularly polarized Microwave field, Phys. Rev. Lett. 64, 511-14.

7. Corkum, P. B., Burnett, N. H. and Brunel F. (1989) Above-threshold ionization in long-wavelength limit, Phys. Rev. Lett. 62, 1259-62.

8. Cooke, W. E. and Shakeshaft, R. (1991) Nonradiative transfer of momentum from an electromagnetic wave to a charged particle, Phys. Rev. A 43, 251-57.

9. Raman, N. K. and Guidotti, C. eds. (1984) Collisions and half-collisions with lasers, Harwood Academic Publishers.

10. Rosenberg, L. (1981) Intermediate- and strong-coupling approximations for scattering in a laser field, Phys. Rev. A 23, 2283-92.

11. (a) Weingartshofer, A., Holmes, J. K., Caudle, G., Clarke, E. M. and Krüger, H. (1977) Direct observation of multiphoton processes in laser-induced free-free transitions, Phys. Rev. Lett. 39, 269-70.
    (b) See also Weingartshofer, A. and Jung, C. (1984) Multiphoton free-free transitions, Ref. 4b pp. 155-87.

12. Krüger, H. and Jung, Ch. (1978) Low-frequency approach to multiphoton free-free transitions induced by realistic laser pulses, Phys. Rev. A 17, 1706-12.

13. Kroll, N. M., and Watson, K. M. (1973) Charged-particle scattering in the presence of a strong electromagnetic wave, Phys. Rev. A 8, 804-09.

14. Becker, W., Long, S. and McIver, J. K. (1990) Short-range potential model for multiphoton detachment of the H^- ion, Phys. Rev. A 42, 4416-19.

15. Dörr, M., Potvliege, R. M., Proulx, D. and Shakeshaft, R. (1990) Multiphoton detachment of H^- and the applicability of the Keldysh approximation, Phys. Rev. A 42, 4138-50.

16. Smith, W. W., Tang, C. Y., Quick, C. R., Bryant, H. C., Harris, P. G., Mohagheghi, A. H., Donahue, J. B., Reeder, R. A., Sharifian, H., Stewart, J. E., Toutouchi, H., Cohen, S., Altman, T. C., Risolve, D. C. (1991) Spectra from multiphoton electron detachment of H^-, J. Opt. Soc. Am. B8, 17-21.

17. Schulz, G. J. (1973) Resonances in electron impact on atoms and diatomic molecules, Rev. Mod. Phys. 45, 378-486.

18. (a) Andrick, D. and Langhans, L. (1978) Measurements of the free-free cross section of electron-Ar scattering, J. Phys. B: At. Mol. Phys. 11, 2355-60.
    (b) Langhans, L. (1978) Resonance structure in the free-free cross section of electron-Ar scattering, J. Phys. B: At. Mol. Phys. 11, 2361-66.
    (c) Andrick, D. (1980) Free-free processes in electron-atom scattering: Experiment, Electronic and Atomic Collisions, N. Oda and K. Takayanagi, eds., North-Holland Publishing Co., 697-704.

19. (a) Andrick, D. and Bader, H. (1984) Resonance structures in the cross section for free-free radiative transitions in e^--He scattering, J. Phys. B: At. Mol. Phys. 17, 4549-55.

(b) Bader, H. (1986) Resonance structures in the cross section for free-free radiative transitions in e⁻-Ne and e⁻-Ar scattering, J. Pys. B: At. Mol. Phys. 19, 2177-88.

20. Brown, L. S. and Kibble, T. W. B. (1964) Interaction of intense laser beams with electrons, Phys. Rev. 133, A705-A719.

21. Frantz, L. M. (1965) Compton scattering of an intense photon beam, Phys. Rev. 139, B 1326-B 1336.

22. (a) Kibble, T. W. B. (1966) Mutual refraction of electrons and photons, Phys. Rev. 150, 1060-69.
(b) Kibble, T. W. B. (1966) Refraction of electron beams by intense electromagnetic waves, Phys. Rev. Lett. 16, 1054-56.

23. Bartell, L. S., Thompson, H. B. and Roskos, R. R. (1965) Phys. Rev. Lett. 14, 851. See Ref. 22a pp. 1061 and 1065.

24. (a) Bucksbaum, P. H., Bashkansky, M. McIlrath, T. J. (1987) Scattering of electrons by intense coherent light, Phys. Rev. Lett. 58, 349-52.
(b) Bucksbaum, P. H., Freeman, R. R., Bashkansky, M. and McIlrath, T. J. (1987) Role of the ponderomotive potential in above-threshold ionization, J. Opt. Soc. Am. B, 4, 760-64.

25. Bucksbaum, P. H., Schumacher, D. W. and Bashkansky, M. (1988) High-intensity Kapitza-Dirac effect, Phys. Rev. Lett. 61, 1182-85.

26. (a) Bucksbaum, P. H. (1988) Above-threshold ionization and quantum mechanics of wiggling electrons, Ref. 3c, pp. 145-55.
(b) Freeman, R. R., Bucksbaum, P. H. and McIlrath, T. J. (1988) The ponderomotive potential of high intensity light and its role in multiphoton ionization of atoms, IEEE J. Quant. Electr., 24, 1461-69.

27. (a) Weingartshofer, A., Clarke, E. M., Holmes, J. K. and Jung, C. (1979) Experiments on multiphoton free-free transitions, Phys. Rev. A 19, 2371-76.
(b) Weingartshofer, A., Holmes, J. K., Sabbagh, J. and Chin, S. L. (1983) Electron scattering in intense laser fields, J. Phys. B: At. Mol. Phys. 16, 1805-17.

28. (a) Wallbank, B., Holmes, J. K., Weingartshofer, A. (1987) Experimental differential cross sections for multiphoton free-free transitions, J. Phys. B: AT. Mol. Phys. 20, 6121-38.
(b) Wallbank, B., Connors, V. W., Holmes, J. K., Weingartshofer, A. (1987) Experimental differential cross sections for one-photon free-free transitions. J. Phys. B: At. Mol. Phys. 20, L833-L838.

29. Krüger, H. (1990) Multiple harmonic generation by laser modulation of Zitterbewegung: the electron - a highly nonlinear electromagnetic medium, THE ELECTRON 1990 Workshop, August 1990.

30. Baldis, H. A., Pépin, H. and Grek, B. (1975) Third harmonic generation from laser-produced plasma, Appl. Phys. Lett., 27, 291-92.

31. Kupersztych, J. (1987) Inverse half-Bremsstrahlung in multiphoton ionization of atoms in intense light beams, Europhys. Lett., 4, 23-28.

32. Mittleman, M. H. (1988) Extended low-frequency approximation for laser-modified electron scattering: Coulomb effects, Phys. Rev. A 38, 82-92.

33. Jung, C. and Krüger, H. (1978) On the accuracy of soft-photon approximations for resonance free-free transitions, Z. Physik A 287, 7-13.

34. Jung, C. and Taylor, H. S. (1981) Possibility of experimental separation of resonances and cusps from background in electron scattering, Phys. Rev. A 23, 1115-26.

35. Wallbank, B., Holmes, J. K., LeBlanc, L. and Weingartshofer, A. (1988) Simultaneous off-shell excitation of He 2^3S by electron and one or more photons, Z. Phys. D - At. Mol. and Clusters, 10, 467-72.

36. Wallbank, B., Holmes, J. K. and Weingartshofer, A. (1989) Simultaneous electron-photon excitation of metastable states of rare-gas atoms, J. Phys. B: At. Mol. Opt. Phys. 22, L615-L619.

37. Wallbank, B., Holmes, J. K. and Weingartshofer, A. (1990) Simultaneous electron-photon excitation of He 2^2S: an experimental investigation of the effects of laser intensity and polarisation, J. At. Phys. B: At. Mol. Phys. 23, 2997-3005.

38. Wallbank, B., Holmes, J. K. and Weingartshofer, A. (1989) Absorption and emission of radiation during electron excitation of the 2^1S and 2^1P states of helium, Phys. Rev. A 40, 5461-63.

39. Geltman, S. and Maquet, A. (1989) Laser-assisted electron-impact excitation in the soft-photon limit, J. Phys. B: AT. Mol. Opt. Phys. 22, L419-L425.

40. Trombetta, F. (1991) Exchange effects in laser-assisted elastic electron-hydrogen scattering. To be published.

41. Sundaram, B. and Armstrong, Jr., L. (1990) Modeling strong-field above-threshold ionization, J. Opt. Soc. Am. B7, 414-24.

42. Kupersztych, J. While I was presenting a seminar at C.E.N. de Saclay, France, Dr. Kupersztych worked out the numerical problem and after de seminar he handed me a piece of paper with the words "here is a possible explanation of your experiment." I express my appreciation to Dr. Kupersztych for the splendid suggestion.

IONIZATION IN LINEARLY AND CIRCULARLY POLARIZED MICROWAVE FIELDS

T. F. Gallagher
Department of Physics
University of Virginia
Charlottesville, VA 22901
U.S.A.

ABSTRACT. Ionization by microwave fields provides a useful bridge between processes normally thought of as being photon and field driven. In addition, it is possible to see clearly that some phenomena are simply free electron phenomena. Investigations of ionization by circularly polarized microwaves and ATI are described to illustrate these points.

1. Introduction

Recent experiments with high intensity lasers have shown that in some cases multiphoton processes can be though of as field driven processes rather than photon absorption.[1] This notion is hardly surprising since higher order processes require lower frequency fields, and at some point the static or quasi static limit is encountered.
With the goal of studying the evolution from photon to field driven processes we have been investigating several aspects of microwave ionization and excitation. Here we discuss two aspects of these studies, ionization of Rydberg Na atoms by circularly polarized microwave fields and above threshold ionization (ATI) of Na and K Rydberg atoms in the low frequency limit.

2. Microwave Ionization With Circular Polarization

In optical experiments it has been observed that substantially higher powers are required to effect multiphoton ionization with circularly, rather than linearly, polarized light.[2] This difference is attributed to the fact that there are more paths to ionization, and hence more potential resonances using circularly than linearly polarized light. Since the difference in the number of available paths increases with the number of photons absorbed, microwave ionization of a Rydberg atom, requiring hundreds of photons could show striking differences. With this in mind we have begun the experimental study of microwave ionization by circularly polarized microwave fields.

In our experiment Na atoms in an atomic beam pass through a Fabry Perot microwave cavity, where they are excited to a Rydberg state using two pulsed tunable dye lasers.[3] The lasers are tuned to the 3s-3p and 3p-Rydberg transitions at 5890Å and ~4140Å respectively. The atoms are excited to the Rydberg states in the presence of the circularly polarized field, and we observe the ions which are produced after the microwaves have been turned off.

The experimental approach to producing the circularly polarized field is to use a Fabry

311

D. Hestenes and A. Weingartshofer (eds.), The Electron, 311–320.

Perot cavity which supports two orthogonally linearly polarized 8.5 GHz fields 90° out of phase. The major experimental problem is to minimize the interaction between these two nominally degenerate cavity modes, as any interaction lifts the degeneracy. To minimize the interaction we feed the cavity with the two orthogonal polarizations from orthogonally polarized waveguides through irises in the Fabry Perot mirrors. In our 8.5 GHz cavity the two modes are offset by 2 MHz, and they have Q's of 2,000 and 2,100, so the linewidths of the modes are 4 MHz. Due to the fact that the Q's are not identical we can achieve circular polarization in steady state, but not as the cavities modes are filled with the microwave field. To ensure that we are using circularly polarized microwaves we excite the atoms to a Rydberg state in the microwave field.

A much higher circularly polarized field is required to ionize the atoms than linearly polarized field, as shown by Fig. 1, a plot of the threshold fields, where 50% ionization occurs, for linear and circular polarization. As shown by Fig. 1 the circularly polarized microwave field required for ionization follows an $E = 1/16n^4$ law. This field is the same as the static field required to ionize a Rydberg Na atom and much higher than the field required for ionization by linearly polarized microwaves, $E = 1/3n^5$.[4]

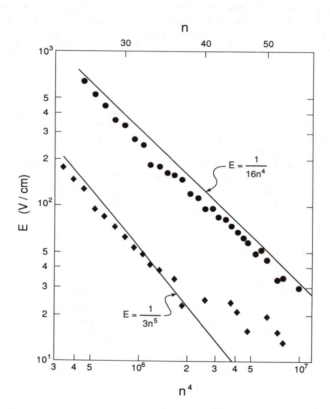

Figure 1. Ionization threshold fields for linear (♦) and circular (●) polarization as a function of n when the atoms are excited in the microwave field.

In addition there is a sharp dependence of the observed ionization signal on the relative phase between the two polarizations in the cavity. This point is shown for n=46 in Fig. 2.

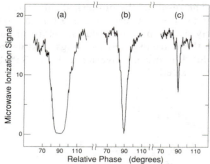

Figure 2. Microwave ionization signal as a function of relative phase between the horizontal and vertical fields when the atoms are excited to the energy of the 46d state and the attentuation is set between the linear and circular polarization ionization thresholds. Specifically, scans of the phase are shown for microwave fields of (a) 34.6 V/cm, (b) 53.5 V/cm, and (c) 61.5 V/cm.

When E is just below $1/16n^4$, so that ionization by a circularly polarized field does not occur, a very small phase deviation from the 90° phase shift required to produce circular polarization leads to a sharp increase in the ionization as shown. As shown in Fig. 2 when the field amplitude is lower the phase is not so sensitive, which is hardly surprising. Apparently a small ellipticity in the polarization produces ionization, while pure circular polarization does not.

Since the threshold field is the same as for a static field, it is tempting to simply call the field quasistatic. However it is not. First the symmetry is different. In a static field azimuthal symmetry is preserved and m is a good quantum number. In a circularly polarized field states of $\ell + m$ even and odd form two mutually exclusive sets of states. Second, the microwave frequency of 8.5 GHz is not always small compared to the Δn separation and not always small compared to the separation of the Stark states. For both these reasons the most simple minded quasi static picture is inappropriate.

As is often the case, a variation of the simplest picture does provide a good description. If we transform the problem to a frame rotating at the microwave frequency, the microwave field is in fact static and cannot induce transitions. The transformation to the rotating frame, often used to describe two level resonance experiments, is discussed Salwen.[5] If we have a microwave field given by

$$\overline{E} = \hat{x}\, E \cos\omega t + \hat{y}\, E \sin\omega t , \qquad (1)$$

and we transform the problem to a frame rotating about the z axis at angular frequency ω, the field becomes a static field in the x (or y) direction in the rotating frame. If we use as eigenstates in the laboratory frame the usual spherical $n\ell m$ states, a rotation through angle ø about the z axis transforms the wave function $\psi_{n\ell m}$ to $e^{im\phi}\,\psi_{n\ell m}$. In the rotating frame ø = ωt and $\psi_{n\ell m} \to e^{-im\omega t}\,\psi_{n\ell m}$, in the energy is shifted by $-m\omega$. Transforming the Na $n\ell m$ wave functions to the rotating frame only adds $-m\omega$ to their energies. In principle, it is then a simple matter to diagonalize the new Hamiltonian matrix containing a field in the x direction to find the energy levels in the rotating frame. In practice, the fact that there are $n^2/2$ levels

for each principal quantum number complicates the problem. To develop an understanding
we have diagonalized the Na Hamiltonian matrix, for n = 5, 6, and 7;, in a frame rotating at
1500 GHz, so that the frequency is   2% of the n = 4-5 interval. A frequency of 8.5 GHz is
2% of the n = 25 to 26 interval. The result of the numerical calculation is shown in Fig. 3 for
ℓ + m even. There are two points to note. At zero field m $\neq$ 0 levels are displaced from their
normal energies, as shown clearly by the p m = ±1 states. More important, when the Stark
manifolds of adjacent n overlap there are avoided crossings, just as in a static field. In the
limit of very low frequency the Stark manifolds overlap at E = $1/3n^5$, but the overlap occurs
at lower fields as the frequency is raised.

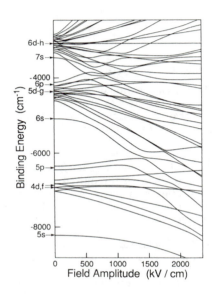

Figure 3. Schematic energy level diagram for Na near n=24 in a frame rotating at 102 GHz
For clarity we have only shown s, p, and d states, which are assumed to have quantum defects
of 1.35,0 and 0 respectively. Their energies in the nonrotating frame are shown by the
arrows. Note that in zero field the p m=±1 and d m=±0 levels are displaced from the non
rotating energies of the p and d states by ∓102 and ∓204 GHz respectively. The lowest field
avoided crossing, between the m=±2 levels, is negligibly small, while those in fields larger
than $1/3n^5$ are substantial due to the admixture of the s state with its non zero quantum
defect. In fields greater than $1/16n^4$ the avoided crossings are with unstable states, and in this
region ionization occurs very rapidly. In the rotating frame the field is static, and there are
no transitions. Ionization occurs if atoms are excited above the classical field for ionization
E=$1/16n^4$.

        We have also done the same calculation for H, and the levels cross, as in a static field
Thus the avoided crossings in Na are due to the non coulomb core coupling not the
transformation to the rotating frame. In static fields $1/3n^5 < E < 1/16n^4$ the core coupling
is manifested in avoided level crossings, and in fields E > $1/16n^4$ the same coupling between
discrete states and the underlying Stark continuum of ionized red Stark states leads to
ionization. [6] In other words field ionization of Na at E > $1/16n^4$ is really autoionization
In the rotating frame the situation is presumably the same. There are avoided crossings fo

$E < 1/16n^4$, presumably due to the m = 0 and 1 parts of the wave function, and ionization when $E > 1/16n^4$. In other words we have almost come back to the original notion; ionization by a circularly polarized microwave field is field ionization in the rotating frame.

It is interesting to consider the pronounced effect of ellipticity in the polarization in terms of the above rotating frame description. When the polarization is elliptical, in the rotating frame there is a field oscillating at 2ω superimposed on the static field. If the static field in the rotating frame exceeds $1/3n^5$, the atom is in the field regime in which there are many level crossings, and even a very small additional oscillating field can drive transitions to higher lying states via these level crossings.

The above description of ionization by a circularly polarized microwave field is clearly a field description. At the other extreme is photoionization, which is also described as autoionization in the rotating frame. For example photoionization of an m = 0 state to the m = 1 continuum appears as autoionization into the degenerate m = 1 continuum with a rate proportional to the light intensity. This notion is easily extended to low order multiphoton processes which occur via virtual intermediate states. An interesting, but outstanding question is how to connect these pictures to the quasi static picture advanced above to describe microwave ionization in a circularly polarized field.

3. Above Threshold Ionization

Above threshold ionization (ATI), the absorption of more photons than necessary for multiphoton ionization was first observed by Agostini et al. in the energy spectrum of electrons produced by multiphoton ionization of Xe by intense 1.06 μm light.[7] Shortly thereafter Kruit et al observed that the lowest energy peaks in the electron spectrum were suppressed as the intensity was raised.[8] This was attributed to the upward shift in the ionization limit by the ponderomotive energy, $W_p = e^2 E^2 / 4mW^2$, in the intense laser field.[9] The ponderomotive energy is simply the average kinetic energy of a free electron of mass m and charge e oscillating in a field E $\cos \omega t$. The fact that the ionization limit and therefore the Rydberg states shift in this manner is not a property of intense light, but is generally true for light which is of frequency low compared to the transitions available to the ground state, but high compared to the frequencies of strong transitions of the Rydberg state in question. For example the room temperature black body radiation induced shift of the ionization limit is equal to the ponderomotive energy.[10] In the laser field the ionization limit is increased, so more photons must be absorbed, the excess energy going into the oscillation of the photoelectron. However when the photoelectron leaves the focus of the laser beam the oscillatory ponderomotive energy is converted to translational kinetic energy. Thus the ponderomotive energy does not alter the energies of the detected electrons, assuming of course that the laser field remains present as the electrons leave the focus. However it does alter which electrons are actually observed. No electrons can be observed with an energy less than the ponderomotive energy.

Examination of optical ATI experiments using linear polarization indicates that there are typically ATI electron peaks at energies as high as three or four times the ponderomotive energy, implying that the ponderomotive energy can be used to crudely estimate the amount of ATI. Since the ponderomotive energy scales as $1/\omega^2$, it is large for even modest microwave fields. The ponderomotive energy for 8 GHz microwaves is 1 eV for an intensity of $10^4 W/cm^4$, a power level that is easily produced. With this in mind we decided to see if microwave ATI existed and, if so, if it was simply understandable.

The experimental approach is to analyze the electrons from Rydberg atoms ionized by an 8.2 GHz microwave field.[11]   The first experiments were carried out using the arrangement shown in Fig. 4.

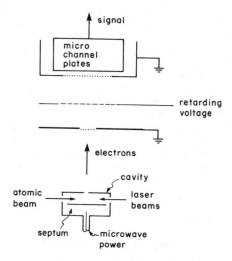

Figure 4. Schematic digram of the apparatus for electron energy spectroscopy.

The microwave cavity is a rectangular cavity with a septum and operates on the TE_{101} mode at 8.2 GHz. The Na atomic and laser beams pass through holes in the sidewalls of the cavity, so that a pencil shaped volume of excited atoms is created by the two dye laser beams which excite the Na atoms from the ground state to a Rydberg state via the 3ps state. The energies of electrons ejected through a hole in the top of the cavity are analyzed either by a retarding potential or by time of flight. The Na atoms are excited are excited to the Rydberg states in the presence of the microwave field. When atoms are excited to the energy of the 40s state electrons are observed at energies as high as 10eV when the computed ponderomotive energies are ~3eV, and the energy distribution is peaked at ~$3W_p$. On the other hand when atoms are excited with the laser to the energy of the 15s state the electron energy is observed to peak at W_p.

The above results represent the energy spectrum of the detected electrons, but are not indicative of the energy spectrum of the electrons as ejected from the atoms, due to the ponderomotive force. Ponderomotive forces were first shown to distort the angular distributions of ATI electrons from laser multiphoton ionization by Freeman et. al,[12] and the effect is, if anything, worse in the microwave case. The ponderomotive force F_p is given by

$$F_p = -\nabla W_p$$
$$= -\frac{e^2}{4m\omega^2}\nabla E^2 \quad ,$$

(2)

In a focused traveling wave laser beam F_p is radially outward. In a standing wave beam there is also a variation in E^2 along the beam propagation direction, leading to ponderomotive forces in the propagation direction, the Kapitza Dirac effect.[13] If we define a rectangular coordinate system with its origin at the center of the cavity, we can write the square of the microwave field amplitude in our TE_{101} cavity as

$$E_z^2 = E_0^2 \cos^2\frac{\pi x}{a} \cos^2\frac{\pi y}{b} , \qquad (3)$$

where a and b are the transverse dimensions of the cavity. In our cavity a ≈ b. Inside the cavity there is evidently ponderomotive force in the x-y plane, which is approximately radially outward, since a ≈ b, and increases approximately linearly with distance from the z axis. There is evidently no z component to the ponderomotive force inside the cavity. However at the exit hole in the top of the cavity the field amplitude changes from E_0 to zero, so there is a ponderomotive force in the z direction, and electrons gain energy W_p in the form of motion in the z direction on leaving the cavity.

Electrons ejected from the atom in the +z direction with large kinetic energies, >W_p, are unaffected by the ponderomotive forces in the cavity, escape from the cavity, and reach the detector. However those electrons ejected from the atom with low, ~0, kinetic energies are virtually certain to be deflected horizontally and will not leave the cavity and be detected. Thus low energy electrons are suppressed in the detected electron spectrum. This distortion is most dramatic for high n states, as has been shown by applying a small bias voltage to the septum.[14] As shown by Fig. 5, when a voltage is applied the electron spectrum at high n is seen to consist of not only a peak at $3W_p$ but a peak at W_p as well.

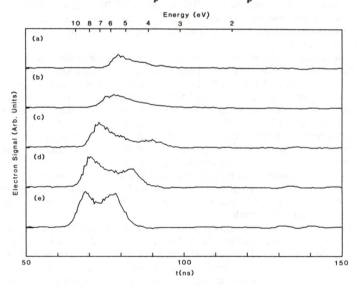

Figure 5. Electron time of flight spectra when Na atoms are excited to n=71 in the presence of a 3.0kV/cm 8.173 GHz microwave field. The voltages on the septum are (a) 0V, (b) – 2.00V, (c) –4.00V, (d) –6.00V, and (d) –8.00V. At higher magnitudes of the septum voltage the later peak, due to the low energy electrons is quite evident. With small voltages the low energy electrons are lost due to ponderomotive forces in the cavity. Note that each additional negative two volts on the septum raises the electron energy by 1 eV.

At low n, where virtually all the electrons have low energy the dominant effect is an overall suppression of the signal. Taking into account the effect of the ponderomotive forces in the cavity, it is apparent that as ejected from the atom the electron energy distributions for low n and high n are respectively singly peaked, at W_p, and doubly peaked at W_p and $3W_p$.

To explain these electron spectra requires only a simple model, sometimes called the "simpleman's theory", developed independently by van Linden van den Heuvell and Muller,[14] Corkum et al [15] and Gallagher [11]. Consider an electron liberated in a classical field $E=E_0\sin\omega t$ at time t_0 with initial velocity v=0. At any later time t the velocity is given by

$$v = \frac{eE}{m\omega}(\cos \omega t - \cos\omega t_0) \tag{4}$$

and the time average velocity by

$$\bar{v} = \frac{eE}{mW}\cos\omega t_0. \tag{5}$$

The average kinetic energy is given by

$$W = W_p (1 + 2\cos^2\omega t_0), \tag{6}$$

which is composed of two parts, an oscillation energy of W_p and a translational energy of 2 $W_p \cos^2\omega t_0$. When the electron leaves the microwave field the energy of oscillation is converted to directed translational motion, so the energy of Eq. (5) is the energy of the electrons when detected.

When the atom is excited to a low n state, the microwave field is too small to directly field ionize the initially excited state, and the atom makes transitions to higher n states until field ionization is allowed, which is most likely to occur near the peak of the field, $\omega t_0 = \pi/2$. For this value of ωt_0 the electron is ejected from the atom with no translational energy, and $W \approx W_p$, in accord with the observations. On the other hand, if a high n state is excited it can be ionized at nearly any phase of the microwave field. The excitation and ionization are therefore simultaneous and computing the energy spectrum is simply a matter of computing $(dW/dt_0)^{-1}$. Since the function $\cos^2\omega t_0$ spends most of its time at its turning points, $\omega t_0=0$ and π, the electron energy spectrum is peaked at W_p and $3W_p$, in agreement with observations.

If the atoms are excited to a Rydberg state and then exposed to a pulsed microwave field we see electrons of energy approximately equal to the ponderomotive potential due to the field necessary to ionize the atom. For low n states this energy can be quite high, but for high n states it is vanishingly small.

The simpleman's theory agrees quite well with electron energies observed at microwave and CO_2 laser frequencies. From Eqs. (4) and (5) it is apparent that there is phase dependence in both $\bar{v}$ and W. The phase dependence has been observed explicitly by phase locking a microwave oscillator to a mode locked laser which excites K atoms with 5ps laser pulses.[16] A phase shifter between the oscillator and the microwave cavity allows the microwave phase, ωt_0, at which the atoms are excited to be varied at will. The electron energies are analyzed by time of flight.

At low n there is no phase dependence, as expected. Irrespective of when ionization occurs several microwave cycles are required to make transitions to higher n states and ionization occurs near the peak field, resulting in electrons of energy ~W_p. For high n states, which are readily ionized by the field, both $\bar{v}$ and W exhibit the predicted phase dependence.

For half the microwave cycle, no electrons are detected, indicating that the electrons are ejected down, away from the detector. In addition the electron energy varies with phase, as shown in Fig. 6.

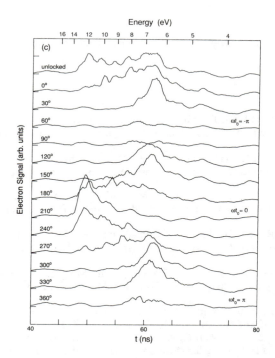

Figure 6. Electron time of flight spectra as a function of phase when the microwaves are phase-locked and the n=42-46 d states are excited. The top trace is unlocked for comparison. The microwave intensity is $33kWcm^{-2}$ ($3.5kVcm^{-1}$;W_p=2.1eV). There is an offset of 4eV on energy scale due to septum voltage of -8.0V.

When ωt_0=0 the electrons have energy 3Wp and when ωt_0=$\pm\pi/2$ the electrons have energy W=Wp, as predicted by Eq. 5. Electrons detected with energy $\sim W_p$ have very low kinetic energy in the microwave field, and irrespective of the direction in which the electrons are ejected from the atom they are detected because of the bias field. On the other hand electrons produced at ωt_0=$\pm\pi$ have kinetic energy $2W_p$ in the cavity but are directed downward, hit the septum, and are not detected.

These experiments clearly validate the simpleman's theory. In addition they demonstrate, that the ionization, as opposed to the excitation en route, occurs in a single cycle, otherwise there would be no phase dependence in the ATI spectrum. Thus ATI in this frequency range is fundamentally different from ATI at optical frequencies, where well resolved ATI peaks are observed, indicating that ionization takes place over many cycles.

4. Conclusion

As shown by these experiments, at low frequencies multiphoton processes are well described as field driven processes, and some of the effects, such as ATI are simply the reflection of the response of a free electron to an oscillating field.

5. Acknowledgements

This work has been supported by the Air Force Office of Scientific Research under grant AFOSR-90-0036.

6. References

1.   Augst, S. D., Strickland, D., Meyerhofer, D. D., Chin, S. L., and Eberly, J. H. (1989) 'Tunneling Ionization of Noble Gases in a high intensity Laser Field', Phys. Rev. Lett. 63, 2212-2215.
2.   Lompre, L. A., Mainfray, G., Manus, C. and Thebault, J. (1990) 'Multiphoton ionization of rare gases by a tunable-wavelength 30-psec laser pulse at 1.06 μm', Phys. Rev. A 15, 1604-1612.
3.   Fu, P. M., Scholz, T. J., Hettema, J. M., and Gallagher, T. F. (1990) 'Ionization of Rydberg Atoms by a Circularly polarized microwave field', Phys. Rev. Lett. 64, 511-514.
4.   Freeman, R. R., McIlrath, T. J., Bucksbaum, P. H., and Bashkansky, M. (1986) 'Ponderomotive Effects on Angular Distributions of Photoelectrons', Phys. Rev. Lett. 57, 3156-3159.
5.   Salwen, H., (1985) 'Resonance Transitions in Molecular Beam Experiments. I. General Theory of Transitions in a Rotating Magnetic Field*', Phys. Rev. 99, 1274-1286.
6.   Littman, M. G., Zimmerman, M. L., and Kleppner, D. (1976) 'Tunneling Rates for Excited States of Sodium in a Static Electric Field*', Phys. Rev. Lett. 37, 486-489.
7.   Agostini, P., Fabre, F., Mainfray, G., Petite, G., and Rahman, N. K. (1979) 'Free-Free Transitions Following Six-Photon Ionization of Xenon Atoms', Phys. Rev. Lett. 42, 1127-1150.
8.   Kruit, P., Kimman, J., Muller, H. G., van der Wiel, M. J. (1983) 'Electron spectra from multiphoton ionization of xenon at 1064, 532, and 355^{nm}', Phys. Rev. A 28, 243-255.
9.   Muller, H. G., Tip, A., and van der Wiel, M. J. (1983) 'Ponderomotive force and AC Stark Shift in multiphoton ionization', J. Phys. B 16, L679-L685.
10.   Gallagher, T. F. and Cooke, W. E. (1979) Interaction of Black Body Radiation with Atoms', Phys. Rev. Lett. 42, 835-838.
11.   Gallagher, T. F. (1988) 'Above Threshold Ionization in the low frequency limit', Phys. Rev. Lett. 61, 2304-2307.
12.   Bucksbaum, P. H., Schumacher, D. W., and Bashkansky, M. (1988) 'High-Intensity Kapitza-Diract Effect', Phys. Rev. Lett. 61, 1182-1185.
13.   Gallagher, T. F. and Scholz, T. J. (1989) 'Above Threshold Ionization at 8 GHz', Phys. Rev. A 40, 2762-2765.
14.   van Linden van den Heuvell, H. B. and Muller, H. G. (1989) 'Limiting cases of excess photon ionization', in S. J. Smith and P. L. Knight (eds.), Multiphoton Processes, Cambrdige University Press, Cambridge, pp. 25-34. S. J. Smith, and P. L. Knight, (eds.) (Cambridge Univ Press., Cambridge, pp. 25-35.
15.   Corkum, P. B., Burnett, N. H., and Brunel, F. (1989) 'Above-Threshold Ionization in the Long-Wavelength Limit', Phys. Rev. Lett. 62, 1259-1262.
16.   Tate, D. A., Papaioannou, and Gallagher, T. F. (1990) 'Phase sensitive above threshold ionization of Rydberg atoms at 8 GHz', Phys. Rev. A 42, xxx.

MULTIPLE HARMONIC GENERATION IN RARE GASES AT HIGH LASER INTENSITY

Anne L'HUILLIER, Louis–André LOMPRE and Gérard MAINFRAY
Service de Physique des Atomes et des Surfaces
C.E.N. Saclay, 91191 Gif sur Yvette, FRANCE

ABSTRACT. We briefly review the main experimental results on high—order harmonic generation in rare gases and their interpretation.

Introduction

Recent experiments show the production of very high harmonics of the laser field. Up to the 17th harmonic of a KrF pump laser (248 nm) has been observed in neon by the group at the university of Illinois at Chicago (McPherson et al.(1987)). In Saclay, we have seen up the 33rd harmonic of a Nd–YAG laser (1064 nm) in a 10 Torr argon gas jet (Li et al.(1989)). The intensities of the harmonics drop fairly quickly for the first harmonics, then exhibit a broad plateau and finally decreases again sharply. The length and width of the plateau depend on the atomic medium and on the laser intensity.

Calculations of single–atom photoemission spectra going beyond perturbation theory reproduce fairly well the experimental results (Kulander and Shore (1989), Potvliege and Shakeshaft (1989), Bandarage et al. (1990), Eberly et al., 1989). This seems to imply that propagation effects play no role or rather that they affect all the harmonics in the same way. However, a calculation of phase matching in the weak field limit shows that phase matching should severely degrade with the order. We present a calculation of harmonic generation in a rare gas medium, involving the resolution of the paraxial propagation equation for a non–perturbative polarization field. It reconciles experimental results and

321

D. Hestenes and A. Weingartshofer (eds.), The Electron, 321–332.
© 1991 *Kluwer Academic Publishers. Printed in the Netherlands.*

calculations of single–atom emission spectra, since it shows that phase matching is the same for all the harmonics of the plateau region.

I– Experimental results with a Nd–YAG laser

An harmonic generation experiment consists in focusing an intense laser radiation into a rather dense rare gas medium (a few Torr) and in analyzing along the propagation axis the VUV light emitted during the interaction. In our case, we use the fundamental frequency of a mode–locked Nd–YAG laser (40 ps pulse width– 1064 nm wavelength), with a maximum energy of 1 GW at a 10 Hz repetition rate. The laser pulse is focused by a 200 mm–focal length lens over a 18 μm focal radius (confocal parameter b = 4 mm). The gaseous medium is provided by a pulsed gas jet producing a well–collimated atomic beam (1 mm FWHM) with a 15 Torr pressure at 0.5 mm from the nozzle of the jet. The VUV light (from 350 nm to 10 nm) is analyzed along the laser axis by using a grating monochromator and detected by photomultipliers or a windowless electron multiplier at wavelengths below 120 nm.

Fig.1 shows the number of harmonic photons obtained in xenon at a 15 torr pressure at several laser intensities, 3×10^{13} W.cm^{-2}, solid line, 1.3×10^{13} W.cm^{-2}, dashed line, 9×10^{12} W.cm^{-2}, dot–dashed line, 7×10^{12} W.cm^{-2}, dotted line, 5×10^{12} W.cm^{-2}, double–dot–dashed line (Lompre et al.,1990). We only observe *odd* harmonics (we did not however look for second order harmonic generation). This is to be expected for harmonic generation in an isotropic gaseous medium, with inversion symmetry. At the lowest laser intensity, the conversion decreases with the order. This is a rather intuitive result : the harmonic signal decreases as the nonlinear order increases. However, at higher laser intensity, a plateau appears. The length of the plateau increases with the intensity up to the point at which the medium gets ionized (above 1.3×10^{13} W.cm^{-2}). The highest harmonic detected is the 21st, which means an energy of 25 eV, a wavelength of 50 nm. The vertical scale is an order–of–magnitude estimate of the number of photons produced at each laser shot. The efficiency for the plateau harmonics at 3×10^{13} W.cm^{-2} is about $10^{-8} - 10^{-9}$. This represents a very high brightness, 10^{17} ph/Ås(mrad)^2.

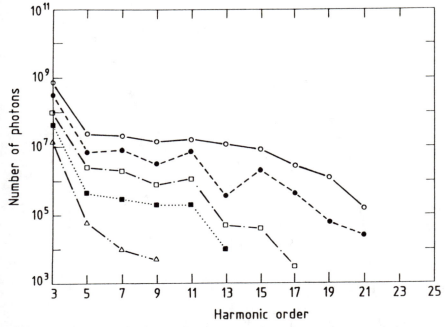

Figure 1 : Number of photons obtained in xenon at a 15 Torr pressure at several intensities (see text). The incident wavelength is 1064 nm, the confocal parameter estimated to 4 mm, the interaction length 1 mm.

Another way of looking at these results is to plot the number of photons as a function of the laser intensity. Figure 2 shows the intensity dependence of the 5th and 17th harmonics. The signal increases first rather rapidly with the laser intensity, before saturating when the medium is ionized. Ionization limits harmonic generation because the main medium responsible for the conversion gets depleted and also because phase matching conditions break down owing to the presence of free electrons (Miyazaki and Kashiwagi, 1978, L'Huillier et al., 1990). Below saturation, the qth harmonic signal varies with the laser intensity as I^p with $p = q$ for the first orders and $p < q$ for the highest orders. Table 1 summarizes the effective orders of nonlinearity determined by a least–square fitting procedure for the different harmonics. Some harmonics (in particular the 13th) present a more complicated behavior than a simple power law, which might be due to the influence of AC Stark shifted resonances.

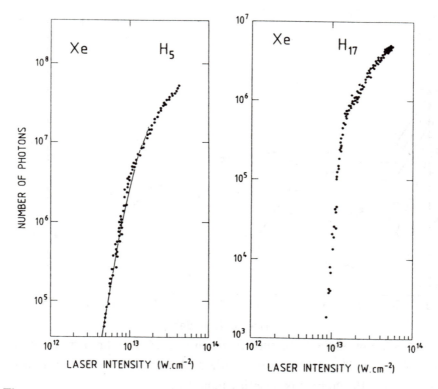

Figure 2 : 5th and 17th harmonics as a function of the laser intensity

Table 1: harmonic power laws below saturation

3	5	7	9	11	13	15	17	19	21
3.3±0.2	6±0.5	7.5±0.7	8.3±0.7	7.7±0.7	5.3±0.5	11.7±0.7	12.5±1	13±1	12.5±1

Similar results have been obtained with the other heavy rare gases (krypton and argon). The distributions obtained in Xe, Kr, Ar at 3×10^{13} W.cm^{-2} are shown in Fig. 3. The conversion efficiency decreases from Xe to Ar, which is not surprising, since xenon is more polarizable than lighter rare gases. However, the maximum order that can be observed increases from 21 in Xe, 29 in Kr to 33 in Ar. The 33rd harmonic (32 nm, 38 eV) is the shortest wavelength radiation that we were able to produce with our Nd—YAG laser system (limited, however, to about 20 mJ in 40 ps). Atoms with higher ionization energies have in general a lower conversion efficiency but they can produce more harmonics. Moreover, they can experience a higher laser intensity without being ionized. The results shown in Fig.3 have been obtained at an intensity of 3×10^{13} W.cm^{-2}, which is higher than the saturation intensities for xenon and krypton. The Ar harmonic intensity distribution presents an anomaly. The 13th harmonic is missing. This may be due to breakdown of phase matching conditions or strong reabsorption in the vicinity of (possibly AC Stark shifted) Rydberg states.

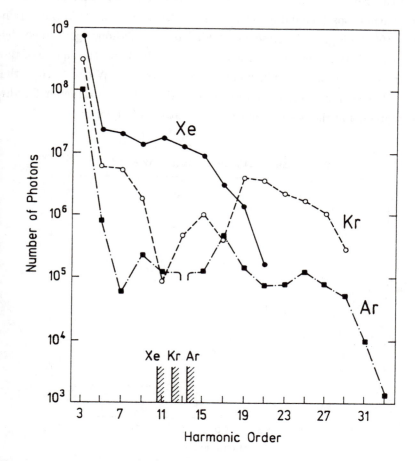

Figure 3. Number of photons obtained in Ar, Kr and Xe at 3×10^{13} W.cm^{-2}, 1064 nm

II— Interpretation of the experimental results

1— Single—atom response

The coherent spectrum emitted by a single atom is the square of the Fourier transform of the dipole moment d(t) induced by the laser field. d(t) can be obtained from the wavefunction of the atomic system, as the expectation value of the dipole operator. This problem has recently received considerable attention. Various approximations have been used, time—dependent methods (Kulander and Shore, 1989, 1990, Eberly et al., 1989), Floquet theory (Potvliege and Shakeshaft, 1989), model calculations (Bandarage et al., 1990, Becker et al., 1990, Sundaram and Milonni, 1990). Figure 4 presents results of calculations performed by Kulander and coworkers in xenon (Krause et al., 1991). It shows the square of the dipole moment as a function of the harmonic order at $3\text{x }10^{13}$ W.cm^{-2}. Note the close similarity between theoretical and experimental results in particular for the length of the plateau and the position of the cut off at high frequency.

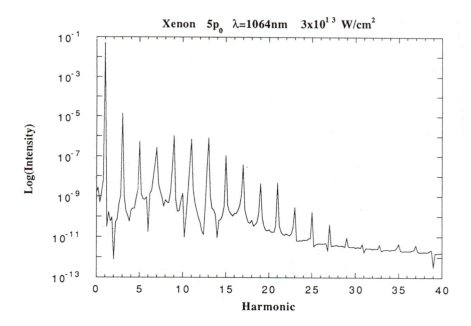

Figure 4. Relative harmonic intensities as a function of the harmonic order q.

Most of non–perturbative calculations, even crude models such as a two–level system (Sundaram and Milonni, 1990), are able to reproduce qualitatively the experimental results, with the decrease in efficiency for the first harmonics, the plateau and the cut off. This is probably a very general property of a strongly driven nonlinear system. As a further illustration, we show in Fig.5 the spectrum emitted by a classical anharmonic oscillator, obtained by solving the following equation : $\ddot{x} + \omega_0^2 x + \Gamma\dot{x} + vx^3 = F\cos\omega t$. The parameters are chosen to be $\omega_0 = 10\omega$, $\Gamma = 1$, $v = F = 500$. Surprisingly, the main features (plateau and cut off) observed in the experiments are reproduced by this simple model.

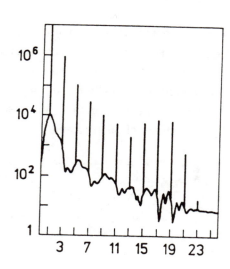

Figure 5. Spectrum emitted by an anharmonic oscillator.

2– Many–atom response

The generation of the qth harmonic field is described by the propagation equation

$$\nabla^2 \mathscr{E}_q + k_q^2\, \mathscr{E}_q = -4\pi(\frac{q\omega}{c})^2\, \mathscr{P}_q$$

where k_q is the wavevector of the qth harmonic field, ω the laser frequency and $\mathscr{P}_q$ the nonlinear polarization field, equal to $2\mathscr{N}d(q\omega)$, where $\mathscr{N}$ denotes the atomic density and $d(q\omega)$ the qth harmonic component of the time–dependent dipole moment. Equivalently, this can be written as (Jackson, 1975)

$$\mathcal{E}_q(\mathbf{r'}) = \left(\tfrac{q\omega}{c}\right)^2 \int \frac{e^{ik_qR}}{R}\, \mathcal{P}_q(\mathbf{r})\, d^3\mathbf{r}$$

with $R=|\mathbf{r}-\mathbf{r'}|$. The calculation of the harmonic profile in the far field is thus reduced to a three dimensional integration over the nonlinear medium (L'Huillier et al., 1990, 1991). The number of photons emitted at a given harmonic frequency is obtained by integrating the temporal and spatial distribution as follows

$$N_q = \frac{c}{4\hbar q\omega} \int \mathbf{r'}\,|\,\mathcal{E}_q(\mathbf{r'},t)\,|^2\, d\mathbf{r'}dt$$

The single–atom contribution is usually factorized out as $N_q = |d(q\omega)|^2\,|F_q|^2$, where $|d(q\omega)|^2$ is the single atom contribution and $|F_q|^2$ the phase matching factor, reflecting the integration of the propagation equation. This factorization can actually be done analytically for an incident Gaussian beam, in the weak field limit. In a more general situation, in a strong field regime, one calculates numerically the number of photons and then derives the phase matching factor.

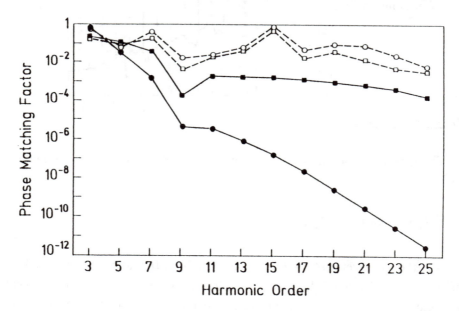

Figure 6. Phase matching factor $|F_q|^2$ for different intensities and geometries (see text)

Figure 6 shows the phase matching factor $|F_q|^2$ as a function of the order q both in the weak field limit (solid lines) and in a strong field regime (3×10^{13} W.cm^{-2}, dashed lines). Two different geometries have been investigated : a collimated geometry where the laser confocal parameter b (4 mm) is larger than the interaction length L = 1 mm (squares in the figure) and a confocal geometry where b (1 mm) is of the order of the interaction length (circles).The difference between the two intensities is striking. In the weak field limit, $|F_q|^2$ decreases rapidly with the process order. In contrast, at 3×10^{13} W.cm^{-2}, $|F_q|^2$ stays approximately constant, independent of the focusing geometry. Note the twelve orders of magnitude difference between the weak and strong field regimes in the tight focused geometry.

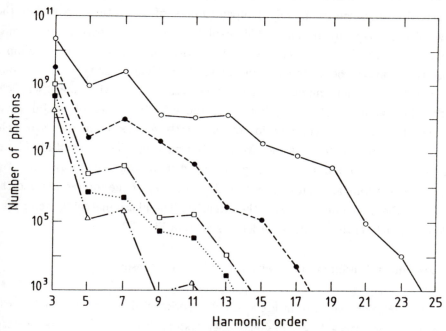

Figure 7. Calculated number of photons for b = 4 mm; L = 1 mm, 15 Torr. Same intensities as in Fig.1.

The numbers of photons N_q, calculated at the same laser intensities as in Fig.1 are shown in Fig.7. The results (L'Huillier et al., 1991) are in good agreement with the experimental data, for the absolute value of number of photons as well as for the general behavior of the harmonic intensity distribution as a function of the incident field.

In the experiments reported in this paper, the pressure is rather small. Consequently, the phase mismatch $\Delta k = k_q - q k_1 = (n_q - n_1) q \omega / c$ is close to zero. It remains much smaller than the effective phase mismatch induced by focusing which originates from the π phase slip across the focus of a Gaussian beam and which is centered around $-2q/b$. Perfect phase matching is not realized. In a weak field picture, increasing the nonlinear order q has about the same effect as decreasing the confocal parameter b. High–order harmonics become more tightly focused. The phase matching function evolves from a sinx/x type of function for the lowest orders to a much more rapidly varying function for the highest orders, such that the wings are completely damped (orders–of–magnitude lower than the average of the sinx/x oscillations). This explains why, in the perturbative limit, the phase matching factor decreases rapidly with the nonlinear order, simply because the geometry becomes more and more of the tight focus type. In a strong field regime, the dipole moment varies much less rapidly with the laser field strength, for example as $|E|^p$, where p is the effective order of nonlinearity. Consequently, the high harmonics do not get tightly focused as in the weak field limit. They are defocused, in some cases exhibiting rings. The magnitude of the phase matching function in the wings of the distribution, i.e. away from the maximum, stays constant, since it reflects the geometry of the interaction and the variation of the dipole moment with the intensity. The phase matching factor does not depend much on the nonlinear order as shown in Figs.6,7.

III– Experimental results in neon with a 1 ps Nd–Glass laser.

In conclusion, we show in Fig.8 recent experimental results obtained in a 15 Torr neon vapor with a 1 ps Nd–Glass laser system developed in our laboratory. The laser wavelength is 1053 nm, the pulse width 1.2 ps, the energy up to 1 J, though only 25 mJ are used in the present experiment for avoiding damaging the UV analysis grating. We use a 1 m focal length in order to increase the conversion efficiency (b^3 scaling, see Lompré et al., 1990). After a rapid decrease for the first harmonics (up to the 13th harmonic) there is a very long plateau, extending up to the 53th harmonic (62 eV, 20 nm). These preliminary results are encouraging because one might expect to reach shorter wavelengths and higher efficiencies by increasing the incident laser intensity.

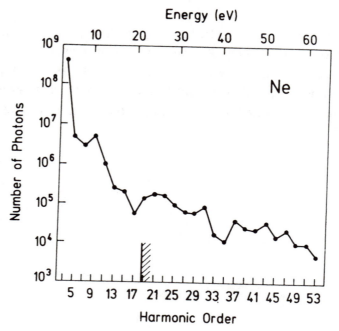

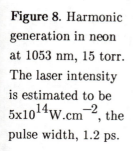

Figure 8. Harmonic generation in neon at 1053 nm, 15 torr. The laser intensity is estimated to be $5 \times 10^{14} \text{W.cm}^{-2}$, the pulse width, 1.2 ps.

REFERENCES

Bandarage G., Maquet A. and Cooper J., Phys. Rev. A 40, 3061 (1989)

Becker W., Long S. and McIver J. K., Phys. Rev. A 41, 1744 (1990)

Eberly J. H., Su Q. and Javanainen J., Phys. Rev. Lett. 62, 524 (1989) ; J. Opt. Soc. Am. B 6, 1289 (1989)

Jackson J. D. (1975) "Classical Electrodynamics", Second Edition, (Wiley, New York).

Krause J. L., Schafer K. J. and Kulander K. C. (1991) to be published.

Kulander K. C. and Shore B. W. Phys.Rev.Lett. 62, 524–526 (1989).

Kulander K. C. and Shore B. W. J. Opt. Soc. Am. B 7, 502–508 (1990).

L'Huillier A., Li X. F. and Lompré L. A. J. Opt. Soc. Am. B 7, 527–536 (1990).

L'Huillier A., Schafer K. J. and Kulander K. C. (1991) to be published.

L'Huillier A., Lompré L. A., Mainfray G. and Manus C., Proc. of the fifth int. conf. on multiphoton processes, Paris, France, 24–28 sept. 1990.

Lompré L. A., L'Huillier A., Monot P., Ferray M., Mainfray G. and Manus C. J. Opt. Soc. Am. 7, 754–761 (1990).

Li X. F., L'Huillier A., Ferray M., Lompré L. A. and Mainfray G. Phys.Rev.A 39, 5751–5761 (1989).

McPherson A., Gibson G., Jara H., Johann U., Luk T. S., McIntyre I., Boyer K. and Rhodes C. K. J.Opt.Soc.Am.B 4, 595–601 (1987).

Miyazaki K. and Kashiwagi H. Phys. Rev. A 18, 635–643 (1978).

Potvliege R. M. and Shakeshaft R. Phys.Rev.A 40, 3061–3079 (1989).

Sundaram B. and Milonni P. W. Phys. Rev. A 41, 6571–6573 (1990).

ABSORPTION AND EMISSION OF RADIATION DURING ELECTRON EXCITATION OF ATOMS

BARRY WALLBANK
Department of Physics
St. Francis Xavier University
Antigonish, Nova Scotia
Canada, B2G 1C0

ABSTRACT.   Recent progress in experiments studying the excitation of
atoms by electron impact in the presence of an intense carbon dioxide
laser is described.   Possible future experiments in what is a new field
in atomic collision physics are also discussed.

1.   INTRODUCTION

Atoms are normally excited into higher states from their ground states
through collisions with electrons or photons.   Recently, however, it has
been   demonstrated   that   excitation   can   also   occur   through   the
"simultaneous" impact of an electron and one [1] or more [2] photons.
This excitation mechanism, first predicted by Göppert-Mayer [3], may be
described by

$$e^-(E_i) \ + \ (n{+}v)\hbar\omega \ + \ A \ \rightarrow \ A^* \ + \ n\hbar\omega \ + \ (E_i \ + \ v\hbar\omega \ - \ E_{ex}) \qquad (1)$$

where an electron of incident energy E_i is scattered from atom A in the
presence of a laser, photon energy $\hbar\omega$, and emerges with energy
$E_i + v\hbar\omega - E_{ex}$ after exciting the atom to a higher state, excitation
energy E_{ex}, and absorbing ($v{>}0$) or emitting ($v{<}0$) v laser photons.   If the
atom were to remain in its ground state then (1) would describe the so-
called free-free transitions that have been studied for several years
[4].
      Experimentally, the simultaneous electron-photon excitation (SEPE)
process described by (1) has been studied for the excitation of the 2^3S
state of helium through the detection of the metastable atoms at electron
energies close to the threshold of excitation ($E_i{-}E_{ex}$) in the field of a
CW CO_2 laser ($I{\sim}10^4$ W cm^{-2}) [1,5].   The much more interesting problem of
excitation involving more than one photon, requiring higher laser
intensities ($I{\sim}10^8$ W cm^{-2}), has been examined by detecting the metastable
atoms produced through the excitation of helium (2^3S and 2^1S), neon ($^3P_2$)
and argon ($^3P_2$), again where $E_i{\sim}E_{ex}$, using a pulsed, CO_2 TEA laser [2,6,7].

D. Hestenes and A. Weingartshofer (eds.), The Electron, 333–339.

These data describe the total excitation cross-sections in the presence
of the laser.

If the scattered electron is detected in experiments examining (1)
then differential scattering cross-sections, which should be much more
sensitive tests of theory than total cross-sections, may be obtained.
For $E_1 \sim E_{ex}$, there are technical difficulties in performing such
experiments but at higher incident energies it has been demonstrated that
differential scattering cross-sections are measurable [8]. The
experimental conditions used to obtain these data ($E_1 \sim 45$ eV, $I \sim 10^8$ W cm^{-2},
excitation to the 2^1P and 2^1S states of helium) may also be more tractable
theoretically.

In order to interpret the total cross-section experiments a simple
extension of the Kroll-Watson treatment [9], used to examine free-free
transitions in the field of a low frequency laser, has been applied
[10,11]. These calculations were successful in reproducing the main
features of the data but are inappropriate for some experimental
conditions. The earliest calculations were primarily confined to cases
where the electrons have incident energies well above the threshold for
excitation of the hydrogen atom to the 2^1S state from its ground state
[12-15]. These calculations should be relevant to experiments where
scattered electrons were detected.

The experimental data and theoretical treatments mentioned above
will be briefly reviewed and possible future progress in this new field
of atomic collision physics will be discussed.

2.   RESULTS AND DISCUSSION

2.1   Total Scattering Cross-Sections

Figure 1 presents data typical of those obtained in these types of
experiment for excitation of helium to the 2^3S state ($E_{ex} = 19.820$ eV) by
the SEPE process. The change in metastable atom yield due to the
presence of a pulsed CO_2 laser, accumulated over 5000 laser pulses, is
shown as a function of incident electron energy around the excitation
energy. The change in yield can be seen to rise to a maximum around
$E_1 = E_{ex}$ before dropping to such an extent that a decrease in excitation
cross-section in the presence of the laser, compared to field free
excitation, is observed. Similar general behaviour has been observed for
excitation of the He 2^3S, He 2^1S, Ne $^3P_2$ and Ar $^3P_2$ metastable states
[2,6,7]. Excitation of the He 2^3S state has been examined as a function
of two of the variables that the laser brings to the collision process,
the polarization and intensity of the laser [7]. It has been observed
that the maximum changes in cross-section occur when the laser
polarization and the incident electron energy are parallel but with
significant changes ($\sim 1/3$ of the maximum increase at $E_1 = E_{ex}$) still
occurring when the laser polarization is perpendicular to the incoming
electron direction.

The main features of these experimental data have been reproduced
using a simple Kroll-Watson Ansätz for He 2^3S [10] excitation and the
closely related Instantaneous Collision Approximation for He 2^3S and 2^1S

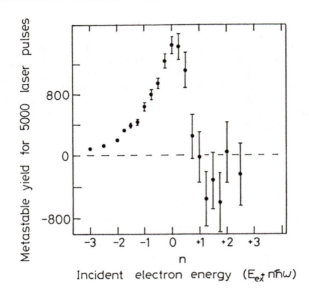

Figure 1. The change in helium metastable yield due to the laser as a function of incident electron energy in units of the photon energy with respect to the 2^3S excitation energy.

[11]. This would indicate that the atom plays a fairly passive role in the process with the laser-electron interaction being most important in determining the cross-section for the SEPE process. However, neither of these treatments is applicable to the situation where the laser polarization and incident electron energy are perpendicular.

Recently, the role of electronic exchange in laser-assisted elastic electron-hydrogen collisions has been reconsidered [16]. It has been predicted that the exchange, under certain circumstances, may largely dominate the differential scattering cross-section, although, in the corresponding field-free situation, it is only a small correction to the direct amplitude. This is expected to occur in some angular regions when only a few photons are exchanged but over the full angular range for many-photon processes. The effect is enhanced when the collision energy is low and should therefore occur at relatively weak laser fields for such collision energies. These findings may be of particular importance to the data discussed above which were obtained for pure exchange transitions. Indeed, this interaction may be playing a dominant role in these collisions and, therefore, this exciting, new aspect of electron-atom collisions in a laser field is certainly worthy of further study.

2.2 Differential Scattering Cross-Sections

The first theoretical treatments of SEPE considered the excitation of the
2^1S state of hydrogen at high incident-electron energies ($E_i \gg E_{ex}$) [12,15].
One can consider SEPE as resulting from two processes: (i) the energy of
the electron is changed by the direct absorption or emission of one or
more photons while scattering from the atomic potential, and (ii) the
atom may be 'dressed' by the laser field with the electron scattering
from this 'prepared' atom.

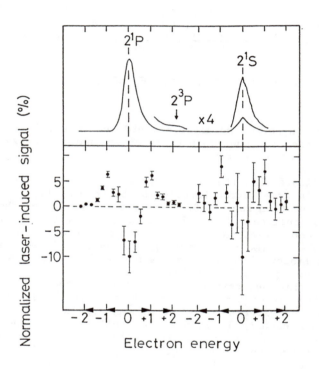

Figure 2. The electron signal detected after scattering 45 eV electrons
through 12° in the energy region of excitation of the 2^1P and 2^1S states
of helium in the absence of the laser (upper) and the change in electron
signal recorded when the laser is present. The change in signal has been
normalized to the field-free signal at the excitation energy of either
the 2^1P or 2^1S state and the electron energy is displayed in units of
laser quanta with respect to the excited states.

The Kroll-Watson treatment ignores the latter process. However, at high incident electron energies and small scattering angles, differences in cross-section between absorption and emission of photons, occurring through the interference, either constructive or destructive, of the scattering amplitudes from mechanisms (i) and (ii) have been predicted [15]. At small scattering angles the two amplitudes may be of similar magnitude making interference effects particularly visible. The actual occurrence of such marked differences between absorption and emission depends on such parameters as the laser intensity and polarization, the atom, the incident electron energy etc. and requires detailed calculation of the scattering amplitudes.

The data shown in Figure 2 was obtained at a scattering angle of 15° in the region of excitation of the 2^1P and 2^1S states of helium. These data are the recorded electron counts in the presence of the laser minus those in its absence normalized to the field-free count rate at the electron energy corresponding to the appropriate excitation, either 2^1P or 2^1S, as a function of scattered electron energy. A decrease in cross-section at the excitation energy and an increase at $\pm 1\hbar\omega$ are clearly seen. The data around the 2^1S excitation show a greater statistical uncertainty because of the much lower field-free excitation cross-section. The peaks at $+ 1\hbar\omega$ represent the change in excitation cross section due to SEPE involving the absorption of one photon and those at $-1\hbar\omega$ involving the emission of one photon. These data are reminiscent of those published for elastic scattering of electrons from atoms [4] which showed a depletion in signal at the elastic peak with this loss in signal being distributed over the peaks due to the absorption and emission of photons, to satisfy an overall sum rule. Such a sum rule appears to be satisfied for the data presented in Figure 2. These experimental results demonstrated that the differential cross-sections for SEPE are, in principle, measurable for the 2^1P and 2^1S states of helium. The interested reader should consult Ref [8] for full experimental details.

In Figure 3, preliminary results are displayed for the differential scattering cross-sections for SEPE to the 2^1P state of helium involving either the absorption or emission of one photon over the angular range 2° to 34° at an incident electron energy of 45 eV with the laser polarization parallel to the incident electron direction and similar laser intensities to that used for the data of Figure 2. There is some indication in these data that at small scattering angles (<20°) the cross-section for emission of one photon is somewhat higher than for absorption. There also appears to be a feature at ~28° where the emission cross-section shows a dramatic increase while that for absorption shows a similar size decrease. However, these differences are almost within one standard deviation and therefore should be considered with some scepticism. We are now involved in trying to improve the uncertainties in the data to enable us to state definitively whether the expected differences in cross-section for emission and absorption arising from interferences of the scattering amplitudes are indeed observed under the conditions of these experiments.

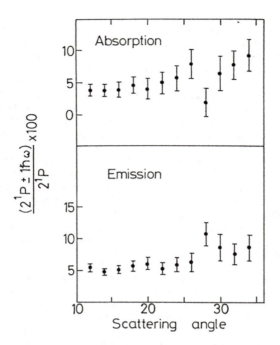

Figure 3. The change in electron signal due to the laser at scattered electron energies corresponding to 1 laser photon above (absorption) and 1 laser photon below (emission) that of excitation of the 2^1P state of helium as a function of scattering angle.   The incident electron energy is 45 eV and the change in electron signal is normalised to the field-free signal at the 2^1P excitation energy.

3.   FUTURE PROSPECTS

The investigation of collisions in laser fields is an exciting new field of research that is of fundamental importance to collision physics as well as to applied fields such as laser heating of plasmas and presents almost unlimited possibilities for future experiments.   The interesting, new, theoretical ideas on the electron proposed at this meeting may reveal even more possibilities since photon-electron interactions may be one of the few ways open to examine some of these properties.

     Our immediate goal at St. Francis Xavier University, however, is to pursue experiments investigating ionizing collisions in a laser field. The motivation behind such experiments, involving electrons, UV photons and metastable atoms as projectiles, have already been discussed by Professor Harald Morgner, University of Witten and the interested reader is directed to his contribution for further details.

4.   ACKNOWLEDGEMENTS

This work was performed with the help and collaboration of Mr. J. K. Holmes and Professor A. Weingartshofer, and with the financial support of the St. Francis Xavier University Council for Research and the Natural Sciences and Engineering Research Council of Canada.

5.   REFERENCES

1.   Mason, N. J. and Newell, W. R., (1987), J. Phys. B, $\underline{20}$, L323.
2.   Wallbank, B., Holmes, J. K., LeBlanc, L. and Weingartshofer, A., (1988), Z. Phys. D, $\underline{10}$, 467.
3.   Göppert-Mayer, M., (1935), Ann. der Physik, $\underline{9}$, 273.
4.   Weingartshofer, A and Jung, C., (1984), 'Multiphoton Free-Free Transitions', in S. L. Chin and P. Lambropoulos (eds), Multiphoton Ionization of Atoms, Academic Press, New York, pp. 155-187.
5.   Mason, N. J. and Newell, W. R., (1989), J. Phys. B, $\underline{22}$, 777.
6.   Wallbank, B., Holmes, J. K. and Weingartshofer, A., (1989), J. Phys. B., $\underline{22}$, L615.
7.   Wallbank, B., Holmes, J. K. and Weingartshofer, A., (1990), J. Phys. B, $\underline{23}$, 2997.
8.   Wallbank, B., Holmes, J. K. and Weingartshofer, A., (1989), Phys. Rev. A. $\underline{40}$, 5461.
9.   Kroll, N. and Watson, K., (1973), Phys. Rev. A, $\underline{8}$, 804.
10. Geltman, S. and Maquet, A., (1989), J. Phys. B. $\underline{22}$, L419.
11. Chichkov, B. N., (1990), J. Phys. B, $\underline{23}$, L333.
12. Rahman, N. K. and Faisal, F. H. M., (1976), J. Phys. B, $\underline{9}$, L275.
13. Rahman, N. K. and Faisal, F. H. M., (1978), J. Phys. B, $\underline{11}$, 2003.
14. Jetzke, S., Faisal, F. H. M., Hippler, R. and Lutz, H. O., (1984), Z. Phys. A, $\underline{315}$, 271.
15. Schwier, H., Jetzke, S., Hippler, R. and Lutz, H. O., (1987), 'Continuum Transitions in Multiphoton Ionization and Electron Scattering', in S. J. Smith and P. L. Knight (eds.), Multiphoton Processes, Cambridge University Press, Cambridge, pp. 43-57.
16. Trombetta, F., (1991), 'Exchange Effects in Laser-assisted Elastic Electron-Hydrogen Scattering' To be published.

PENNING IONIZATION IN INTENSE LASER FIELDS

HARALD MORGNER
Department of Science
University Witten/Herdecke
Stockumerstr.10
D-5810 Witten , F.R.G.

ABSTRACT. The essential features of Penning ionization are reviewed in order to provide the basis for the discussion of laser modified Penning ionization. Expected experimentally observable effects are described. The relationship to free- free transitions is discussed.

1. Introduction

The purpose of this contribution is to describe expected effects that an intense laser has on the the spontaneous process of Penning ionization

$$A^* + B \longrightarrow A + B^+ + e^- \tag{1}$$

where A^* denotes an excited rare gas atom and B a target atom or molecule, usually in its ground state, but not necessarily so. A noteworthy feature of Penning ionization as opposed to other collisional ionization events is the fact that this process occurs even in the limit of zero kinetic energy. Thus, Penning ionization can be viewed as autoionization of the collision complex A^*B.

The main emphasis of this contribution is not a rigorous theoretical formulation, but rather a discussion of how the laser modification can be identified experimentally.

Several theoretical studies on laser- assisted or laser- modified Penning ionization have been published over the last decade, e.g. [1],[2],[3],[4]. They all have in common that they treat the field-matter interaction by the dipole approximation. The present considerations lead us to the conclusion that the most important experimental features are not accounted for in this way.

The paper is organized as follows: section 2. is devoted to the presentation of our knowledge on field free Penning ionization. The necessary extension of the Hamilton operator to the incorporation of the laser field and the qualification of the term 'intense' in the present context is discussed in the next section. The last two sections

341

D. Hestenes and A. Weingartshofer (eds.), The Electron, 341–351.
© 1991 Kluwer Academic Publishers. Printed in the Netherlands.

deal with the qualitative description of the observable effects and how field-modified Penning ionization can be viewed as part of a broader topic, namely free-free transitions in the long range Coulomb field.

2. Field free Penning ionization

We will concentrate our discussion on the prototype reaction

$$He^*(1s2s) + Ar(3p^6) \longrightarrow He(1s^2) + Ar^+(5p^5) + e^- \tag{2}$$

since a large body of experimental and theoretical information is available for this system. The main source of experimental information is the spectroscopy of the emitted electron, i.e. the registration of its kinetic energy and its direction.

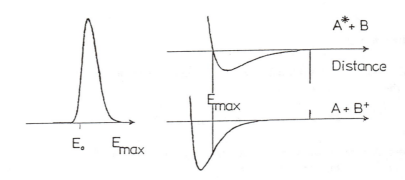

Fig.1 Schematical potential curves for initial (He*+Ar) and final state (He+Ar+).
Electrons with maximum energy E_{max} are emitted predominantly when the
nuclei are close to the classical turning point. The width of the electron
energy distribution (left panel) reflects the variation of the energy separation between the potential curves.

If the ionization takes place at large internuclear separations, i.e. with essentially unperturbed atomic states, the energy of the electron is given by

$$E_0 = E[He^*] - IP[Ar]$$

which is roughly 5eV. However, since ionization is enabled only as a consequence of close encounter most ionizing events take place at small internuclear separations where the energy levels of both atoms are disturbed. Fig.1 shows how this leads to electron energies other than only E_0. The electron energy distribution is roughly 100 meV broad. Since the energy resolution of electron spectrometers is much better than that, it is possible to distinguish experimentally between electrons that have different energies. This has interesting consequences with respect to the measurement of the emission direction. It is possible to derive angular distributions not only as an average over all emitted electrons, but also differentially for electrons out of a finite energy interval within the energy distribution. Fig.2 (upper panels) shows the different angular distributions obtained for different electron energies from the system

$$He^*(2^1S) \quad + \quad Ar(3p^6) \quad \longrightarrow \quad He(1s^2) \quad + \quad Ar^+(3p^5) \quad + \quad e^- \qquad (3)$$

These results show convincingly that electrons with different energies can be in different states even though they originate from one and the same process. If one calculates from the laboratory fixed frame back into the molecular fixed frame the differences between the angular distributions become even more drastic [5], [6], [7] as shown in fig.2 (lower panels). According to [8] it is helpful to distinguish three steps in the process of Penning ionization:

1. the approach of A^* and B, a typical half collision in the context of molecular dynamics

2. the electronic transition, i.e. the switch from the bound configuration into one with one electron in the continuum

3. the nuclear motion of $A+B^+$.

The rigorous treatment of step 1 requires the use of a nonlocal and energy dependent potential operator which is formally complicated and computationally time consuming. However, in a recent study [9] it has been shown that most Penning ionizing systems - including reaction(2) - can be treated by means of a local and energy independent potential without loosing accuracy beyond experimental uncertainty. From the results of ref.[9] it becomes evident that this situation is unchanged if the interaction with the laser field is added. General characteristic features in Penning ionization exist only for step 2 as pointed out in ref.[8]. In the following we list the characteristic properties of the electronic transition according to the discussion of ref.[10]:

Verticality

- step 2 is a vertical transition in the sense that switching from initial to final continuum configuration is fast compared to nuclear motion. A statement in the literature [11] that the Franck-Condon approximation does not apply to Penning ionization is not in contradiction with this. The reason being that in [11] the above mentioned formal nonlocality of the potential operator (with impact on the calculation of the initial state wave function of nuclear motion) has been shifted to the transition operator. The thus introduced non- local transition is formally nonvertical, of course, but the physical meaning of verticality, namely the above mentioned different time scales for electronic and nuclear motion, is obviously not affected.

A consequence of verticality in the above sense is the conservation of total spin during step 2: the time for the change of configuration is too short to allow spin orbit interaction of valence orbitals to be effective. This statement is backed by much experimental evidence [8],[10] . We note, however, that spin conservation does not hold through the whole process including step 1 and step 3.

Exchange mechanism for A* being a metastable

the minimum number of electrons that must move during a Penning reaction like (2) is two . In principle the final state can be reached in two possible ways which are visualized by arrows in the following. For the direct mechanism we have

$$\text{He}^*(1s2s) + \text{Ar}(3p^6) \qquad\qquad (3a)$$

with arrows indicating $\longrightarrow e^-$

$$\text{He}^*(1s2s) + \text{Ar}(3p^6) \qquad\qquad (3b)$$

with arrows indicating $\longrightarrow e^-$

As long as the projectile is in a metastable state the process is dominated by the exchange mechanism. A particular striking piece of experimental evidence has been presented by Ebding and Niehaus [12] . They had the projectile He* moving fast and the target Ar being at rest in the laboratory frame. Doppler shift measurements allowed them to identify the fast moving He atom as the source of electron emission. It is obvious that the direct mechanism would not give rise to a Doppler shift. We note here that this may have some impact onto the design of an experiment on laser- modified Penning ionization: if the presence of the electromagnetic field should enable the direct mechanism this could easily be traced experimentally by Doppler shift measurements. We will come back to this consideration in section 4 of this paper.

The Independent Particle model

various pieces of evidence prove that the two electrons that are considered to move during the Penning ionization process do not exchange angular momentum [10]. To illustrate this further we write out the two electron transition matrix element as

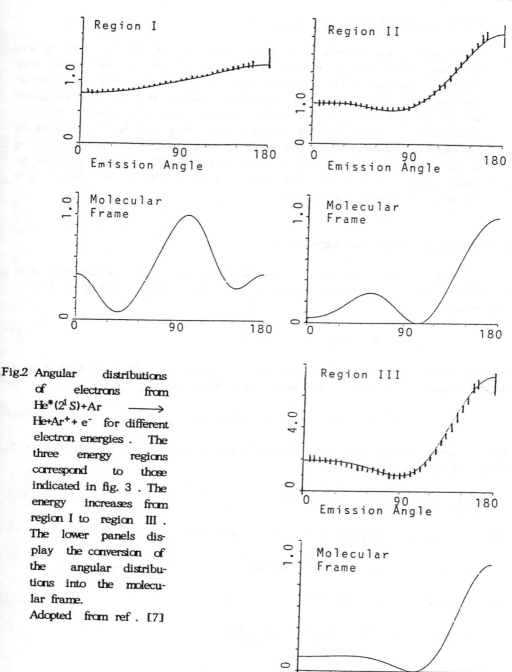

Fig.2 Angular distributions of electrons from He*(2^1 S)+Ar $\longrightarrow$ He+Ar$^+$+ e$^-$ for different electron energies . The three energy regions correspond to those indicated in fig. 3 . The energy increases from region I to region III . The lower panels display the conversion of the angular distributions into the molecular frame.

Adopted from ref . [7]

$$\langle X(1) \cdot \varphi_{He1s}(2)|U(1,2)|\varphi_{Ar3p}(2) \cdot \varphi_{He2s}(1)\rangle \qquad (4)$$

where χ denotes the scattering state of the emitted electron and the subscripts label the atomic orbitals that the electrons would populate if the collision partners were separated adiabatically. The appropriate transition operators U need not be specified further in the present context. The above mentioned rule has the consequence that χ (1) as well as φ_{Ar3p} (2) must have sigma-symmetry with respect to the internuclear axis. A second interesting feature is the finding that the two electrons do not exchange momentum [10]. This means in particular that the ionization process is subject to the restriction that the electron labelled (1) in eq.(4) preserves its instantaneous momentum. This has impact on the details of the Doppler shift measurements mentioned above [10].

3. Coupling of the electromagnetic field

Let H_{el} be the electronic Hamiltonian of the field free system. The electromagnetic field is introduced into the Schrödinger equation in the standard way by adding the interaction operator

$$H_{int} = e(mc)^{-1}A_p + e^2(2mc^2)^{-1}A^2 \qquad (5)$$

where A denotes the vector potential in the three dimensional space. A shall describe linearly polarized light with the polarization vector in z-direction

$$A = A_0 \sin(\omega t - kr) = (0,0,A_{Oz}) \sin(\omega t - \mathbf{kr}) \qquad (6)$$

For any two eigenstates $|a\rangle, |b\rangle$ of the field free Hamiltonian H_{el} the transition matrix element is given as

$$\langle b|H_{int}|a\rangle = e(mc)^{-1} \{\langle b|AP|a\rangle + \langle b|A^2 e/(2c)|a\rangle\} \qquad (7)$$

The first term which is linear in A can be rewritten as

$$T_{lin} = i(m/\hbar)A_{Oz} \langle b|\sin(\omega t - \mathbf{kr})V(\mathbf{r}) z|a\rangle \qquad (8)$$

and the quadratic term yields

$$T_{quad} = (1/2) e(2c)^{-1} A_{Oz}^2 \langle b|\cos(2\omega t - 2\mathbf{kr})|a\rangle \qquad (9)$$

where we have skipped $\langle b|a\rangle$ since $|a\rangle$, $|b\rangle$ are considered orthogonal. The electron described by $|a\rangle$, $|b\rangle$ moves in the potential $V(r)$ which can be considered as Coulomb potential $V(r)=Q/r$ for our purposes. With $z=r\,Y_{10}(\Theta)$ this allows to transform eq.(8) into

$$T_{lin} = i(m/\hbar)A_{0z}\,Q\,\langle\,b|\sin(\omega t-\mathbf{kr})\quad Y_{10}(\Theta)|a\,\rangle \qquad (8a)$$

Comparing (8a) and (9) shows that the brackets have the same order of magnitude. Thus for accessing the relative strength of the quadratic term it suffices to compare the respective prefactors. The break even point is determined by

$$(m/\hbar)A_{0z}\,e = (1/2)\,e(2c)^{-1}\,A_{0z}{}^2$$

which corresponds to an intensity of more than 10^{13}Watt/cm^2 for the wavelength of a CO_2 -laser. Such high intensities are totally unsuitable to study Penning ionization. In this intensity regime multiphoton ionization sets in. This would mean that He* and Ar would hardly have a chance to collide before being ionized by the laser field. Indeed, a planned experiment on laser modified Penning ionization is designed with an intensity below 10^9 Watt/cm^2 [13]. As a consequence we content ourselves to consider only that part of the transition amplitude that is linear in the vector field $\mathbf{A}$.

The second, likewise reasonable, approximation consists in restricting the laser interaction onto the most loosely bound electron of either configuration, He*(1s2s)+Ar or He + Ar$^+$+e$^-$. In case of the initial configuration this is the electron in the excited state that asymptotically starts as the 2s orbital of helium. In the final state we have to consider the electron in the continuum state. Consequently we can treat the states $|a\rangle$, $|b\rangle$ as one electron states with respect to the interaction operator H as long as we consider transitions within either of the configurations (cf. cases 1 and 3 below).

We have to distinguish three cases:

case 1: both states $|a\rangle$, $|b\rangle$ belong to the subspace of H_{el} that contains the excited (bound) states of the helium atom. If only the lower lying Rydberg states are effectively populated than we are allowed to make use of the dipole approximation. The effect thus described is the formation of dressed states of He*(1s2s)+Ar

case 2: one state, say $|a\rangle$, belongs to the subspace of the preceeding case whereas $|b\rangle$ describes an electron that moves in the field of the molecular ion He Ar$^+$. The limited spacial extension of $|a\rangle$ again justifies the dipole approximation

case 3: both states are part of the subspace that contains the scattering states of He+Ar$^+$. The long range interaction between this ion and the electron is a Coulomb potential. Thus, the formulation (8a) of the transition matrix element is applicable. Taking into account that $|a\rangle$, $|b\rangle$ are scattering states we see that the integration over spacial coordinates does not converge. The range that contributes to this integral extends formally to infinity as long as the laser field is described as a plane wave. It is obvious that in this situation the exact shape of the laser beam profile must be of influence. Further, one may expect that this type of interaction with the electromagnetic radiation is felt much more strongly than the other cases. It is evident that the dipole approximation is necessarily invalid for this matrix element.

Existing literature [1], [2], [3], [4] is essentially concerned with the coupling according to case 2. Some authors [1] do not specify the final state in terms of eq.(3). Other authors, e.g. ref.[3] state explicitly that they employ the radiative mechanism which can be visualised by eq.(3a) with the additional remark that the transition is not spontaneous but driven by the external laser field. We do not follow, however, the statement of ref.[3] that the radiative mechanism should be active for $He^* 1s2s, 2^3S$.

Our reasoning is that according to (3a) the 2s electron had to return to its ground state which would require a spin flip. The electromagnetic radiation cannot do that directly (under circumstances that are considered to justify the electric dipole approximation). An indirect effect via spin-orbit interaction may be conceivable in atoms other than helium which is known to follow the LS-coupling scheme up to very high main quantum numbers. Our own analysis of case 2 indicates that the external laser field can lead to both final states described by the right sides of eq.(3a) and eq.(3b). However, at intensities considered here, i.e. below $10^9 W/cm^2$, the corresponding transition matrix elements are much smaller than those of the spontaneous process [14].

Laser action according to case 1 can modify the character of the orbital of the excited electron. This can be expressed by the expansion:

$$|\varphi_{He2s}> \longrightarrow \Sigma\, c_{nl}\, |nl> \tag{10}$$

with $|nl>$ being Rydberg states of helium. This modification may be noticeable in particular for contributions with $l=1$ which subsequently enable the spontaneous direct process (3a).

The most important influence is to be expected from the matrix element of case 3. It corresponds to a free-free transition in the field of $HeAr^+$. The long range character of the $HeAr^+ - e^-$ interaction should make this process even more important than free-free transitions mediated by the short range potential between Ar and e^-. The latter are active at CO_2 laser intensities well below $10^9 W/cm^2$ as shown by Weingartshofer and colleagues [15],[16]. To our knowledge, the importance of case 3 has not been recognized before in the literature.

4. Observable effects

The main expected effect of the laser field on Penning ionization is the change in the electron energy distribution. As visualised schematically in fig.3 the energy spectrum familiar from the field free case is augmented by peaks which are displaced from the original peak by integer multiples of the photon energy.

The net change in energy allows to judge the minimum order to which the laser has interacted with the collision system. However, the real order of interaction may be much higher than can be read off the energy displacement since multiples of the photon energy can be absorbed as well as emitted.

Thus, the inspection of the angular distribution gives a much more detailed handle on the process. The role of the angle between the polarization vector and the internuclear axis has been discussed before by Dahler [1] for case 2.

The experimental distinction between the cases 1, 2 and 3 is not unique. Still, if the direct process (3a) is isolated via Doppler shift measurements as mentioned at the end of section 2., it is obvious that matrix elements according to case 1 (initial state) or according to case 2 must have played a role. Free-free transitions alone cannot switch the emission origin from Helium to Argon.

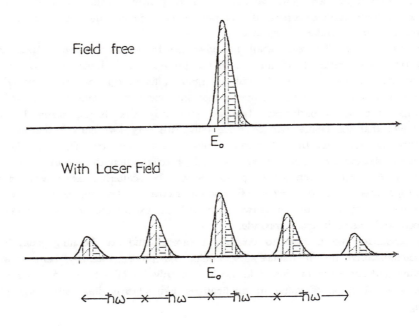

Fig.3 upper panel: field free electron energy spectrum lower panel: field-modified electron energy spectrum

5. Synopsis with other free-free transitions

As discussed above the main effect is to be expected from the interaction of the laser
field with the emitted Penning electron. Thus laser modified Penning ionization can
provide interesting information on the free-free transitions of electrons in the
long range Coulomb field. The charm of this approach to free-free transitions consists in
the fact that the starting conditions of the electronic state can be controlled to quite
some detail and varied with ease: as demonstrated in section 2. different electron energies in the field free situation lead to very different angular distributions indicative
of different states (cf. fig.2 and ref. [6], [7]). Different states in the field free case
have to be understood as different starting conditions for the laser driven free-free process.
This type of varying a parameter is obtained almost automatically in a given experiment.
Another interesting experiment is the study of VUV photoionization in a laser
field. Again the laser is not the cause of the ionization, but interacts only with
the emitted electron. The laser intensity has to be kept in the same range as discussed
for Penning ionization. Since the angular distribution of photoelectrons without the laser is
very well known we again have a situation where the starting conditions of the electrons
are known and different from all conditions encountered in Penning ionization. This type
of experiment appears quite naturally as accompanying Penning ionization in a laser field
since in laser free experiments the comparison between Penning ionization and VUV photoionization is a standard thing to do.
Of course, this discussion shall not give the impression that an experiment in which
electrons are scattered off an ion in the presence of a laser were superfluous. In contrast, this arrangement would result in again different starting conditions of the electron.
It is necessary to mention, however, that this experment suffers from two shortcomings
compared to the experiments discussed previously. One is experimental and is due to
the fact that ion beams can be made only with very low number densities compared to
neutral beams. Thus, the count rate would be extremely low. The other is of theoretical
nature: whereas in Penning ionization and photoionization the electron is set into a state
with a very limited number of partial waves the scattering experiment cannot avoid to
average over a large number of angular momenta. The question is then either how
detailed information can be drawn from such an experiment or how stringent a test of a
theoretical result it could provide.
In conclusion we propose to conduct a careful study of Penning ionization and VUV
photoionization in an intense laser field, the proper intensity being discussed above.
Better understanding of this field would not only be of interest to basic science but at
the same time be of value in applications like plasma diagnostics and plasma modelling.

Acknowledgment

The author wishes to express his gratitude to Prof. A. Weingartshofer for making the workshop on "The Electron" a reality.

References

[1]   Saha,H.P. Dahler,J.S. and Nielsen,S.E. (1983) Phys.Rev. A28, 1487-1502

[2]   Saha,H.P. and Dahler,J.S. (1983) Phys.Rev. A28, 2859-67

[3]   Bellum,J.C. and George,T.F. (1979) J.Chem.Phys. 70, 5059-71

[4]   Lam,K.S. and George,T.F. (1985) Phys. Rev. A32, 1650-6

[5]   Morgner,H. (1978) J.Phys.B11, 269-80 Hoffmann, V. and Morgner,H. (1979) J.Phys.B12, 2857-74

[6]   Hertzner,A. Morgner,H. Roth,K. and Zimmermann,G. (1985) XIVth ICPEAC, Book of Abstracts, Palo Alto 1985, eds.M.J.Coggiola, D.L. Huestis and R.P.Saxon

[7]   Hertzner,A. (1986) Diplomthesis, Freiburg

[8]   Morgner,H. (1981) Comm.Atom.Mol.Phys. 11, 271-285

[9]   Morgner,H. (1990) Chem.Phys. 145, 239

[10]   Morgner,H. (1988) Comm.Atom.Mol.Phys. 21, 195-215

[11]   George,T.F. (1983) J.Phys.Chem. 87, 2799

[12]   Niehaus, A. and Ebding, T.
       ICPEAC, Book of Abstracts, eds. J.S. Risley and R. Geballe (Univ. Washington Press, Seattle, 1975)
       Ebding, T. (1976)
       Ph. D. Thesis, Freiburg

[13]   Wallbank, B., Weingartshofer, A. and Morgner, H.

[14]   Hertzner, A. and Morgner, H. 1990 to be published

[15]   Weingartshofer, A., Holmes, J.K., Sabagh, J. and Chin, S.L. (1983), J. Phys. B16, 1805-17

[16]   Wallbank, B., Holmes, J.K. and Weingartshofer, A. (1967) J. Phys. B20, 6121-38

MICROWAVE IONIZATION OF H ATOMS: EXPERIMENTS IN CLASSICAL AND QUANTAL DYNAMICS

L. MOORMAN
Department of Physics
State University of New York
Stony Brook, NY 11794-3800
U.S.A.

ABSTRACT. Experiments on microwave ionization of hydrogen atoms at various frequencies compared with theoretical calculations have shown that the problem may be divided into different regions of distinct behavior depending on the precise experimental situation. Some regions may be described up to a certain level of accuracy using classical atomic dynamics while others require quantum atomic dynamics. At a more detailed level, the latest comparisons with theoretical calculations indicate that experimentally we can measure atomic quantal interferences in an otherwise "classical region" and make links to classical atomic dynamics in an otherwise "quantal region". In the latter case, "scars" promise to play a particularly important role in the understanding of how quantal and classical dynamics merge for a system whose classical dynamics is at least partly chaotic.

1. INTRODUCTION

This chapter deals with the recent advances in experiments and theory of microwave ionization of hydrogen atoms. We will mainly discuss the experiments performed in the laboratory at Stony Brook. Our theoretical discussion will focus on one of the most recently proposed interpretations of strong local stabilities observed in a region that, as far as the treatment of the atomic dynamics is concerned, is of intermediate classical-quantal nature. The intertwining of quantal and classical dynamics in this region and the fact that the latter has an irregular character makes

353

D. Hestenes and A. Weingartshofer (eds.), The Electron, 353–390.
© 1991 *Kluwer Academic Publishers. Printed in the Netherlands.*

354 L. MOORMAN

this microscopic problem of the simplest atom placed in a microwave field extremely rich and interesting.

To the experimentalist the hydrogen atom in a microwave field is a particularly attractive system to study because it allows for careful control over the key parameters. Our experiments expose hydrogen atoms in a Rydberg state with initial principal quantum number in the range from $n_0 = 24$ to 90 to a carefully controlled microwave field. Depending on the microwave frequency, w (atomic units are used throughout au) in order to ionize a H atom for some n_0 values as many as 300 photon energies are necessary to bridge the energy gap to a final free-electron (continuum) state.

Theoretically this system can be studied from two very different points of view: "Classical" theories regard both the atom and the field as classical quantities. "Quantal" theories describe the atom quantum mechanically, but the field and its coupling to the atom are still considered classically. This combined with the "irregularity" of the classical problem therefore represents a (time dependent) example of a family of problems that may be grouped under the name "quantum chaos".

Classically, the non-linearity of the harmonic drive of the microwave field combined with the Coulomb force may give the possibility for chaotic dynamics and quantitative tests in which Liapunov exponents were calculated have proven this to be so [1,2,3]. The chaotic dynamics in its turn provides the irregular trajectories through which the atom may ionize, the effects of which were observed experimentally [4]. For certain conditions – higher scaled frequencies – this is sometimes called diffusive ionization and a theoretical prediction that in that case quantum mechanics suppresses the classical diffusive ionization process [5] has also been verified by at least two experimental groups [6,7]. The confirmation of this purely quantal effect in the hydrogen microwave ionization experiments initiated a lively discussion among theorists as to what this quantal suppression of classical microwave ionization originates from and whether such a behavior could successfully be described by simplified, truncated theories or (classical) map approximations that may be quantized in some way.

From the comparison of experimental results with theoretical calculations that closely model the experimental conditions a picture has emerged in which the microwave ionization of hydrogen may be divided in different regions [8,10,9] for the parameters that control the experiment. In some regions the experimental data compare very well to calculations based on classical atomic dynamics while in others quantal atomic dynamical calculations are necessary. But, if one looks in more detail (e.g. with higher resolution) one finds indications for quantized-atom interferences within the "classical" region, and links with classical atomic dynamics in the "quantal" region. The proven accuracy [11,12] with which experiments can be done and the wealth of information they have yielded so far make ionization by microwaves a good probe system to study the transition of quantal to classical dynamics, which some may conjecture to be the region where a correspondence principle would be applicable. It is therefore appropriate to include these investigations under the heading of Quantum Chaology [13,14].

2. THE HYDROGEN ATOM IN A MICROWAVE FIELD

The hydrogen atom is prepared in a Rydberg state with principal quantum number in the range ($24 < n_0 < 90$). The method by which this is done will be discussed below. The free field energy spectrum of these states becomes denser with increasing principal quantum number and their energies are approximately given by $E_{n_0} = -1/(2n_0^2)$ (au). The external electro-magnetic field used is so strong that the atom ionizes, or is brought on the verge of ionization. But because the ratio of the photon energy to the free field energy splitting is of order unity or less, the atom has to absorb at least the order of 10^2 times the energy of a single microwave photon in order to be eventually ionized. Of course, although the final energy balance can be made up this way, the process of ionization can not be considered in this way, where the atom is decoupled from the field, because in the real experiment the atom ionizes in the presence of the field.

The relevant parameters in microwave ionization of hydrogen, or single electron systems in general, are the classically scaled frequency, $n_0^3\omega$, and the classically scaled amplitude, n_0^4F. The scaled frequency is the ratio of the "external" frequency ω (au) to the "internal" Kepler frequency, $1/n_0^3(au)$. In terms of quantum mechanical quantities this represents the ratio of the photon energy to the energy splitting of free field n states at $n = n_0$. The scaled amplitude is the ratio of the external electric field amplitude, $F(au)$, to the mean internal field strength of the nucleus – averaged over one classical Bohr orbit around the nucleus.

Under the large variety of experimental conditions investigated so far, we have found not only quantum behavior but also clear indications of classical behavior in the ionization process. Table 1 summarizes what the comparison of experimental and theoretical investigations on the microwave ionization problem of hydrogen has taught us. Experimental ionization curves were always obtained by measuring the ionization signal as a function of field amplitude at a fixed frequency. These ionization curves together with the results of theoretical calculations have indicated up to now the existence of (at least) six regions the characteristics of which were explained in Ref [8,9,10]. We emphasize that with respect to scaled frequency the six regions are not disjoint; the boundaries overlap. The frequencies at which the experiments were performed are also indicated in the second column of the table because the quantum dynamics does not depend only upon classical scaled quantities as is required by classical dynamics. The abbreviations under 'Exper. type' are explained in the next section. The third column, labeled 'Exper. character.,' listing the occurence of experimentally anomalies is based on a comparison of

Region	Cl.sc.freq. $n_0^3\omega$	Freq. (GHz)	Exper. type	Exper. character. [19]	model characteristic	Remarks Exp ↔ Th.
static	0	0			Tunneling through static barrier [20]	
I tunneling	≲0.05	9.92	DIP	Absence of NMS in ionization curves	Tunneling through slowly moving barrier	QM lower than classical threshold
II low frequency	≈ 0.05 to 0.3	7.58 9.92 11.9	DIP+SA	NMS + changes of slope in ionization curves.	3D/1D Cl. MC; QM S.eq solution; QM adiabatic basis	Thresholds ≈ Cl.; NMS are QM that may *lower* or *raise* thresholds.
III semi classical	≈ 0.1 to 2	7.58 9.92 11.9 26.4 36.0	SA DIP+SA SA SA DIE+SA	Classical scaling of thresholds; Local stabilities in thresholds at $n_0^3\omega = 1/q$	3D/1D Class.MC; 1D Q.approx. on adiabatic states; 1D S.eq.	Classical 1D islands in Poincaré sections correspond to local stabilities in thresholds; Quantal scars?
IV transition	≈ 1 to 2	26.4 36.0	SA DIE+SA	Local structures over 50% of n_0^4F values. Is there classical scaling?	3D/1D Class.MC; 1D QM S.eq; QRCE basis	QM increases (up to 100%) or decreases (few %) Cl. thresholds
V high frequency	≳2	26.4 36.0	SA DIE+SA	Large structures superimposed on rising thresholds	1D S.eq; Delocalization; Cantorus flux quantization; QKM; QRCE-basis	QM rise on average about 100% above Cl. threshold; QRCE calc. show stabilization as well as destabilization; Delocalization and QKM not good for $n \gg n_0$. Quantal Scar.
VI Photo electric effect	$> \frac{1}{2}n_0$				QM perturbation theory	QM expected to describe one photon ionization + ATI

Table 1: *Six regions of Hydrogen microwave ionization are distinguished by comparing Stony Brook experiments to several theories. This chapter contains more detailed information about the regions II-V. Abbreviations used in the table are: 1D(3D) = 1(3) dimensional; Cl=Classical; DIP, DIE: see section 3; MC=Monte Carlo; NMS=non monotonic structures; QKM = quantized Kepler map; QM = quantum mechanical; QRCE-basis: basis of 'quasi resonant compensated energy' states; sc.=scaled; SA: see section*

different experimental data only, *i.e.* no theoretical arguments, except scaling, are used. The fourth column lists which types of theoretical models have been used. The last column briefly notes general conclusions that follow from the comparison between experiment and theory. In this paper we report on the regions II through V. We will not discuss region I in which only preliminary measurements have been made and for which the border with II in terms of scaled parameters is not well investigated yet. Region VI is largely theoretical, combining results of calculations with what is known from experimental strong field (pulsed laser) interactions with atoms, which use frequencies far above the microwave region. Experiments were also performed with two simultaneous microwave fields but these will not be discussed in this contribution [15,16].

3. EXPERIMENTAL METHOD

A very simplified picture of the apparatus is given in Fig 1 [17,18,21]. A fast beam of hydrogen atoms ($\approx$ 15 keV) is produced by charge exchange of protons with Xe gas. In two separate field regions with static fields F_1, F_3 two $^{12}C^{16}O_2$ laser transitions are made using a double resonance laser excitation technique. Typical field values for F_1 were about 30 kV/cm and $F_3 \approx 1 - 500$ V/cm. Since each laser polarization is parallel to the direction of the static fields, only $\Delta m = 0$ transitions are driven. A typical scheme involves excitation of the extremal Stark state with parabolic quantum numbers $(n_0, n_1, |m|) = (7, 0, 0)$ to the state $(10, 0, 0)$ with the first laser, and from there to $(n_0, 0, 0)$ with the second laser. Spectroscopic scans [17] made as a function of F_3 showed that available laser lines and experimental spectroscopic resolution allowed us to populate and resolve states with principal

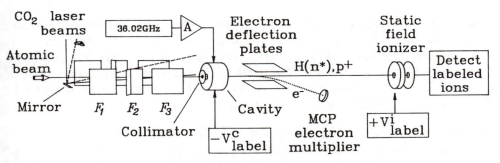

Figure 1. *Schematic view of the apparatus used in Stony Brook for the detection of microwave ionization. Atomic beam particles enter from the left and are, after preparation in highly excited states n_0, ionized in the cavity due to the interaction with a strong microwave field. For various detection schemes see text.*

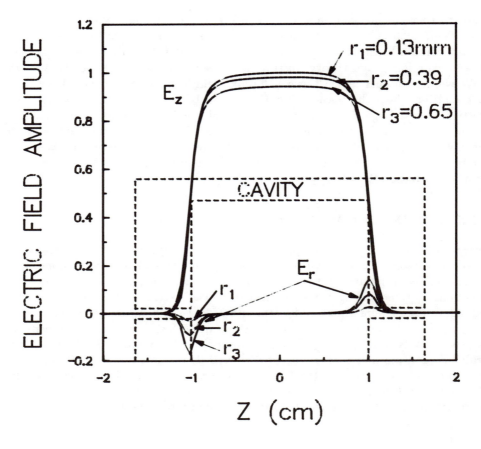

Figure 2. *Longitudinal (E_z) and radial (E_r) components (with regard to the direction of the beam axis) of the electric field amplitude inside the cavity and beam holes. The calculation by 'Superfish' solves Maxwell's equations for cylindrical geometries for many radii. Three solutions close to the axis and normalized to the center of the cavity are shown. Part of the cavity contour, scaled to the z axis, is shown by the dashed line. $\nu = 36.02$ GHz for the TM_{040} mode*

quantum numbers in the range $24 \leq n_0 \leq 90$ for hydrogen [16,4].

Though a unique substate was produced in F_3, it could not be maintained all the way into the cavity. A nominally "zero field" region dominated by stray fields before the atoms entered the cavity led to a statistical distribution of all substates [4] of the given n_0-value, or classically an ensemble of orbits with a microcanonical distribution of orbital planes and eccentricities ($\epsilon \leq 1$). Such an ensemble (or distribution) will be called 3D. We will later refer to 1D models, in which the

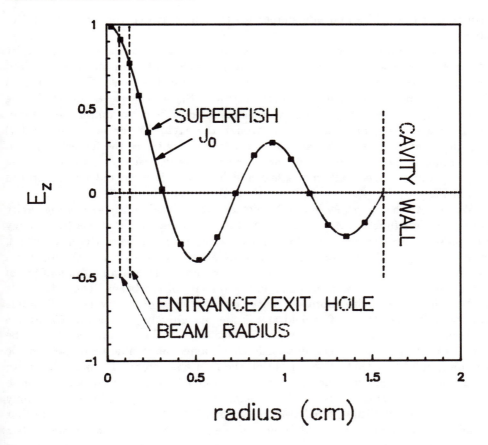

Figure 3. *Longitudinal component (E_z) of the electric field amplitude on the mid-plane inside the same cavity as Fig 2. The result of the calculation with 'Superfish' is compared to an 'ideal cavity' solution. The latter, which has no beamholes, is a J_0-Besselfunction with the fourth node at the cavity wall for this TM_{040} mode.*

electron bounces back and forth on a line between the nucleus and the classical turning point of the Coulomb potential, roughly corresponding to what would be in quantum mechanics an extremal Stark state [22,23,24,25].) This 3D atom is the initial state with which we perform actual experiments.

The beam of Rydberg atoms was narrowed by a collimator before entering a circular cylindrical copper cavity with beam holes in the center of the two endcaps. The cavity was resonantly driven in one of the modes TM_{0p0} with p=2, 3 or 4 such that the atom on its way through the cavity experiences several hundred oscillations of the field. Table 2 summarizes the atomic state properties, characteristics of the cavities and the field amplitude shapes for the different experiments done in our laboratory.

The field amplitude in the cavity was specified as accurately as possible by solving Maxwell's equations with a standard finite element computer program 'Superfish' [26] which exploits the cylindrical symmetry of the cavity. The calculated electric field amplitudes along the beam direction (E_z) seen by the atoms traversing the cavity are given in Fig 2. The various curves represent the amplitude components along straight line trajectories through the cavity at various distances (r_i) from the symmetry axis. The atoms on axis of the cavity see the largest maximal field amplitude. A cross section of the cylindrical cavity, rescaled to the horizontal axis, including correct wall thicknesses and hole sizes is indicated by the broken lines in Fig 2.

We refer to the curves E_z in Fig 2 which are normalized on the axis of the cavity as the envelope function $A(z)$. Seen from the rest frame of the atoms (in analogy with pulsed laser experiments) this represents the pulse shape of the field, A(at), where $a^{-1} = \frac{L}{V}$ is the characteristic timescale for the atom to travel the length of the cavity. The turn on and off of the pulse therefore occur over a known number of oscillations and its duration depends on the velocity of the beam. The field inside the beamholes was well suppressed since the holes were of relatively small radius and the walls were thick. The field amplitude was already reduced to 50% at the inside of the cavity wall and virtually vanished outside the beamholes. As can be seen in Fig. 2, the radial field amplitude E_r became comparable to E_z only inside the holes, indicating that at that position the field was rotated 45° with respect to the beam axis. However, such a large angle would be experienced only by those particles at the extreme outside of the beam at $r = 0.7$ mm, and in addition only there where the amplitude is reduced to less than 20%. It is important to note that at the inside wall where the field amplitude is reduced to 50%, this rotation is less than 25°. The field is to a good approximation longitudinal near the position where the total field is maximal, i.e. where we expect ionization to be most probable. We have done many tests on the accuracy of these calculations of the amplitude envelope including in situ measurements by passing a small dielectric 'bead' through the cavity and determining the local field amplitude from the resonance frequency shift. These tests demonstrate that for the cylindrically symmetric modes used in the experiments reported on here, the pulse shape is known very precisely (at the percent level) and can be used reliably in numerical calculations of the atomic dynamics.

With a collimator of typical radius 0.7 mm (see Table 2) for the high frequencies, the total beam contained atoms that experienced different field amplitudes, F_{max}, ranging over 4 and 7% for the two modes used (ΔF_{max}; Table 2). Fig. 3 shows the field, E_z (TM$_{040}$) in the midplane of the cavity calculated by 'superfish' with the beamholes included, compared to an analytic expression of an ideal mode for a cavity without beam holes (a J_0 Bessel function). The F_0 values given throughout in this paper will always be the geometric average of F_{max} over the cross section of the beam. In some cases, not reported on here [11,12], the beam was collimated down until ΔF_{max} was 0.2%, but a special calibration technique described in [27] estimated the absolute accuracy in the value F_0 generally to be 5%.

In response to the electric field, which is maximal on the axis and polarized along the beam, atoms may be ionized depending on the field amplitude. To detect the ionzation we employ three different methods with the following objectives:

A) *Direct Ionization, Proton* (DIP) detection method. In this method the protons produced inside the cavity by the interaction with the microwaves are detected in a particle multiplier downstream from the cavity.

B) *Survived Atom* (SA) detection method. The atoms that survive the microwave field in a Rydberg state are ionized in a separate (longitudinal) static field ionizer downstream from the cavity. Those ions are subsequently detected in the same particle multiplier that is used in the previous method.

C) *Direct Ionization, Electron* (DIE) detection method. The electrons produced in the cavity by the interaction with the microwaves are extracted from the beam immediately after the cavity and then detected with a microchannel plate.

The names given to these methods are meant to indicate which object, electron, proton, neutral atom, leaving the cavity we intended to detect. All three detection methods made use of a label voltage on that part of the apparatus where the charged particles of interest were produced. (For DIP the cavity body was held at ≈ 100 V, for DIE it was at ≈ -6 V and for SA the cavity body was grounded and the static field ionizer was at about 150 V). This 'energy label' allowed for velocity selection downstream from the cavity and the static field ionizer in a static field lens system (not shown in Fig 1). In addition all measurements were performed by phase sensitive detection on the first laser transition.

A notable side effect of the DIP method was that an atom existing in an extremely high Rydberg state after the cavity (say 10 times n_0) could be ionized by the small field along its flight to the detector. This ion (or electron) would also be energy labeled by the field in which it was formed and therefore pick up extra velocity.

Another complication was that even a minute stray transverse component of the magnetic field could produce a Lorentz force (or equivalently a 'motional' electric force in the rest frame of the atom) strong enough to ionize extremely high Rydberg states. The combination of these two ionization mechanisms sets the limits, or cutoff values, n_c, indicated in the 4th line of Table 2. Notice the large values of n_c, 160-190, for the DIE experiment compared to the initial values n_0. In the experiments reported on here the cutoff values were completely determined by effects outside the microwave region and therefore the atom-microwave interaction dynamics inside the cavity is completely free from static field interaction effects. This is different from experiments done elsewhere [4,7], in which static fields were intentionally maintained in the interaction region.

type of experiment	H SA or DIP	H SA or DIP	H SA or DI[
n_0 range	24-90	43-49 and 62-77	45-80
n_c	90-95	90-95	86-92 or 160-
kinetic energy (keV)	14.0	14.0	14.6
frequency (GHz)	9.9233	7.582 / 11.889	26.40 / 36.(
nominal cavity mode	TM_{020}	TM_{020} / TM_{030}	TM_{030} / TM
cavity length (cm)	4.928	4.501	2.007
cavity radius (cm)	2.658	3.498	1.565
collimator radius (cm)	0.251 / 0.050	0.251	0.070
turn-on (5%-95%)(osc)	79	50 / 80	82
flat max. (95%-95%)(osc)	231	200 / 300	334
turn-off (95%-5%)(osc)	95	50 / 80	82
calc. radial variation ΔF_{max} of F	7% / 0.2%	4%/10%	4% / 7%
absolute accuracy in F_0	5%	5%	5%
$n_0^3 \omega$	0.02-1.1	$\approx$ 0.14-0.22 and 0.43-0.55	0.5 - 2.8

Table 2:*Experimental apparatus data for H microwave ionization and excitation experiments reported in this contribution. The turn-on, -off, and flat maximum describe in number of field oscillations the field pulse-shape that the atoms experienced in their rest frame. The absolute accuracy for the microwave field amplitude on the beam axis is based on an extensive calibration procedure [27]. Relative accuracy within a given experimental curve, e.g., Fig. 5, is better than one or two percent, but the field drops quadratically away from the beam axis by the amount shown in the second-to-last line of the Table.*

Direct ionization experiments were taken by slowly increasing the amplitude F_0 from a value where only a background signal was measured to a value where ionization occurred. The signal was observed to increase and finally saturate to a maximum, defining the 0 to 100% levels. This increasing signal, recorded in conjunction with the microwave power, produces a curve whose shape is characteristic of those of the DIE experiments as well. In contrast, the type of curve measured in a survival experiment, SA, represents a decreasing signal from 100% down to 0%. If one takes the inverse of the SA signal, a Q curve is obtained (Q=100%-SA). This curve represents the amount of signal quenched by the microwaves and is similar in appearance to that of DIP and DIE curves.

All experimental curves DIP, DIE and Q record the microwave field dependence of the probability for hydrogen atoms to be excited to a final state with an energy larger than that of a state with principal quantum number n_c, a cutoff which is much larger than n_0 in most cases. The contributing processes include true ionization, *i.e.* excitation to the continuum by microwaves, as well as excitation to bound states above n_c. The characteristic which distinguishes the three experimental modes is the value of n_c. Another difference is that the SA method is sensitive to extreme microwave de-excitation. Although its occurance is unlikely, (except in some special cases) it can be diagnosed by varying the voltage in the static field ionizer which determines the minimum n-state that will ionize. Provided that the value of the static field voltage is kept high enough there should be no difference in behavior among the detection methods. This was in fact experimentally verified.

Although complete curves were always recorded, a useful way to characterize them is to extract the field amplitude at which 10% of the Rydberg atoms were ionized which we will call *the 10% electric field amplitude threshold* $(F_0(10\%))$ or simply *the 10% threshold*. Following this procedure, many comparisons carried out for hydrogen at 9.92 GHz with n_0 ranging from 32 to approximately 70, showed that quench and ionization curves gave the same information – even if they contained structure(s). When n_0 approached n_c, not surprisingly, differences became apparent. Experiments done at 36 GHz also revealed differences between the DIP and DIE methods. These will be discussed later.

4. CLASSICAL SCALING

In Fig 4A, five curves are shown representing measurements made on atoms in three different initial states, $n_0 = 67$, 48, and 43, and at three different frequencies 9.9, 26.4, and 36.0 GHz respectively. The electric field amplitudes necessary for ionization seem to be very different for the three cases. Moreover, from a quantum mechanical point of view one might not expect them to have much in common each requires a very different number of photon energies to reach the free field continuum (*c.q.* the binding energy of the initial states is 73, 53, and 47 $\hbar\omega$). It may come as a surprise then that by rescaling the electric field amplitude (F, expressed in atomic units) with the fourth power of the initial principal quantum number (n_0^4) these curves become nearly congruent, see Fig 4B. A key point in the magical degeneracy of these seemingly unrelated curves may be realized by rescaling the frequencies ($\times 2\pi = \omega(au)$) with the third power of the initial principal quantum number (n_0^3). One finds that the three situations correspond to values which are extremely close to each other, namely 0.45, 0.44, and 0.43 au respectively. This hints at the existence of a relation between the various curves.

We will show that this type of scaling has a physical significance and is in fact required under classical conditions. By writing the classical Hamiltonian for the fully three dimensional classical hydrogen atom interacting with a linearly polarized external field pulse, with electric amplitude $FA(at)$, as the sum of the free kinetic energy of the electron, the nuclear Coulomb potential energy and the potential

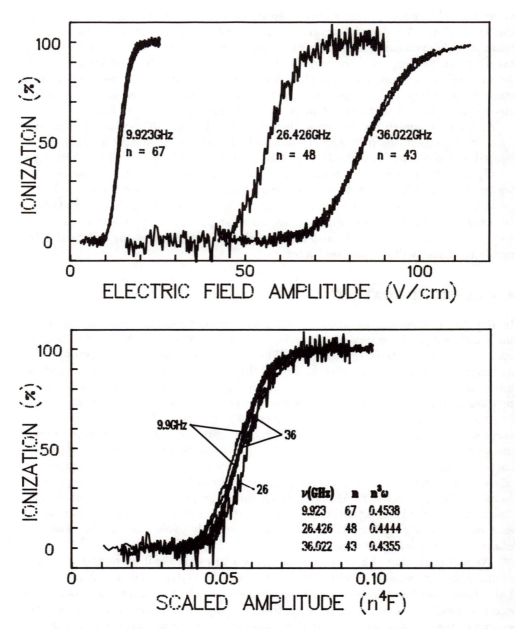

Figure 4. *Example of five ionization curves in region III for three different frequencies that show classical scaling (see text). A:(top) absolute amplitude units, B:(bottom) scaled amplitude units.*

energy of the unbound electron in the external field:

$$H(\vec{r}, \vec{p}, t; F, \omega, a, \phi) = \frac{1}{2} |\vec{p}|^2 - \frac{1}{|\vec{r}|} + FA(at)z\cos(\omega t + \phi). \tag{1}$$

In this $A(at)$ is the normalized shape of the amplitude seen in the restframe of the atom and a^{-1} is the total time the atom is inside the cavity.

It is easy to check that this Hamiltonian has the following exact classical scaling property with arbitrary constant α (ϕ is invariant):

$$H'(\vec{r}', \vec{p}', t'; F', \omega', a') = H(\alpha^{-2}\vec{r}, \alpha\vec{p}, \alpha^{-3}t; \alpha^4 F, \alpha^3\omega, \alpha^3 a) = \alpha^2 H(\vec{r}, \vec{p}, t; F, \omega, a). \tag{2}$$

Due to the complete freedom of the value α one may rewrite any solution $(\vec{r}(t), \vec{p}(t))$ of H (say e.g. describing a Kepler orbit) as a solution $(\vec{r}'(t'), \vec{p}'(t'))$ of H' [28,29]. We may for instance choose to define the primed system to be the system in which $F' = 1$ (unit of amplitude), or we may choose $\omega' = 1$ (unit of angular frequency). Instead however we make another, more natural choice: We will choose to define the primed system such that the initial classical principal action variable of the orbit $(I'_n)_0$ [31] is $\hbar$, i.e. one unit of action (1 au), when the field is not yet turned on. We can use dimensional arguments as in [24]: The principal action variable has a dimension $rp_r = \frac{Ml^2}{T}$ [31], and therefore the principal action variable of the initial orbit scales as:

$$(I_n)_0 = \alpha_0(I'_n)_0 = \alpha_0\hbar \tag{3}$$

The other two action variables for the initial conditions $(I_l)_0$ and $(I_m)_0$ scale with the same factor because they have the same dimensions [31].

In classical considerations later $(I_n)_0$ will be abbreviated by I_0. Using Bohr-Sommerfeld-Wilson quantization we may replace I_0 by $n_0\hbar$ to translate expressions from classical into quantal language (or n_0 if au are chosen for all quantities), and using Eq. (3) we may replace α_0 by n_0 (both dimensionless) in scaling relations.

The important parameters which characterize the dynamics are now $(F', \omega', a') = (n_0^4 F, n_0^3\omega, n_0^3 a)$ in which F, ω and a are in principle in arbitrary units but particular convenient physical interpretations for the values of these scaled parameters can be given if they are expressed in atomic units (see below). It is for the reason described above that these parameters are called *classically scaled variables*. For a field pulse that rises slowly enough we assume that we may leave $n_0^3 a$ unspecified. As for our experiments one can in principle extract this information from Table 2.

The goal of the explanation after Eq. (2) above was to show that the scaled parameters are a natural choice to label the solutions rescaled to a system where $I'_0 = \hbar$ (or 1 au). There is another way of interpreting these parameters. In the initial state with action I_0 there are two time scales in the classical dynamics: The internal Kepler frequency ω_K and the external driving frequency ω. (There would be more frequencies for instance if more electrons would be present). Therefore, because one is free to scale the time parameter (see Eq. (2)) these two frequencies can be reduced to one; namely their ratio. Therefore the classical dynamics for hydrogen and one-electron ions depends only on one frequency ratio:

$$\Omega = \frac{\omega}{\omega_K} = \frac{\omega}{\frac{1}{I_0^3}} = I_0^3 \omega = n_0^3 \omega. \tag{4}$$

Although the ratio is dimensionless and thus does not depend on the system of units, we clearly choose atomic units in the second ratio and therefore ω in the right hand side should be expressed in au also. Since time scales are important one should expect the classical dynamics to behave differently for different values of $n_0^3 \omega$ ($\ll 1$; ≈ 1 ; $\gg 1$). Of course as can be seen from Eq. (1) we have introduced a third timescale a^{-1}. However since from Eq. (2) one is free to scale only one parameter we have to require that the ratio a/ω is preserved under scaling.

Quantum mechanically the ratio $n_0^3 \omega$ has a very simple meaning: It represents the ratio of the energy of the microwave photon to the average energy splitting of the n states for the initial quantum number.

$$\frac{\Delta E_{ph}}{\Delta E_{n_0}} = \frac{\hbar \omega}{(\hbar/n_0^3)} = n_0^3 \omega. \tag{5}$$

Similarly, we may interpret the classically scaled field amplitude as the dimensionless ratio of the electric field amplitude that the electron experiences from the external field and the mean electric field of the nucleus averaged over one classical Bohr orbit about the nucleus:

$$\frac{F_{ext}}{F_{nucl}} = \frac{F_0}{<1/r^2>} \approx \frac{F_0}{1/I_0^4} = n_0^4 F_0. \tag{6}$$

In general, tested for the three frequencies as shown in our example in Fig. 4, we can say that the 10% threshold amplitudes scale very well within a few percent; these tests were done over a wide range of scaled frequencies ($0.2 \leq n_0^3 \omega \leq 0.7$). This indicate that classical theory might be able to describe ionization curves in this region and also seems to justify that the exact scaling of the field pulse (*i.e.* including scaling of the switch- on and -off of the amplitude at the entrance and exit of the cavity) was not neccesary for these scaled frequencies. In addition to the scaling results just described there is much more evidence for classical behavior in this region. Individual curves can be calculated by classical methods using Monte Carlo simulations and directly compared to the experiments. This will be discussed in section 6.

5. ON THE CLASSICAL-QUANTAL CORRESPONDENCE OF THE AVERAGE COULOMB FORCE ON THE ELECTRON

In the previous section we emphasized that the classically scaled amplitude can be understood very easily by taking the ratio of amplitude and the averaged Coulomb force $F_c = Zr^{-2}$ (in the initial, unperturbed state) on the electron in a classical Bohr orbit around a nucleus of charge Z and used $< r^{-2} >_c = I_0^{-4}$. The latter is trivial by realizing that a Bohr orbit is a circular orbit in which the radius has a fixed value during the motion say $|r| = R = I_0^2 a_0$ ($a_0 = 1$ in atomic units). This is the reason we may rewrite the average:

$$< r^{-2} >_c \overset{Bohr}{=} < r >_c^{-2} = I_0^{-4}. \tag{7}$$

The first step is in general not allowed, neither for elliptic Kepler orbits nor for quantal states as we will see below. We will investigate the correspondence between the average force on electrons in quantum states and classical orbits a bit further.

A. QUANTAL

Using the orthonormality of the spherical harmonics, the average strength of the Coulomb force of the nucleus on the electron in a given quantum mechanical substate $|n_0 lm >$ is found to be given by [34,35,36]:

$$< F >_Q = < n_0 lm | Zr^{-2} | n_0 lm > = \int_0^\infty R_{n_0 l}^* \frac{Z}{r^2} R_{n_0 l} r^2 dr = \frac{1}{(l + \frac{1}{2})n_0^3} \left(\frac{Z^3}{a_0^2} \right), \tag{8}$$

in which R_{nl} are the radial wavefunctions and a_0 is the Bohr radius. It is clear from this that for different states of fixed angular momentum l and arbitrary substate m, for example an s state, $< F >_Q$ decreases $\sim n_0^{-3}$ and not as the classical situation for a Bohr orbit described above . This would be relevant for instance for He (and other many-electron atoms), where the initial state for experiments may be a S state for different n_0 [11,12].

The quantal states that correspond with Bohr's circular electron orbits are the states with no radial nodes in the wavefunction when $l = n - 1$. The $m = \pm l$ states are easiest to visualize as circular states, as they have their maximum electron density in the azimuthal plane perpendicular to the quantization axis, and there is a direction of charge current. The state with $m = 0$ is also easy to visualize as its maximal electron density is aligned along the quantization axis, approaching closest to a 1 dimensional state (Another way to view this is that it corresponds to an ensemble of classical circular orbits lying in all orbital planes that contain the quantization axis). In that case, with $l = n - 1$, from Eq. (8) we find indeed $< r^{-2} > \sim n_0^{-4}$ for large n_0 which corresponds to Eq. (7).

For the case of a statistical distribution of substates l and m, we find the following dependence on n_0:

$$<< r^{-2} >_Q >_{Stat.distr.} = \frac{\sum_{l=0}^{n_0-1} \sum_{m=-l}^{l} < n_0 lm | Zr^{-2} | n_0 lm >}{\sum_{l=0}^{n_0-1} \sum_{m=-l}^{l} 1} = \frac{2}{n_0^4} \left(\frac{Z^3}{a_0^2} \right). \tag{9}$$

This has the same n_0 dependence again as for the classical circular orbit except that it is larger by a factor 2 due to all other states in the distribution that are bound stronger to the nucleus than the $l = n_0 - 1$ (Bohr) state. Of course for a coherent superposition of initial substates within one n_0 value ('shell') which are observed in modern atomic physics experiments the dependence on n_0 may be very different between the boundaries $\sim n_0^{-3}$ and $\sim n_0^{-4}$.

The averaged energy splitting between neighboring states can be found from the total energy $E_{n_0} = -\frac{1}{2n_0^2}\left(\frac{Z^2}{a_0}\right)$ which is, due to the total degeneracy of the state with principal quantum number n_0, the same for a statistical distribution of l and m states. The average energy splitting around the initial state is therefore:

$$\Delta E_{n_0} = (1/2)\{(E_{n_0} - E_{n_0-1}) + (E_{n_0+1} - E_{n_0})\} = \frac{1}{n_0^3}\frac{1}{(1 - 1/n_0^2)}. \qquad (10)$$

This is in fact closely related to Bohr's correspondence principle and from it we see that the value n_0^{-3} in Eq. (5) is a good approximation. The first correction exists of a term that is smaller by a factor n_0^{-2} which is typically in the order of 10^{-3} in our experiments. The first relevant correction for a spinless particle from a different origin is a relativistic correction that is smaller by a factor $\frac{3}{2}\frac{\alpha_s^2}{n_0} \lesssim 3 \cdot 10^{-6}$ (fine structure constant α_s). Both corrections can be omitted for our purpose.

B. CLASSICAL

Now we will treat the classical derivation of the mean force $< F >_C$ for an arbitrary (3D) Kepler orbit of a 'classical' electron of mass $m = 1$ au orbiting around the nucleus in a potential $V = -G/r$. Although this can essentially be found in Born (1924) [30] (for a later reference see [25]), we prefer to reintroduce a few concepts and parameters that label Kepler orbits as it is not common to view an atom as a classical object nowadays. Also, giving some details may help to understand what we mean when we compare our experiments later in sections 6 and 7 with numerical calculations that used a microcanonical distribution of classical initial states.

By using spherical coordinates we can define a set of angle-action coordinates $(r, \theta, \phi, I_r, I_\theta, I_\phi)$ and from this 3 new action variables which are linear combinations of the old ones can be defined that have more physical significance: I_n is the principal action variable, meaning the action for which the frequency is "nondegenerate and different from zero" (If no static fields are applied it is the only one that is nonzero). I_l is the total angular momentum and I_m is the component of the angular momentum onto the polar axis [31]; consistent with modern notation we use $2\pi I_i$ [32,24] for J_i in [30,25].

The first two new actions, which are constants of the motion, can be related to the energy of the motion, eccentricity and semimajor and -minor axis of the classical elliptic Kepler orbit:

$$E = -\frac{G^2}{2I_n^2} \qquad (11)$$

$$\epsilon = \sqrt{1 - \left(\frac{I_l}{I_n}\right)^2} \tag{12}$$

$$a = -\frac{G}{2E} = G^{-1}I_n^2 \tag{13}$$

$$b = a\sqrt{1 - \epsilon^2} = G^{-1}I_lI_n. \tag{14}$$

Since in our case with $n_0 \geq 24$ (for a spinless particle as well as for the spin-orbit interaction [36]) the first relativistic corrections are smaller by a factor $\alpha_S^2/n_0 \lesssim 2 \cdot 10^{-6}$ it is sufficient to use the non- relativistic approximation here. The orbit itself in the orbital plane is described by: $r = a(1 - \epsilon \cos u)$, in which u is the eccentric anomaly defined by $\omega\tau = u - \epsilon \sin u$, where ω is the fundamental frequency and τ a time or epoch measured from an instant when the particle was at the perihelion. Born shows that for every elliptic orbit $< r^{-2} > = \frac{1}{ab}$. This can directly be derived without performing the integration from the second law of Kepler. We find that the mean force on the electron in a Kepler orbit is inverse proportional to the area (πab) enclosed by the ellipse:

$$< F >_c = \frac{G}{ab} = \frac{G^3}{I_l I_n^3}. \tag{15}$$

This shows corresponding to the quantal treatment that for fixed angular momentum I_l and increasing I_n the mean force on a classical electron goes down as I_n^{-3}.

The first two action variables may be used for imposing Bohr-Sommerfeld-Wilson rules to quantize the atomic motion by replacing $I_n = n\hbar$ with $n = 1, 2, \cdots$ and $I_l = k\hbar$. In the old quantum theory k was $1, 2, \cdots, n$. Modern semi-classical quantization methods (Einstein-Brillouin-Keller) prescribe $k = l + \frac{1}{2}$ with the orbital angular momentum quantum number $l = 0, 1, 2 \cdots, n-1$ [31]. One should not be mislead to identify the $\frac{1}{2}$ as originating from the spin, as in both quantal and classical treatments in this section the intrinsic spin is not taken into account. If the spin was taken into account the quantization rule would instead be $k = j + \frac{1}{2}$ with the total angular momentum quantum number $j = \frac{1}{2}, \frac{3}{2}, \cdots, n - \frac{1}{2}$ that combines the electron spin and the orbital angular momentum. In that case $k = 1, 2, \cdots, n$ is integer valued as in the old quantum theory, which is the combined effect of the two contributions just described: semi-classical quantization and spin. Each shifts the energy of the quantized states, the first conceptually theoretically, the latter through the spin- orbit interaction, which is a relativistic effect. The *combined result* however was that it did not shift the (free field) energy levels of the hydrogen atom, although there were twice as many (doubly degenerate) states due to the spin – now labeled by (n, j) instead of (n, l) – but allowed for different selection rules consistent with the presence (and intensity ratios) of certain fine structure lines in the observed Balmer spectrum of hydrogen (For instance the absence of the b'- line, and presence of the c-line by G. Hansen in 1925) . For an introduction, references and the continuation of this facinating story on the hydrogen spectrum and of the development of the new quantum theory see [33].

Returning to the spinless electron we see that with $I_l = (l + \frac{1}{2})\hbar$ we find an exact correspondence between the mean force in the quantized classical Eq. (15) and the purely quantal situation of Eq. (8).

For a circular-Bohr orbit ($\epsilon = 0$; $a = b$) we obtain $I_n = I_l$ leading to the familiar result that was used in eq. (6), see also (7) ($G = 1\ au$; $I_n = I_0$):

$$< F >_C \overset{Kepler}{=} \frac{1}{I_l I_0^3} \overset{Bohr}{=} \frac{1}{I_n^4}. \tag{16}$$

The mean Coulomb force has therefore, classically, a pole for orbits that are a line through the nucleus ($I_l = 0$; $\epsilon = 1$), (where the "$\frac{1}{2}$" due to the semi-classical quantization rule does not allow for a corresponding quantal state). Averaging over all classical Kepler orbits for a fixed energy means averaging over a microcanonical distribution as in [32] (here a uniform uniform distribution of ϵ^2) resulting in:

$$<< F >_C >_{Microc.distr} = \int_0^1 \left(1 - \epsilon^2\right)^{-1/2} (I_0)^{-4} d(\epsilon^2) = \frac{2}{I_0^4}. \tag{17}$$

This pure classical result corresponds indeed to the quantal statistical average calculated in Eq. (9). Finally, the classical Kepler frequency can be derived from the energy in Eq. (11) $\omega_K = I_n^{-3}$ corresponding to Eq. (10) and used in Eq. (5)

From Eq. (8) and Eq. (15) we may conclude that the mean Coulomb force on the electron, averaged over its orbit, is smallest for the circular- Bohr orbit, classically among all Kepler orbits of the same energy, and quantally among all substates of the same energy. We saw also that the mean force from the nucleus is independent of m (or I_m) in the absence of a field. The reader is reminded here that these statements about relative strengths do not hold for the total binding energy for those states, as the latter is the sum of potential and kinetic ($e.g.$ centrifugal) energy. The goal in this paragraph, however, was to provide us with an internal atomic measure for electric field amplitudes and frequencies (or internal clock).

6. LOW SCALED FREQUENCY EXPERIMENTS

Several H atom DIE, DIP and SA experiments have now covered the wide range of scaled frequencies $0.05 \leq n_0^3 \omega \leq 2.8$. The curves were recorded at various fixed microwave frequencies between 7.58–36.02 GHz for a large number of n_0-values (Table 2). From these curves we extracted the classically scaled amplitude threshold $n_0^4 F(10\%)$. Because of finite signal-to-noise ratios, the actual $0^+\%$ threshold field amplitude could not precisely be located. However because the majority of curves increased quite abruptly, except for occasional structures near threshold, the 10% threshold provided a faithful representation of the onset of microwave ionization.

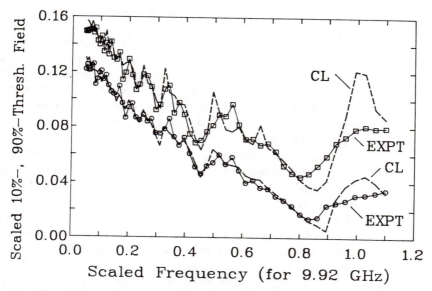

Figure 5. *Classically scaled ionization threshold amplitudes versus classically scaled frequency for 10% (O) and 90% (□) experimental ionization data taken at 9.92 GHz compared to the results of the 3D classical theory (broken line).*

In Sects. 6.1 and 6.2 we will refer only to experiments and classical theory with $n_0^3 \omega \leq 1.1$, which spans the low frequency (II) and the semi- classical (III) regions. In Sect. 6.3, at higher scaled frequencies, $n_0^3 \omega \gtrsim 1$, we will see a very different behavior.

6.1. CLASSICAL IONIZATION THRESHOLDS

Fig. 5 shows 9.9 GHz scaled amplitude threshold data $(n_0^3 \omega)$ for 10% (circles) and 90% (squares) ionization probability as a function of scaled frequency. Each point in the graph is an average of individual thresholds extracted from many curves measured with both DIP and SA methods. Each set of curves yields two data points corresponding to the two threshold levels. Connecting the data points for the same threshold levels in Fig. 5 produces two curves which in the average decrease with increasing scaled frequency while also displaying many local structures. The structures appear as a series of local maxima. The scaled field values at which they occur are found to be near $1, 1/2, 1/3, 1/4, 1/5, 1/6$. There is also a maximum which is most prominent in the 90% curve just above $1/2$ (0.56), and another which is most prominent in the 10% curve just below $2/5$. The figure compares these experimental data with 3D classical Monte Carlo calculations that closely model the experiments [4,6,28,37,38,29,39,40,41]. The model employs a classical microcanonical 3D-distribution of orbits that correspond to the uniform distribution of quantal substates [32]. It also uses the experimental envelope function, and the cutoff value n_c. The results of this model follow the general trend in the curve very well. Below $n_0^3\omega = 0.8$, apart from occasional exceptions, there is agreement to within a few percent. We emphasize the quality of this agreement by noting that the scaled values are a *linear* (not logarithmic) and *absolute* (not relative) function of the threshold field amplitude F_0. Moreover, the classical calculation reproduces the experimentally observed local maxima as well. Looking in even greater detail, it appears that the classically computed thresholds agree more closely with the measurements for scaled frequencies between the local maxima than for those precisely at the maxima.

It has been shown that these local maxima can be interpreted in a very direct way by a much simpler classical model in one dimension. The Poincaré sections of the phase space [38] for such a 1D model exhibit resonant islands and the overlap of adjacent resonant islands is known to accompany the onset of ionization. Overlapping resonant islands (or "islets") can be linked to the experimentally observed local stabilities as can be seen from Sanders *et al.* [42] who have done numerical calculations for the 10% thresholds and compared their results to experiments done in our laboratory. They find the distinct peaks at $1/p$ ($p = 1, 2, 3, 4, 5$) as in the experiments (Fig. 5), but also find 2/3 to be locally more stable. Similarly a calculation based on the half width of the $1/p$ primary islands by Blümel *et al.* reaching the $2/(2p - 1)$ resonances reproduced the measured thresholds to better than 30% [43] and an interpolation based on higher order perturbation theory, leading to solving an implicit equation, was reasonably successful (their figure 2) in calculating the relative threshold variation around these resonances. They showed that this result can also be obtained from quantal perturbation theory. A very different theory, the 1D adiabatic low frequency quantal theory of Richards *et al.* [44] which takes only two adiabatic states into account found resonances near the frequencies $\frac{n(n+\frac{1}{2})}{(n+1)^2 p}$ (p a small integer). This result agrees also well with that of the 1D classical theory.

6.2. FULL IONIZATION CURVES

While the 1D-classical dynamics is capable of reproducing quite well the experimental 10%-threshold fields for $n_0^3\omega \simeq 0.05$–1.2, it cannot reproduce the amplitude dependence of the actual (3D) direct-ionization or survived-atom experimental curves. Classical 3D Monte Carlo calculations, in contrast, are able to do this over large parts of this scaled frequency range. But there are two exceptions: (i) Agreement with experiment can break down for scaled frequencies on or near the classical resonances of low order rational values. (ii) The classical 3D calculations can not reproduce the distinct structures found in more than a dozen ionization or quench curves for $0.05 \lesssim n_0^3\omega \lesssim 0.3$ (the low frequency Regime (II)) [45,46,47], including non-monotonic structures (not shown here), steps, or changes in slope. Both (i) and (ii) indicate the importance of quantal effects.

Quantal (1D) calculations [45] done for the same range of scaled frequencies as

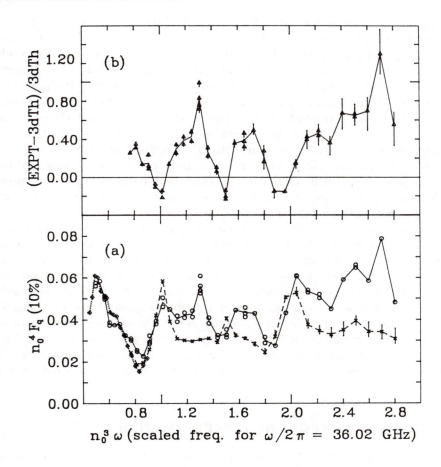

Figure 6. *Curves have been drawn to guide the eye. A: (bottom) scaled 10%-threshold amplitudes for (O) experimental survived atom (SA) and (×, ◇) 3-D classical calculations vs. scaled frequency. B: (top) △, fractional differences between O and × from (A). For vertical line segments in (A) and (B) see ref. [6].*

above reproduce the experimental 10% threshold field strengths quite well. Two sets of quantal 1D calculations have explained the structures mentioned in (ii): Blümel and Smilansky [45] linked them to the effect of occasional clustering of many avoided crossings between Floquet (=quasi-energy) curves. Some avoided crossings served to mediate transport to the continuum via so-called window states. Richards et al. [44] associated the experimental structures in (ii) with resonances between a small number of states in an adiabatic basis. That both sets of quantal 1D results are able to explain the existence of structures in an experiment with 3D atoms, *c.q.* statistical mixture of substates, indicates that the 3D dynamics is linked to that of the much more restricted 1D problem in important ways. It may be expected that such a link only exists for dynamics in relatively weak fields and is a consequence of the hydrogen atom being degenerate, i.e. only one frequency being present, and will probably not work for other systems ([52]).

The local breakdown of classical atomic dynamics near the resonances in (i) is somewhat more subtle. It would be wrong to expect classical behavior based only on the presence of large values of n_0 in the experiments. Islands and other structures in the Poincaré sections of the classical phase space are of finite size; one might conjecture that the structures in the classical Poincaré section having an area less than $\hbar$ would play an inferior role in the quantal atomic dynamics. Climbing the hierarchy of higher order resonances in classical perturbation theory leads to structures in the classical Poincaré sections of ever smaller volumes. Thus, one expects that only the largest, lowest order classical phase space structures could correspond to experimental features. This is consistent with the observed behavior in Fig. 5.

7. HIGH SCALED FREQUENCY EXPERIMENTS

For experiments with scaled frequencies rising beyond 1 the SA (Fig 6A bottom) and the DIE (Fig 7A bottom) methods were employed. We found that the measured 10%-threshold amplitudes (open circles) were no longer well reproduced by classical 3D calculations ($\times$ and diamond). However, the experimental data continued to show the effect of local stability, near *e.g.*, $\frac{1}{1}$, $\frac{4}{3}$, $\frac{5}{3}$, $\frac{2}{1}$, $\frac{5}{2}$, $\frac{8}{3}$, and, perhaps, others. Some of this behavior can be captured by the classical calculations, *e.g.*, for $n_0^3\omega$ near $\frac{1}{1}$, $\frac{2}{1}$, and $\frac{5}{2}$, but not all. The experimental dip near $\frac{3}{2}$ in Fig 6A is a clear counter-example to the classical local maxima there. The reverse is also true: the experiment exhibits strong local stability near 1.30, where there is little or none classically. The reason is that the island associated with the relatively high order $\frac{4}{3}$ classical resonance is too small to have an important effect on the classical dynamics. Even so, recent calculations do show an important relation to the quantal dynamics there (see Section 8).

For $n_0^3\omega > 2$, classical dynamics breaks down completely by systematically underestimating the field required for ionization. Our experimental observation of

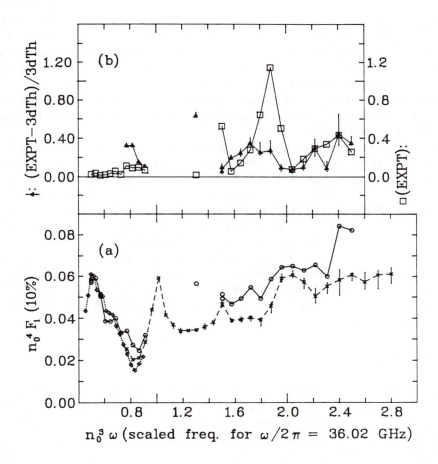

Figure 7. *Curves have been drawn to guide the eye. A: (bottom) same as Fig. 6A except ○ are experimental 'direct ionization, electron' (DIE) results. B: (top) △, fractional differences (with scale on the right vertical axis) between ○ and × from (A); □, fractional differences between DIE (Fig. 7A) and SA (Fig. 6A) 10% threshold fields. For vertical line segments in (A) and (B) see ref. [6].*

the enhanced stability of the quantal atom in the "high frequency" region was the first confirmation of theoretical predictions of this effect initiated by Casati *et al.* [5,48,49,50,51] This will be discussed in detail in Section 8. The triangles Figs. 6B (top) and 7B (top) denote the relative difference between the experimental and theoretical thresholds (both averaged over the cross section of the beam). The fluctuating behavior seen in the SA data (Fig. 6A) indicates that the real (quantal) experiment shows very strong stabilisation which is highly significant with respect to experimental uncertainties.

In Fig. 7B (top) the open squares compare the two experimental methods in a relative way: $(DIE - SA)/SA$. We remind the reader that the only difference between DIE and SA is due to the cutoff. The SA (Fig 6A) has a relatively low cutoff, $n_c = 86 - 92$, whereas the DIE (Fig. 7A) has a relatively high cutoff, $n_c = 160 - 190$. We see from the squares that the relative difference of the thresholds obtained with the two methods fluctuates between 0 and 1.20. We interpret the small relative threshold differences occuring for $n_0^3 \omega \lesssim 0.9$, and $n_0^3 \omega = 1.3; 1.6; 2.06$ as the lack of final state population with n-values between the two cutoffs. On the other hand the appreciable differences occuring at $n_0^3 \omega = 1.5$, and $n_0^3 \omega = 1.9$ indicate an appreciable population of the final states between the two cutoffs. We will see later in Section 8 that this provides important additional information on the dynamics.

8. THEORIES

We have already seen some of the results of a comparisons of theoretical calculations with our experiments. Before continuing our discussions of specific theories that have been applied to our experimental situation further, some general remarks on possible theoretical approaches are becoming. There seem to be three relevant questions here: The first is whether it is necessary for theoretical models to quantize the microwave field (as in QED). The second question is whether in the semi-classical approximation in which only the atom is quantized, a perturbation series expansion converges for all orders in some perturbation parameter. The third question is if one may use a completely classical model in which the atomic motion is also considered classically.

With regard to the first question, about quantizing the microwave field, consider the photon density per unit volume λ^3 [53]. If this is much larger than 1 the description of physical phenomena based on classical electrodynamics is reliable, and we may say that we are in the semi-classical region. (Except for specific coherent photonic state effects as for instance squeezing of light, bunching and anti-bunching of photon statistics *etc.* which we will not consider here). For our cavities typical values for this entity $\frac{N}{V}\lambda^3$ range from 10^9 to 10^{16} per volume of wavelength. At these photon densities the difference in the effect of a creation operator resulting in a factor $\sqrt{N+1}$ and anihilation operator resulting in $\sqrt{N}$ even for 100 photons absorbed or emitted out of the field, would not be detectable compared to the spontaneous quantum fluctuations of the field [53].

The second question has been answered in part by a recent proof [54]. It states that for N-body atomic systems the Rayleigh-Schrödinger perturbation expansion converges to the resonances of the AC-LoSurdo-Stark effect, defined as eigenvalues of the complex scaled Floquet Hamiltonian. It is further mentioned that this is valid for a "weak" external AC electric field and very much in contrast to the DC field case, where such a perturbation expansion is known to be asymptotic divergent, $i.e.$ the sum diverges for fixed perturbation parameter if evaluated to increasing orders. The relevant question seems to be if this finite convergence radius is large enough to be applicable to our problem.

The third question, the relevance of classical models, is supported by having high n_0 Rydberg states in our experiments and using the correspondence argument. In the previous section we already saw the success of completely classical calculations for the experiments discussed in the low scaled frequency regions (II, III). However even where there is classical behavior for certain experimental conditions one would expect that by improving the experimental situation to higher resolution and increasing its selectivity of for instance the prepared and detected state, quantal dynamics is necessary to describe the experiments. It is in this sense that the correspondence argument is not a mathematical theorem. For the other regions accurate comparisons have to be made to find an answer whether classical dynamics is sufficient to describe the experimental results at the level of accuracy (and selectivity) at which the experiments are done at this moment.

Having concluded that quantization of the microwave field is not approprite for the description of our experiments and that semiclassical perturbation theory and fully classical methods might work under specific circumstances, we begin our discussion of theoretical calculations which were directly compared to our experimental data.

The experimental data was previously compared with the classical Monte Carlo calculations of Richards following the regularization methods of Ref [29]. As discussed in section 6.1 and 6.2., which we will not repeat here in detail, the agreement was found to be very good at low scaled frequencies ($0.05 \lesssim n_0^3 \omega \lesssim 1.1$).

Blümel and Smilansky [23] have also investigated this scaled frequency region with a 1D-quantal model. They used a basis of the unperturbed Hamiltonian for the bound and continuum part of the spectrum, and neglected only continuum-continuum transitions. The resulting set of integro-differential equations were solved and resulted in two basic mechanisms responsible for ionization. The picture is a two step process consisting of excitation to states inside a "window" followed by transfer of probability to the continuum as a decay process. In the first mechanism the population probability is transferred via Floquet (quasi-energy) states to so called window states, which are highly excited states that decay with appreciable rates to the continuum. The second mechanism explained the subthreshold non-monotonic structures as the clustering of avoided crossings of pairs of quasi energies. Each coherent superposition of the Floquet states overlaps with the unperturbed initial states and final (window) states allowing for an efficient transfer of population to the window states.

Richards [55] proposed an expansion in 1D- adiabatic states in which ionization was represented phenomenologically by the addition of decay terms to the differential equations. Two, four and eight state approximations were discussed in ref. [44]. They were quite successful in calculating positions where the 3D experiment showed non-monotonic resonances. The theory also suggests a link between these non-monotonic (quantal) structures on individual ionization curves and the local stabilities observed for the 10% thresholds that were previously always explained as indications of resonance islands in Poincare sections of the phase space of classical models.

For the high scaled frequency several theories have been proposed. Predictions were made [5] that quantized atom effects would suppress ionization and therefore raise the thresholds compared to those given by classical atomic dynamics calculations. One such theory is the localization theory for the sinusoidally driven Kepler problem. The basis for this theory lies in the numerical observation that wavefunctions did not spread indefinitely in the action parameter, but instead they localize around the initial state after some time scale, *i.e.* their distribution function decays exponentially away from n_0. This is in contrast to the continual diffusion that occurs in classical dynamics for high scaled frequencies.

Localization theory represents a dynamical analogue of what is known as Anderson Localization in the field of condensed matter physics. The analogy stems from a formal equivalence with the tightbinding model equations [56,5,49,50,51,57,2] (originally done for the so-called "kicked rotator" [58]) which attempt to describe the effects of extrinsic disorder on electronic wavefunctions in spatially periodic potentials. Localization theories, however, are theories for average behavior and do not encompass mechanisms which would produce the rich structures observed in our experiments.

A numerical comparison with our data shows that for the SA experiments with $n_c = 89 - 92$, the analytical curve resulting from Localization theory only crudely reproduces the average of the experimental data [59,2] (not shown here). Indeed, the experimental data display many robust deviations of up to 50%. For the DIE experiments with $n_c = 160$, localization strongly overestimates the average 10% thresholds.

Further improvements designed to produce simplified theories which precisely model the structures observed in the experiment were made using other approximate treatments of the quantal atomic dynamics via maps. In particular, the quantum Kepler map has been developed in both 1D- and 2D-versions [51,60]. The 1D-version has been used to model our 36.02 GHz survived atom experiment, including the cutoff n_c and the envelope function. It reproduces surprisingly well most of the structures in the $n_0^3\omega$ -dependence of the 10% threshold fields $n_0^4 F$ (10%) for 3D atoms. However, as $n_0^3\omega$ is varied by miniscule amounts, e.g., $\Delta n_0^3\omega = 10^{-7}$, significant fluctuations of the thresholds are calculated. These were interpreted as mesoscopic effects such as in solid state physics for so-called universal conductance fluctuations. Averaging the Quantum Kepler Map over the relative accuracy with which the atom can determine the frequency (10^{-3}) suppresses the effect of the microscopic oscillations of the calculated threshold amplitudes.

In strong contrast with the above is the explanation offered by Leopold and Richards [61]. They conclude from a comparison of their own calculations with their quasi-resonant compensated-energy (QRCE) basis calculations that the QKM is a reasonable approximation provided that the field is insufficiently strong to excite states far from the initial state. On the other hand, they find that the QKM treats the very highly excited states incorrectly which leads to an overestimation of the ionization probability and to wild fluctuations with very small variation in the field frequency whenever one of the quasi-resonant states lies very close to the continuum. They claim that these fluctuations are therefore spurious in the sense that they are a consequence of the approximation and not due to a connection between classical chaos and quantum mechanics in the hydrogen microwave problem.

Jensen et al. [62] have made large quantal 1D numerical calculations on a CRAY that included both the cutoff, n_c, and an approximate form of the envelope function used in the 3D quench experiment. Their result of (5% ionization thresholds) agrees quite well with the experimental 10% threshold fields obtained with 3D atoms.

In another paper, Jensen et al. [63] have argued that for high scaled frequencies $n_0^3\omega > 2$ the main reason for the disagreement between classical and quantal atomic dynamics is that a dynamical selection rule picks out a few quantal states which dominate the motion. In Richards et al [64] it is shown, numerically, that these states provide a good representation of the quantal dynamics. Instead of an infinite number of quasi-resonant states between the initial level and the continuum, they confine their interest to a finite number of quasi-resonant states which play an important role in reaching the continuum. A crude estimate of this number is given by the binding energy of the initial state divided by the photon energy, $(2n_0^2)^{-1}/\omega$. This may be re-expressed as $n_0/(2n_0^3\omega)$, that is the number of photons required to reach the continuum.

Instead of working in the unperturbed basis, $<z|n>$, Leopold and Richards use a time dependent basis of states that diagonalizes the compensated energy [65]. The domination of these QRCE states leads to strong selection rules. These are stronger than those of the unperturbed basis, because the coupling between the QRCE states is stronger than that of the virtual resonant states in the unperturbed basis. This makes the description on the QRCE basis very accurate even for a small number of states. For H($n_0 = 80$) atoms in a 36.02 GHz field, the number of quasi-resonant states is only about 15. Leopold and Richards found their basis-state truncation method to work reasonably well when $n_0^3\omega \gtrsim 2$. They [66,67,64] also considered the manner in which the classical limit of the quantal behavior might be reached in the limit of large initial principal quantum numbers n_0 for fixed scaled field strength and frequency. It was concluded that our experimental 10% threshold fields for $n_0^3\omega \leq 2.8$ and n_0-values up to 80 were not within the classical atomic dynamics region.

The selection of theoretical models briefly mentioned above is not at all an exhaustive list of the work that has been done in this field. While we have been forced to limit our discussion to only a few of these, we would like to mention that alternative approaches exist. One of these employs the language of classical dynamics using Cantori, which are KAM like boundaries with Cantor set properties. Classical stochasiticity will be suppressed when the phase space area escaping through classical cantori each period of the electric field is small compared to Planck's constant [68,69]. Others use quantum atomic dynamical approaches such as Floquet (quasi-energy) calculations to show various aspects of the microwave excitation and ionization problem [70,71,72,73]. In the next section we will focus on still another very different and quite recent explanation that may well play a key role in bringing us new insights on the connection between classical and quantal atomic dynamics.

9. QUANTAL SCARS OF CLASSICAL ORBITS

In recent years new links between quantal and classical dynamics have been suggested by comparing certain transformations from QM eigenstates with unstable periodic orbits in the phase space of the classical problem. Such a quantal state is said to exhibit a scar of the classical unstable periodic orbit [74]. 'Scars' have now been identified in several theoretical problems. First recognized in the Bunimovitch stadium [74,75] they have been identified in other time independent problems as the hydrogen atom in a uniform magnetic field [76], the problem of two coupled anharmonic oscillators [77] and others [78,79]. Recently also the time dependent problems such as the quantum kicked rotor [80] and the hydrogen atom in strong fields [81,82,83] have exhibited scars.

We will concentrate here on a recently proposed explanation for some of the local stabilities in our threshold curves on the microwave ionization of hydrogen [81,82]. Calculating coarse-grained Wigner (or Husimi) transformations of quasi-energy states (QES) in the time dependent H-microwave problem, Jensen et al. was able to compare certain characteristics of a wavefunction $\Psi(t)$ with the Poincaré section of a classical phase space representation for the same control parameters. Similar comparisons had been done by Stevens et al. [84] who used coherent states of the driven surface-state electron on a basis of zero field states and showed the results in the (x, p) phase space representation. They found that for larger frequencies and a fixed field strength, the quantum phase space evolution is restricted or localized in contrast with diffusive motion present in the classical evolution. They found that tunneling is a better characterization of the quantal evolution at larger frequencies than diffusion.

Jensen *et al.* found that the inhibition of (quantal) transport is due to excitation of individual quasi-energy states whose Husimi transformation is highly localized near the classical unstable periodic orbits.

Using the action angle representation of the Hamiltonian, the coherent states were chosen to be "Poisson-like" on a basis of unperturbed states with a complex periodic dependence on the classical angle:

$$|I, \theta > = \sum_{n=0}^{\infty} [A_n(\alpha) I^{\alpha n} e^{-\alpha I}]^{1/2} e^{i2\pi n\theta} |n > \tag{18}$$

with squeezing parameter α which determines relative width of the wavepacket in θ versus I as is consistent with the uncertainty principle. These can be seen as the coarse grained "probe" function in the procedure. A state $\rho_n(I, \theta) = < n|I, \theta >$ would be peaked at action $I = n$ as a 'squeezed' Poisson distribution with width $\Delta I = \sqrt{(\alpha n + 1)/\alpha^2}$ and uniformly distributed in θ. For a general coherent state the Husimi distribution is now simply defined by a projection of the wave function on the probe functions:

$$\rho(I, \theta) = | < I, \theta|\Psi > |^2 \tag{19}$$

with

$$\Psi = \sum_{n=0}^{\infty} a_n |n > \tag{20}$$

Using this prescription they performed numerical experiments. They noticed that for certain scaled frequencies for the (relatively) slow switch on of the field only one QES state is excited inside the interaction region, *e.g.* at scaled frequency $n_0^3 \omega = 1.30$, 97.8% in a single QES. They investigated how these generalized Husimi transformations of QES compare to a phase space Poincaré section. They overlayed the contour lines of the generalized Husimi distribution with the Poincaré section using certain initial conditions. As can be seen from Fig. 8 there are no "contour" lines above $n = 72$ for this coherent state which means a strong localization at low actions. They found that scar like objects are more than exponentially "localized."

Moreover the generalized Husimi distribution seems to be linked to the geometric structures which are maximum near $n_0^3 \omega = 1.3$ ($n_0 = 62$ at 36.02 GHz). This represents a quantal local stability when compared to the 3D classical calculation. Another feature that the scarring phenomena explains is the insensitivity of the 10% threshold field strength to the cutoff value n_c that was observed in the experiments [6]. Most intriguing is the seeming contradiction that quantum dynamics is stabilized by scarring a subset of orbits that are classically unstable. Moreover for the classical dynamics this subset of classical orbits does not seem to be very important (it is of measure "zero") since there is no sharp local minimum near $n_0^3 \omega = 1.3$ in the classical calculations of Fig 6A (bottom) and 7A (bottom). It is presently not understood why the scarred wavefunctions are concentrated on

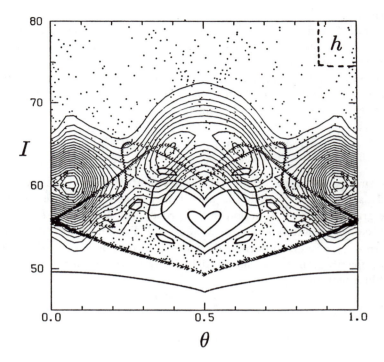

Figure 8. *Contour lines of the Husimi transformed QES that is predominantly excited for* $n_0 = 62$ *with* $n_0^3\omega = 1.3$ *and* $n_0^4 F = 0.05$. *It is shown here overlayed on a classical Poincaré section. The heart shapes, a resonance island, surround a classical stable resonance for* $n_0^3\omega = 1$.

the classical unstable periodic orbits, but the observation seems to promise to play a key role in the understanding how quantal and classical dynamics are linked together for a system for which the classical dynamics is known to be exponentially unstable.

ACKNOWLEDGEMENTS

The review on this research would not have been possible without the dedicated cooperation of many colleagues involved with the experiments in the laboratory in Stony Brook. In particular I would like to thank M. Bellerman, Dr. E.J. Galvez, A.F. Haffmans, Prof. P.M. Koch, B.E. Sauer, and S. Yoakum. I would also like to thank many theoretical colleagues for stimulating discussions, especially

D. Richards, J. Leopold, B. Sundaram, R. Jensen, R. Blümel, D.L. Shepelyansky, B. Meerson, and A. Rabinovitch. In addition I thank D. Richards and S. Yoakum for their readiness to carefully read the manuscript, their comments and many text improvements, and R.V. Jensen for providing us with Fig. 8. This research was supported by grants from the Atomic, Molecular, and Plasma Physics Program of the US National Science Foundation.

References

[1] "Irregular" here means chaotic in the sense of exponentially unstable for finite times, as opposed to infinite times (see e.g., T. Tel, 'Transient chaos', to be published in 'Directions in Chaos, Vol. 3', Bai-Lin Hao (ed.), World Scientific, Singapore). This is numerically proven for the system discussed in this chapter by calculating the Liapunov exponents in [3] for the Kepler map that is believed to be a good approximation of the 1D-ordinary differential equation model of a hydrogen atom in a microwave field.

[2] Moorman L., Koch P.M. (1991) 'Microwave ionization of Rydberg atoms', Bai-Lin Hao, Da Hsuan Feng, Jian-Min Yuan (eds.), Directions in Chaos, Vol. 4, World Scientific, Singapore, Ch 2, to appear.

[3] Haffmans, A.F., Moorman, L., Rabinovitch, A., and Koch, P.M., 'Initial condition phase space stability pictures of two dimensional area preserving maps', in preperation

[4] Leeuwen K.A.H. van, Oppen G. v., Renwick S., Bowlin J.B., Koch P.M., Jensen R.V., Rath O., Richards D., and Leopold J.G. (1985) 'Microwave ionization of hydrogen atoms: Experiment versus classical dynamics', Phys. Rev. Lett. 55, 2231-4

[5] Casati G., Chirikov B.V., Shepelyansky D.L. (1984) 'Quantum limitation for chaotic excitation of the hydrogen atom in a monochromatic field', Phys. Rev. Lett. 53, 2525-28

[6] Galvez E.J., Sauer B.E., Moorman L., Koch P.M., and Richards D. (1988) 'Microwave ionization of hydrogen atoms: Breakdown of classical dynamics for high frequencies', Phys. Rev. Lett. 61, 2011-14

[7] Bayfield J.E., Casati G., Guarneri I., Sokol D.W. (1989) 'Localization of classically chaotic diffusion for hydrogen atoms in microwave fields' Phys. Rev. Lett. 63, 364-67

[8] Koch P.M. (1990) 'Microwave ionization of excited hydrogen atoms: What we do and do not understand', in D.K. Campbell (ed.) Soviet American conference on Chaos, Woods Hole, AIP, pp. 441-475

[9] Richards, D., Leopold, J.G. (1990) 'Classical ghosts in quantal microwave ionisation', in 'The physics of electronic and atomic collisions, XVI', A. Dalgarno, et al. (eds.), AIP conference proceedings, 205, New York, p. 492-8

[10] Koch P.M., (1990) 'Microwave excitation and ionization of excited hydrogen atoms', in S. Krasner (ed.) "Chaos" perspecitves on nonlinear science', AAAS, Washington, p 75-97

[11] Water W. van de, Leeuwen K.A.H. van, Yoakum S., Galvez E.J., Moorman L., Bergeman T., Sauer B.E., and Koch P.M. (1989) 'Microwave multiphoton excitation of helium Rydberg atoms: The analogy with atomic collisions', Phys. Rev. Lett. 63, 762-65

[12] Water W. van de, Leeuwen K.A.H. van, Yoakum S., Galvez E.J., Moorman L. Sauer B.E., and Koch P.M. (1989) 'Microwave multiphoton ionization and excitation of helium Rydberg atoms', Phys. Rev. A 42, 572-91

[13] Berry M.V. (1989) 'Quantum Chaology, not Quantum Chaos', Phys. Script. 40, 335-336

[14] Gutzwiller, M.C. (1990) 'Chaos in classical and quantum mechanics' Springer Verlag, Interdiscipl. Appl. Math., Vol 1, New York

[15] Moorman L., Galvez E.J., Sauer B.E., Mortazawi-M A., Leeuwen K.A.H. van, Oppen G. v., and Koch P.M. (1989) 'Two-frequency microwave quenching of highly excited hydrogen atoms'in Phys. Rev. Lett. 61, 771-74

[16] Moorman L., Galvez E.J., Sauer B.E., Mortazawi-M A., Leeuwen K.A.H. van, Oppen G. v., and Koch P.M. (1989) 'Two-freqeuncy microwave quenching of highly excited hydrogen atoms' in 'Atomic Spectra and Collisions in External Fields 2', eds. K.T. Taylor, M.H. Nayfeh, and C.W. Clark, Plenum Press, 343-57

[17] Koch P.M., Moorman L., Sauer B.E., Galvez E.J., and Leeuwen K.A.H. van (1989) 'Experiments in quantum chaos: Microwave ionization of highly excited hydrogen atoms' Phys. Script. T26, 51-57

[18] Koch P.M. (1988) 'Microwave ioniztion of highly excited hydrogen atoms: A driven quantal system in the classical chaotic regime' in H.B. Gilbody et al. (eds.), Electronic and Atomic Collisions, North-Holland, Amsterdam, p 501-16

[19] Koch, P.M., Mariani, D.R. (1981) 'Precise measurement of the static electric field ionization rate for resolved hydrogen Stark substates ' in Phys. Rev. Lett. 1275-78

[20] Banks, D., Leopold, J.G., (1978) in 'Ionization of highly-excited atoms by electric fields: (I) Classical theory of the critical electric field for hydrogenic ions' J. Phys. B: Atom. Molec. Phys. 11 37-46; and in (1978) '(II) Classical theory of the Stark effect' J. Phys. B: At. Molec. Phys. 11, 2833-43

[21] Koch, P.M., (1983) 'Rydberg studies using fast beams' in Rydberg States of Atoms and Molecules, R.F. Stebbings and F.B. Dunning (eds.), Cambridge University Press, New York, p 473-512

[22] Kleppner, D., Littman, M.G., Zimmerman M.L. (1983) Rydberg atoms in strong fields', in R.F. Stebbings and F.B. Dunning (eds.) Cambridge University Press, New York, p 73-116

[23] Blümel R. and Smilansky U. (1987) 'Microwave ionization of highly excited hydrogen atoms', Z. Phys. D: Atoms, Molec. and Clusters 6, 83-105

[24] Landau, L.D., and Lifshitz, E.M. (1976) 'Course of theoretical physics, Vol. 1: Mechanics' 3rd edition, Pergamon Press, Oxford, Ch. 3

[25] Goldstein, H.(1977) 'Classical Mechanics', Addison Wesley Inc., Reading, 12th edition, Ch. 3.6

[26] Halbach K. and Holsinger R.F. (1976) 'SUPERFISH, a computer program for the evaluation of RF cavities with cylindrical symmetries', in Part. Accel. 7, 213-22

[27] Sauer B.E., Leeuwen K.A.H. van, Mortazawi-M A., and Koch P.M. (1991) 'Precise calibration of a microwave cavity with a nonideal waveguide system', Rev. Scient. Instr. 62, 189-97

[28] Leopold J.G., and Percival I.C. (1978) 'Microwave ionization and excitation of Rydberg atoms', Phys. Rev. Lett. 41, 944-7

[29] Leopold J.G. and Richards D. (1985) , 'The effect of a resonant electric field on a one-dimensional classical hydrogen atom' J. Phys. B: Atom. Mol. Phys. 18, 3369-94

[30] Born M., 'The mechanics of the atom' (1924), Republished by Frederick Ungar publishing co., New York (1960).

[31] The principal action variable is $I_n = I_r + I_\theta + I_\phi$, the total angular momentum is $I_l = I_\theta + I_\phi$, and the component of the angular momentum onto the

polar axis is $I_m = I_\phi$. with $2\pi I_i = \oint p_i dr_i$ for $i = r$, θ, and ϕ (and no sum convention) [30,32]. The quantization conditions become $I_n = n\hbar$, $I_l = k\hbar$, and in a magnetic field $I_m = m\hbar$. In the old quantum theory $n = 1, 2, \cdots$ is the principal quantum number and k is the subsidiary (or azimuthal) quantum number [30]. This would be $k = 1, 2, \cdots, n$, however modern quantization methods (Einstein-Brillouin-Keller) require $k = l + 1/2$ with $l = 0, 1, 2, \cdots, n$, in which the $1/2$ does not refer to the spin but to the Maslov (or Morse) index in semi-classical quantization. This index counts the number of classical turning points α encountered by a closed trajectory in the classical phase space (for a more general description in terms of caustics see [79]). At each turning point a phase-loss of $\pi/2$ (equivalent to $1/4^{th}$ of a wave) has to be taken into account. Continuity of the phase of the wave function then leads to $I = (n + \frac{1}{4}\alpha)\hbar$. For example the 1 dimensional harmonic oscillator encountering two turning points per period obtains for the same reason (but named after Wentzel- Kramers-Brillouin) the well known 'zero-point' energy, as given in $E_v = (v + \frac{1}{2})\hbar$ for $v = 0, 1, \cdots$. For further reading and how this can be understood from conditions to be satisfied under a coordinate transformation of a pathintegral from Cartesian into spherical coordinates, requiring a new term $\sim \frac{1}{4}\hbar^2$ in the classical Hamiltonian, which adds to the angular momentum part $|L|^2 = l(l + 1)\hbar^2$ giving $(l + \frac{1}{2})^2\hbar^2$, see p 203 and 212 etc. of [14].

[32] Percival I.C., and Richards D. (1975) 'The theory of collisions between charged particles and highly excited atoms', in 'Advances in atomic and molecular physics', Vol 11, p1-82

[33] Series, G.W. (1988), 'The spectrum of the hydrogen atom: Advances', World Scientific Publishing Co, (or the 1957 original version of part I, by Oxford University Press) p20-24.

[34] Woodgate, G.K. (1980) 'Elementary Atomic Structure', Clarendon Press, Oxford, $2^n d$ ed., p 22

[35] Vol 3 of [24] (1977) 'Quantum Mechanics' 3rd edition, Pergamon Press, Oxford, p. 120

[36] Bethe, H.A., Salpeter, E.E. (1977) 'Quantum mechanics of one and two electron atoms', Plenum/Rosetta, Oxford, p.17; p.58

[37] Meerson B.I., Oks E.A., and Sasorov P.V. (1982) 'A highly excited atom in a field of intense resonant electromagnetic radiation: I Classical Motion', J. Phys. B: Atom. Mol. Phys. 15, 3599

[38] Jensen R.V. (1984) 'Stochastic ionization of surface-state electrons: Classical theory', Phys. Rev. A30, 386-97

[39] Leopold J.G. and Richards D. (1986) 'The effect of a resonant electric field on a classical hydrogen atom' J. Phys. B: Atom. Mol. Phys. 19, 1125-42

[40] Jensen, R.V. (1987) 'Effects of classical resonances on the chaotic microwave ionization of highly excited hydrogen atoms', Physica Scripta 35, 668

[41] Richards, D. (1990) 'The Coulomb potential and microwave ionization', International conference on the physcis of electronic and atomic collisions, New York 1989, AIP, 54-64

[42] Sanders M.M., Jensen R.V., Koch P.M. and Leeuwen K.A.H. van (1987) 'Chaotic ionization of highly excited hydrogen atoms', Nucl. Phys. B (Proc. suppl.) 2, 578-579

[43] Blümel R. and Smilansky U. (1989) 'Ionization of excited hydrogen atoms by microwave fields: a test case for quantum chaos', Physica Scripta 40, 386-93.

[44] Richards D., Leopold J.G., Koch P.M., Galvez E.J., Leeuwen K.A.H. van, Moorman L., Sauer B.E., and Jensen R.V. (1989) 'Structure in low frequency microwave ionization of excited hydrogen atoms', J. Phys. B 22, 1307

[45] Blümel R. and Smilansky U. (1987) 'Localization of Floquet states in the rf excitation of Rydberg atoms', Phys. Rev. Lett. 58, 2531-4

[46] Blümel R., Goldberg J., and Smilansky U. (1988) 'Features of hte quasienergy spectrum of the hydrogen atom in a microwave field' Z. Phys. D: Atoms, Molec., and Clusters 9, 95 and Blümel, R., Hillermeier, R.C., and Smilansky, U. (1990) Z. Phys. D 15, 267

[47] Koch P.M. (1982) 'Interactions of intense fields with microwave atoms', Journal de Physique Colloque 43, C2-187

[48] Casati G., Chirikov B.V., Shepelyansky D.L., and Guarneri I. (1987) 'Localization of diffusive excitation in the two- dimensional hydrogen atom in a monochromatic field', Phys. Rev. Lett. 59, 2927

[49] Casati G., Chirikov B.V., Shepelyansky D.L., and Guarneri I. (1987) 'Relevance of classical chaos in quantum mechanics: The hydrogen atom in a monochromatic field', Phys. Rep. 154, 77

[50] Casati G., I. Guarneri I., Shepelyansky D.L. (1987) 'Exponential photonic localization for the hydrogen atom in a monochromatic field', Phys. Rev. A 36, 3501

[51] Casati G., I. Guarneri I., Shepelyansky D.L. (1988) 'Hydrogen atom in monochromatic field: Chaos and dynamical photonic localization' I.E.E.E.: J. Quantum Electron. 24, 1420

[52] D. Richards (private communication).

[53] Sakurai, J.J. (1978) in 'Advanced quantum mechanics' in series in advanced physics, Addison-Wesley, 7th printing, p.35 and p.38

[54] Graffi, S., Grecchi, V., Silverstone, H.J. (1985) in 'Annales de l'institute de Henry Poincaré – Physique theorique', Vol 42, p. 215-234

[55] Richards D., (1987) J. of Phys. B At. Mol. Phys. 20, 2171-92

[56] Casati, G., Guarneri, I., Shepelyansky, D.L. (1990) 'Classical chaos, quantum localization and fluctuations: A unified view' Physica A 163, 205

[57] Grempel D.R., Prange R.E., and Fishman S. (1984) 'Quantum dynamics of a nonintegral system', Phys. Rev. A 29, 1639-47

[58] Fishman S., Grempel D., Prange R.E. (1982) 'Chaos, Quantum recurrences and Anderson localization' Phys. Rev. Lett. 49, 509-12

[59] Koch P.M., Moorman L., Sauer B.E. (1990) 'Microwave ionization of excited hydrogen atoms: experiments versus theories for high scaled frequencies' in a special issue on Quantum Chaos of 'Comments on Atomi- and Molecular Physics', Vol.25, pp. 165-183

[60] Brivio , Casati G., Guarneri I., and Perotti L. (1988) 'Quantum suppression of chaotic diffusion: theory and experiment' Physica 33D, 51-57

[61] Richards, D., Leopold, J.G. (1990), 'On the Quantum Kepler Map' J. Phys. B: At. Mol. Opt. Phys. 23, 2911-2927

[62] Jensen R.V., Susskind S.M., Sanders M.M. (1989) 'Microwave ionization of highly excited hydrogen atoms: A test of the correspondence principle', Phys. Rev. Lett. 62, 1476-79

[63] Jensen, R.V., Leopold, J., Richards, D. (1988) 'High- frequency microwave ionization of hydrogen atoms' J. Phys. B: At. Mol. Phys. 21, L527-31

[64] Richards D., Leopold J.G., and Jensen R.V. (1988) 'Classical and quantum dynamics in high frequency fields', J. Phys. B: At. Mol. Phys. 22, 417-33

[65] Leopold, J.G., Richards, D. (1989) 'Quasi-Resonances for high frequency perturbations' J. Phys. B 22, 1931

[66] Leopold J.G. and Richards D. (1988) 'A study of quantum dynamics in the classically chaotic regime' J. Phys. B: At. Mol. Phys. 21, 2179-2204

[67] Leopold,J.G., and Richards, D. (1988) 'Quantal localization and the uncertainty principle', Phys. Rev. A 38, 2660-3

MICROWAVE IONIZATION OF H ATOMS

[68] Mackay, R.S., and Meiss, J.D. (1988) 'Relation between quantum and classical thresholds for multiphoton ionization of excited atoms', Phys. Rev. A 37, 4702-7

[69] Meiss, J.D., (1989) 'Comment on "Microwave ionization of H- atoms: breakdown of classical dynamics for high frequencies" by E.J. Galvez *et al.*' in Phys. Rev. Lett. 62, 1576

[70] Breuer, H.P., Dietz, K., Holthaus, M. (1988) 'The role of avoided crossings in the dynamics of strong laser field-matter interactions' Z.Phys. D: Atoms, Molec. and Clusters 8, 349

[71] Breuer, H.P., Dietz, K., Holthaus, H. (1988) in Z. Phys. D 10, 12; and (1989) in J. Phys. B 22, 3187

[72] Breuer, H.P., Holthaus, M.,'Adiabatic processes in the ionization of highly excited hydrogen atoms' (3rd paper)

[73] Wang, K., Chu, S-I. (1989) 'Dynamics of multiphoton excitation and quantum diffusion in Rydberg atoms', Phys. Rev. A 39, 1800-1808

[74] Heller E.J. (1984) 'Bound state eigenfunctions of classically chaotic Hamiltonian systems: Scars of periodic orbits', Phys. Rev. Lett. 53, 1515-8

[75] O'Conor P.W., Gehlen J.N., Heller E.J. (1987) 'Properties of random superpositions of plane waves', Phys. Rev. Lett. 58, 1296-9

[76] Wintgen, D., and Hönig, A. (1989) 'Irregular wave functions of a hydrogen atom in a uniform magnetic field', Phys. Rev. Lett. 63, 1467-70

[77] Waterland, R.L., Jian-Min Yuan, Martens, C.C , Gillilan, E., and Reinhardt, W.P. (1989) Phys. Rev. Lett. 61, 2733-6

[78] Feingold M., Littlejohn R.G., Solina S.B., Pehling J.S. (1990) 'Scars in billiards: The phase space approach', Phys. Lett. A 146, 199- 203

[79] Berry M.V. (1989) 'Quantum scars of classical closed orbits in phase space', Proc. Roy. Soc. London, A 423, 219-231; and in (1982) "Chaotic behavior of deterministic systems', G. Iooss, R.H.G. Helleman, and R. Stora (eds.), Proceedings of the Les Houches Summer Institute, North-Holland, Amsterdam, p 172

[80] Radons, G., Prange, R.E. (1988) 'Wave function at the critical Kolmogorov-Arnol'd-Moser surface', Phys. Rev. Lett. 61, 1691-4

[81] Jensen, R.V., Sanders, M.M., Saraceno, M., Sundaram, B., (1989) 'Inhibition of quantum transport due to "Scars" of unstable periodic orbits', Phys. Rev. Lett. 63, 272-5 and preprint NSF-ITP-89-1281 (Yale).

[82] Jensen, R.V., Susskind, S.M., Sanders, M.M. (1991) submitted to Phys. Reports

[83] Jensen, R.V., Sundaram, B., (1990) 'On the role of "Scars" in the suppression of ionization in intense high-frequency fields', Phys. Rev. Lett. 65, 1964-7

[84] Stevens, M.J., and Sundaram, B. (1989) 'Quantal phase space analysis of the driven surface-state electron', Phys. Rev. A 39, 286-277

The Electron 1990 Workshop
List of Participants

BANDRAUK, A. D.	Université de Sherbrooke Sherbrooke, Québec	Canada
BARUT, A. O.	University of Colorado Boulder, Colorado	USA
BOUDET, R.	Université de Province Marseille Cedex 3	France
BRUNEL, F.	National Research Council Laser and Plasma Physics Ottawa, Ontario	Canada
BURNETT, N. H.	National Research Council Laser and Plasma Physics Ottawa, Ontario	Canada
CAPRI, A. Z.	University of Alberta Edmonton, Alberta	Canada
CHIN, S. L.	Université Laval Québec, Québec	Canada
COOPERSTOCK, F. I.	University of Victoria Victoria, British Columbia	Canada
CORKUM, P.	National Research Council Laser and Plasma Physics Ottawa, Ontario	Canada
FREEMAN, R. R.	AT&T Bell Laboratories Electronics Research Department Holmdel, New Jersey	USA
GALLAGHER, T. F.	University of Virginia Charlottesville, Virginia	USA
GRANDY, Jr., W. T.	University of Wyoming Laramie, Wyoming	USA
GULL, S.	Cavendish Laboratory Cambridge	England
HAWTON, M.	Lakehead University Thunder Bay, Ontario	Canada

HESTENES, D.   Arizona State University   USA
                                Tempe, Arizona

HOLLEBONE, B. R.   Carleton University   Canada
                                Ottawa, Ontario

JAYNES, E. T.   Washington University   USA
                                Saint Louis, Missouri

JUNG, Ch.   Universität Bremen   Germany
                                Bremen

KERWIN, L.   Canadian Space Agency   Canada
                                Ottawa, Ontario

KRUGER, H.   Universität Kaiserslautern   Germany
                                Kaiserslautern

L'HUILLIER, A.   Centre d'Etudes Nuclaires de Saclay   France
                                Service de Physique des Atomes et
                                des Surfaces
                                Gif-Sur-Yvette Cedex 3

MARMET, P.   Herzberg Institute of Astrophysics   Canada
                                Ottawa, Ontario

McEWAN, J.   The University of Kent   England
                                Canterbury, Kent

MOORMAN, L.   State University of New York   USA
                                Stony Brook, New York

MORGNER, H.   Universität Witten/Herdecke   Germany
                                Witten-Annen

VAN DYCK, Jr., R. S.   University of Washington   USA
                                Seattle, Washington

WALLBANK, B.   St. Francis Xavier University   Canada
                                Antigonish, Nova Scotia

WEINGARTSHOFER, A.   St. Francis Xavier University   Canada
                                Antigonish, Nova Scotia

WILLMANN, K.   Buchenbach-Unteribental   Germany

This is both a subject and name index, it follows the alphabetical order except for the entries under the topics: electron theory and QED and self-field QED, for clarity we follow the page sequence as presented in these articles.

Fundamental Theories of Physics

Series Editor: Alwyn van der Merwe, *University of Denver, USA*

1. M. Sachs: *General Relativity and Matter.* A Spinor Field Theory from Fermis to Light-Years. With a Foreword by C. Kilmister. 1982   ISBN 90-277-1381-2
2. G.H. Duffey: *A Development of Quantum Mechanics.* Based on Symmetry Considerations. 1985   ISBN 90-277-1587-4
3. S. Diner, D. Fargue, G. Lochak and F. Selleri (eds.): *The Wave-Particle Dualism.* A Tribute to Louis de Broglie on his 90th Birthday. 1984   ISBN 90-277-1664-1
4. E. Prugovečki: *Stochastic Quantum Mechanics and Quantum Species.* A Consistent Unification of Relativity and Quantum Theory based on Stochastic Spaces. 1984; 2nd printing 1986   ISBN 90-277-1617-X
5. D. Hestenes and G. Sobczyk: *Clifford Algebra to Geometric Calculus.* A Unified Language for Mathematics and Physics. 1984
    ISBN 90-277-1673-0; Pb (1987) 90-277-2561-6
6. P. Exner: *Open Quantum Systems and Feynman Integrals.* 1985   ISBN 90-277-1678-1
7. L. Mayants: *The Enigma of Probability and Physics.* 1984   ISBN 90-277-1674-9
8. E. Tocaci: *Relativistic Mechanics, Time and Inertia.* Translated from Romanian. Edited and with a Foreword by C.W. Kilmister. 1985   ISBN 90-277-1769-9
9. B. Bertotti, F. de Felice and A. Pascolini (eds.): *General Relativity and Gravitation.* Proceedings of the 10th International Conference (Padova, Italy, 1983). 1984
    ISBN 90-277-1819-9
10. G. Tarozzi and A. van der Merwe (eds.): *Open Questions in Quantum Physics.* 1985
    ISBN 90-277-1853-9
11. J.V. Narlikar and T. Padmanabhan: *Gravity, Gauge Theories and Quantum Cosmology.* 1986   ISBN 90-277-1948-9
12. G.S. Asanov: *Finsler Geometry, Relativity and Gauge Theories.* 1985
    ISBN 90-277-1960-8
13. K. Namsrai: *Nonlocal Quantum Field Theory and Stochastic Quantum Mechanics.* 1986   ISBN 90-277-2001-0
14. C. Ray Smith and W.T. Grandy, Jr. (eds.): *Maximum-Entropy and Bayesian Methods in Inverse Problems.* Proceedings of the 1st and 2nd International Workshop (Laramie, Wyoming, USA). 1985   ISBN 90-277-2074-6
15. D. Hestenes: *New Foundations for Classical Mechanics.* 1986
    ISBN 90-277-2090-8; Pb (1987) 90-277-2526-8
16. S.J. Prokhovnik: *Light in Einstein's Universe.* The Role of Energy in Cosmology and Relativity. 1985   ISBN 90-277-2093-2
17. Y.S. Kim and M.E. Noz: *Theory and Applications of the Poincaré Group.* 1986
    ISBN 90-277-2141-6
18. M. Sachs: *Quantum Mechanics from General Relativity.* An Approximation for a Theory of Inertia. 1986   ISBN 90-277-2247-1
19. W.T. Grandy, Jr.: *Foundations of Statistical Mechanics.*
    Vol. I: *Equilibrium Theory.* 1987   ISBN 90-277-2489-X
20. H.-H von Borzeszkowski and H.-J. Treder: *The Meaning of Quantum Gravity.* 1988
    ISBN 90-277-2518-7
21. C. Ray Smith and G.J. Erickson (eds.): *Maximum-Entropy and Bayesian Spectral Analysis and Estimation Problems.* Proceedings of the 3rd International Workshop (Laramie, Wyoming, USA, 1983). 1987   ISBN 90-277-2579-9

Fundamental Theories of Physics

22. A.O. Barut and A. van der Merwe (eds.): *Selected Scientific Papers of Alfred Landé.*
    [*1888-1975*]. 1988   ISBN 90-277-2594-2
23. W.T. Grandy, Jr.: *Foundations of Statistical Mechanics.*
    Vol. II: *Nonequilibrium Phenomena.* 1988   ISBN 90-277-2649-3
24. E.I. Bitsakis and C.A. Nicolaides (eds.): *The Concept of Probability.* Proceedings of the
    Delphi Conference (Delphi, Greece, 1987). 1989   ISBN 90-277-2679-5
25. A. van der Merwe, F. Selleri and G. Tarozzi (eds.): *Microphysical Reality and Quantum
    Formalism, Vol. 1.* Proceedings of the International Conference (Urbino, Italy, 1985).
    1988   ISBN 90-277-2683-3
26. A. van der Merwe, F. Selleri and G. Tarozzi (eds.): *Microphysical Reality and Quantum
    Formalism, Vol. 2.* Proceedings of the International Conference (Urbino, Italy, 1985).
    1988   ISBN 90-277-2684-1
27. I.D. Novikov and V.P. Frolov: *Physics of Black Holes.* 1989   ISBN 90-277-2685-X
28. G. Tarozzi and A. van der Merwe (eds.): *The Nature of Quantum Paradoxes.* Italian
    Studies in the Foundations and Philosophy of Modern Physics. 1988
                                                                       ISBN 90-277-2703-1
29. B.R. Iyer, N. Mukunda and C.V. Vishveshwara (eds.): *Gravitation, Gauge Theories
    and the Early Universe.* 1989   ISBN 90-277-2710-4
30. H. Mark and L. Wood (eds.): *Energy in Physics, War and Peace.* A Festschrift
    celebrating Edward Teller's 80th Birthday. 1988   ISBN 90-277-2775-9
31. G.J. Erickson and C.R. Smith (eds.): *Maximum-Entropy and Bayesian Methods in
    Science and Engineering.*
    Vol. I: *Foundations.* 1988   ISBN 90-277-2793-7
32. G.J. Erickson and C.R. Smith (eds.): *Maximum-Entropy and Bayesian Methods in
    Science and Engineering.*
    Vol. II: *Applications.* 1988   ISBN 90-277-2794-5
33. M.E. Noz and Y.S. Kim (eds.): *Special Relativity and Quantum Theory.* A Collection of
    Papers on the Poincaré Group. 1988   ISBN 90-277-2799-6
34. I.Yu. Kobzarev and Yu.I. Manin (eds.): *Elementary Particles. Mathematics, Physics
    and Philosophy.* 1989   ISBN 0-7923-0098-X
35. F. Selleri (ed.): *Quantum Paradoxes and Physical Reality.* 1990   ISBN 0-7923-0253-2
36. J. Skilling (ed.): *Maximum-Entropy and Bayesian Methods.* Proceedings of the 8th
    International Workshop (Cambridge, UK, 1988). 1989   ISBN 0-7923-0224-9
37. M. Kafatos: *Bell's Theorem, Quantum Theory and Conceptions of the Universe.* 1989
                                                                       ISBN 0-7923-0496-9
38. Yu.A. Izyumov and V.N. Syromyatnikov: *Phase Transitions and Crystal Symmetry.*
    1990   ISBN 0-7923-0542-6
39. P.F. Fougère (ed.): *Maximum-Entropy and Bayesian Methods.* Proceedings of the 9th
    International Workshop (Dartmouth, Massachusetts, USA, 1989). 1990
                                                                       ISBN 0-7923-0928-6
40. L. de Broglie: *Heisenberg's Uncertainties and the Probabilistic Interpretation of Wave
    Mechanics.* With Critical Notes of the Author. 1990   ISBN 0-7923-0929-4
41. W.T. Grandy, Jr. (ed.): *Relativistic Quantum Mechanics of Leptons and Fields.* 1991
                                                                       ISBN 0-7923-1049-7
42. Yu.L. Klimontovich: *Turbulent Motion and the Structure of Chaos.* A New Approach
    to the Statistical Theory of Open Systems. (forthcoming 1991)   ISBN 0-7923-1114-0

43. W.T. Grandy, Jr. and L.H. Schick (eds.): *Maximum-Entropy and Bayesian Methods.*
Proceedings of the 10th International Workshop (Laramie, Wyoming, USA, 1990).
1991   ISBN 0-7923-1140-X
44. P.Pták and S. Pulmannová: *Orthomodular Structures as Quantum Logics.* Intrinsic
Properties, State Space and Probabilistic Topics. 1991   ISBN 0-7923-1207-4
45. D. Hestenes and A. Weingartshofer (eds.): *The Electron.* New Theory and Experiment.
1991   ISBN 0-7923-1356-9

KLUWER ACADEMIC PUBLISHERS – DORDRECHT / BOSTON / LONDON